Lecture Notes in Networks and Systems   842

The series "Lecture Notes in Networks and Systems" publishes the latest developments in Networks and Systems—quickly, informally and with high quality. Original research reported in proceedings and post-proceedings represents the core of LNNS.

Volumes published in LNNS embrace all aspects and subfields of, as well as new challenges in, Networks and Systems.

The series contains proceedings and edited volumes in systems and networks, spanning the areas of Cyber-Physical Systems, Autonomous Systems, Sensor Networks, Control Systems, Energy Systems, Automotive Systems, Biological Systems, Vehicular Networking and Connected Vehicles, Aerospace Systems, Automation, Manufacturing, Smart Grids, Nonlinear Systems, Power Systems, Robotics, Social Systems, Economic Systems and other. Of particular value to both the contributors and the readership are the short publication timeframe and the worldwide distribution and exposure which enable both a wide and rapid dissemination of research output.

The series covers the theory, applications, and perspectives on the state of the art and future developments relevant to systems and networks, decision making, control, complex processes and related areas, as embedded in the fields of interdisciplinary and applied sciences, engineering, computer science, physics, economics, social, and life sciences, as well as the paradigms and methodologies behind them.

Indexed by SCOPUS, INSPEC, WTI Frankfurt eG, zbMATH, SCImago.

All books published in the series are submitted for consideration in Web of Science.

For proposals from Asia please contact Aninda Bose (aninda.bose@springer.com).

José Bravo · Gabriel Urzáiz

Editors

Proceedings of the 15th International Conference on Ubiquitous Computing & Ambient Intelligence (UCAmI 2023)

Volume 2

 Springer

Editors
José Bravo
Ciudad Real, Spain

Gabriel Urzáiz
Mérida, Yucatán, Mexico

ISSN 2367-3370   ISSN 2367-3389 (electronic)
Lecture Notes in Networks and Systems
ISBN 978-3-031-48641-8   ISBN 978-3-031-48642-5 (eBook)
https://doi.org/10.1007/978-3-031-48642-5

This Springer imprint is published by the registered company Springer Nature Switzerland AG
The registered company address is: Gewerbestrasse 11, 6330 Cham, Switzerland

Paper in this product is recyclable.

Preface

Ubiquitous computing (UC) is a paradigm that allows software applications to obtain and process environmental information in order to make users feel that changes in the environment do not affect the functionality provided to them, thus aiming at making the technology invisible. UC is made possible by the confluence of computing, communication and control technologies, and it involves measuring and considering context variables (e.g., time and location of the users) to appropriately adapt the functionality of software applications to user needs.

On the other hand, ambient intelligence (AmI) represents functionality embedded into the environment that allows systematic and unattended sensing, as well as proactive acting, to provide smart services. Thus, AmI solutions frequently appear in domains of "being helped," e.g., social, psychological, healthcare and instrumental scenarios.

The UCAmI Conference presents advances in both AmI application scenarios and their corresponding technical support through UC. In doing this, this forum covers a broad spectrum of contributions from the conception of technical solutions to the assessment of the benefits in particular populations. The aim is for users of these smart environments to be unaware of the underlying technology, while reaping the benefits of the services it provides. Devices embedded within the environment are aware of the people's presence and subsequently react to their behaviors, gestures, actions and context.

During the last years, the interest in ubiquitous computing and ambient intelligence has grown considerably, due to new challenges posed by society, demanding highly innovative services for several application domains, such as vehicular ad hoc networks, ambient-assisted living, e-health, remote sensing, home automation and personal security. The COVID-19 pandemic has made us not only more aware of the ubiquity of information technologies in our daily life but also of their need to seamlessly support our everyday activities.

We are concerned with the sound development of UC and AmI as the only way to properly satisfy the expectations around this exciting intersection of information, communications and control technologies. Therefore, this UCAmI edition involves research work in five tracks: AmI for health and A3L, Internet of everything and sensors, smart environments, human–computer interaction and data science.

We received 107 submissions for this 15th edition of UCAmI authored by 207 researchers from 18 countries. A total of 297 reviews were performed, reaching the high average of 2.19 reviews per submission. We would like to thank all the authors who submitted their work for consideration, as well as the reviewers who provided their detailed and constructive reviews. Many thanks also to the track chairs for the great commitment shown in organization and execution of the papers reviewing process.

Finally, we are happy to return to the same place of the 2011 edition of the conference: Riviera Maya (Mexico).

November 2023   José Bravo
Gabriel Urzáiz

Organization

Our Staff

José Bravo (General Chair)   University of Castilla-La Mancha, Spain
Gabriel Urzaiz (Local Chair)   Anáhuac Mayab University, Mexico

Steering Committee

José Bravo, Spain
Pino Caballero, Spain
Macarena Espinilla, Spain
Jesús Favela, Mexico
Diego López-De-Ipiña, Spain
Chris Nugent, UK
Sergio F. Ochoa, Chile
Ramón Hervás, Spain
Gabriel Urzaiz, Mexico
Vladimir Villareal, Panama
Jesús Fontecha, Spain
Iván González, Spain

Organization Committee

Cosmin Dobrescu, Spain
David Carneros, Spain
Laura Villa, Spain
Luis Cabañero, Spain
Tania Mondéjar, Spain
Esperanza Johnson, Spain
Alejandro Pérez, Spain
Brigitte Nielsen, Panama
Paloma Bravo, Spain

Track Chairs

AmI for Health & (A3L) (Ambient, Active & Assisted Living)

Jesús Fontecha, Spain
Ian Cleland, UK

Smart Environment

Macarena Espinilla, Spain
Kåre Synnes, Sweden
Chris Nugent, UK

Internet of Everything (IoT + People + Processes) and Sensors

Joaquín Ballesteros, Spain
Cristina Santos, Portugal

Data Science

Marcela Rodríguez, Mexico
Alberto Morá, Mexico

Human–Computer Interaction

Gustavo López, Costa Rica
Sruti Subramanian, Norway

Satellite Events

International Workshop on Energy Aware Systems, Communications and Security

Mauro Migliardi, Italy
Francesco Palmieri, Italy

Program Committee

Adrian Lara	UCR
Adrian Sánchez-Miguel Ortega	Universidad de Castilla-La Mancha
Alberto Morán	UABC

Alejandro Pérez Vereda	Universidad de Castilla-La Mancha
Alessio Merlo	University of Genova
Alireza Souri	Haliç University
Allan Berrocal	Universidad de Costa Rica
Andres Diaz Toro	UNAD
Andrés Oliva	Anáhuac Mayab University
Antonio Robles-Gomez	UNED
Antonio Albín Rodríguez	Universidad de Jaén
Arcangelo Castiglione	University of Salerno
Arfat Ahmad Khan	Department of Computer Science, College of Computing, Khon Kaen University
Beatriz Garcia-Martinez	Universidad de Castilla-La Mancha
Borja Bordel	Universidad Politécnica de Madrid
Bruno Carpentieri	University of Salerno
Carlo Ferrari	University of Padova
Carlos Rovetto	Universidad Tecnológica de Panamá
Carlos Aguilar Avelar	UABC
Carlos E. Galván	Universidad Autónoma de Zacatecas
Carmelo Militello	Italian National Research Council (CNR)
Carmen Martinez Cruz	University of Jaen
Chris Nugent	Ulster University
Colin Shewell	Ulster University
Constantin Cosmin Dobrescu	Universidad de Castilla-La Mancha
Cristiana Pinheiro	University of Minho
Cristina Santos	University of Minho
Cristina Ramirez-Fernandez	TecNM/I.T. de Ensenada
David Carneros-Prado	Universidad de Castilla-La Mancha
David Gil	University of Alicante
Davide Zuccarello	Politecnico di Milano
Dionicio Neira Rodado	Universidad de la Costa
Eduardo Barbará	Anáhuac Mayab University
Elena Navarro	Universidad de Castilla-La Mancha
Ernesto Lozano	CICESE
Ernesto Vera	UABC
Esperanza Johnson	Høgskolen i Innlandet
Fabio Lopes	Universidade Presbiteriana Mackenzie
Fabio Salice	Politecnico di Milano
Federico Cruciani	Ulster University
Federico Botella	UMH
Francesco Palmieri	University of Salerno
Francisco Flórez-Revuelta	University of Alicante
Francisco Javier Cabrerizo	University of Granada

Gabriel Urzaiz	Anahuac Mayab University
Gerardo Alonzo	Anahuac Mayab University
Gianni D'Angelo	University of Salerno
Gilberto Borrego Soto	Instituto Tecnológico de Sonora
Gustavo Lopez	University of Costa Rica
Hadi Moradi	School of ECE, UT
Helena Gómez Adorno	IIMAS-UNAM
Higinio Mora	University of Alicante
Ian McChesney	Ulster University, School of Computing
Ian Cleland	Ulster University
Idongesit Ekerete	Ulster University
Ignacio Diaz-Oreiro	Universidad de Costa Rica
Iker Pastor López	University of Deusto
Inmaculada Ayala	Universidad de Málaga
Irvin Hussein Lopez-Nava	Centro de Investigación Científica y de Educación Superior de Ensenada
Iván González Díaz	Castilla-La Mancha University
Jan Havlík	Czech Technical University in Prague
Javier Medina Quero	University of Granada
Javier Ferrandez	University of Alicante
Javier Sanchez-Galan	Universidad Tecnologica de Panama
Jessica Beltrán	Universidad Autónoma de Coahuila
Jesus Fontecha	University of Castilla-La Mancha
Jesus Favela	CICESE
Jesús Peral	University of Alicante
Joana Figueiredo	University of Minho
João Lopes	University of Minho
Joaquin Ballesteros	Universidad de Malaga
Jordan Vincent	Ulster University
Jorge Eduardo Ibarra Esquer	Universidad Autonoma de Baja California
Jorge Rivera	Anáhuac Mayab University
Jose Bravo	University of Castilla-La Mancha
José Luis López Ruiz	University of Jaén
Josef Hallberg	Luleå University of Technology
Karan Mitra	Luleå University of Technology
Kåres Synnes	Luleå University of Technology
Kristina Yordanova	University of Rostock
Kryscia Ramírez-Benavides	Universidad de Costa Rica
Laura Villa	University of Castilla-La Mancha
Lilia Muñoz	Universidad Tecnológica de Panamá
Long-Hao Yang	Fuzhou University
Luigi Benedicenti	University of New Brunswick

Luis Cabañero	University of Castilla-La Mancha
Luis Quesada	Universidad de Costa Rica
Luis Pellegrin	UABC
Luís Moreira	University of Minho
Luis A. Castro	Instituto Tecnologico de Sonora (ITSON)
Macarena Espinilla	University of Jaén
Manuel Fernández Carmona	Universidad de Málaga
Marcela Rodriguez	UABC
María Martínez Pérez	University of A Coruña
Maria del Pilar Angeles	IIMAS, UNAM
Matias Garcia-Constantino	Ulster University
Matteo Venturelli	Politecnico di Milano
Matthew Burns	Ulster University
Mauro Migliardi	Universita' degli Studi di Padova
Mauro Iacono	Campania University L. Vanvitelli
Mercedes Amor	Universidad de Málaga
Michele Mastroianni	University of Salerno
Muhammad Asif Razzaq	Fatima Jinnah Women University
Nuno Ferrete Ribeiro	University of Minho
Oihane Gómez-Carmona	University of Deusto
Rafael Pastor-Vargas	UNED
Ramón Hervás	University of Castilla-La Mancha
Rene Navarro	University of Sonora
Rosa Arriaga	Georgia Tech
Rubén Domínguez	Anáhuac Mayab University
Saguna Saguna	Luleå University of Technology
Salvatore Source	Università degli Studi di Enna "Kore"
Sandra Nava-Munoz	UASLP
Sara Comai	Politecnico di Milano
Sara Cerqueira	University of Minho
Sergio Ochoa	University of Chile
Solomon Sunday Oyelere	Luleå University of Technology
Sruti Subramanian	Norwegian University of Science and Technology
Tania Mondéjar	University of Castilla-La Mancha
Tatsuo Nakajima	Waseda University
Unai Sainz Lugarezaresti	University of Deusto
Vaidy Sunderam	Emory University
Vincenzo Conti	University of Enna KORE
Vladimir Villarreal	Universidad Tecnológica de Panamá
Wing W. Y. Ng	South China University of Technology
Yarisol Castillo	Universidad Tecnológica de Panamá
Zaheer Khan	University of the West of England, Bristol

Contents

Data Science

About the Editors

Dr. José Bravo is Full Professor in Computer Science in the Department of Technologies and Information Systems at Castilla-La Mancha University, Spain, and Head of the Modelling Ambient Intelligence Research Group (MAmI, mami- lab.eu). He is involved in several research areas such as ubiquitous computing, ambient intelligence, ambient assisted living, context-awareness, Internet of things, mobile computing and m-Health. He is an author of over 37 JCR articles and the main researcher on several projects. H-Index (Scopus). - 22, H-Index (Google Scholar). – 30. Dr. Bravo supervised 7 PhD and over 45 Computer Science undergraduate theses. Since 2003, José Bravo has been the organizer of the International Conference on Ubiquitous Computing & Ambient Intelligence (UCAmI).

Dr. Gabriel Urzáiz received his BS in Computer Engineering at the National Autonomous University of Mexico and his PhD in Advanced Computer Technologies at the Castilla-La Mancha University. His research activity is focused on computer networks, mainly for the integration of heterogeneous networks and their application in ambient intelligence and ubiquitous computing. He is the author of several conference and journal papers, and he has also participated as a reviewer and guest editor. He has been the director of the Computer Science School of the Anahuac Mayab University in Mexico, a research professor and a postgraduate academic coordinator. His current position is as a full-time professor in the Engineering and Exact Sciences Division, primarily focused on teaching and student mentoring.

Smart Environment

Reinforcement Learning Model in Automated Greenhouse Control

F. Javier Ferrández-Pastor[1]([✉]), José M. Cámara-Zapata[2],
Sara Alcañiz-Lucas[1], Sofía Pardo[2], and Jose A. Brenes[3]

[1] I2RC, Universty of Alicante, 03690 San Vicente, Spain
fjferran@ua.es
[2] CIAGRO, Miguel Hernandez University, 03312 Orihuela, Spain
jm.camara@umh.es
[3] CITIC-PPCI-ECCI, University of Costa Rica, 11501 San José, Costa Rica
joseantonio.brenes@ucr.ac.cr

Abstract. Automated systems, controlled with programmed reactive rules and set-point values for feedback regulation, require supervision and adjustment by experienced technicians. These technicians must be familiar with the scenario where the controlled processes are carried out. In automated greenhouses, achieving optimal environmental values requires the expertise of a specialist technician. This introduces the need for an expert in the installation and the problem of depending on them.

To reduce these inconveniences, the integration of three paradigms is proposed: user-centered design, deployment of data capture technology based on IoT protocols, and a reinforcement learning model. The objective of the reinforcement learning model is to make decisions in the programming of set-points for the climate control of a greenhouse. In this way, the need for manual and repetitive supervision of the specialized technician is reduced; meanwhile, the control is optimized.

The design, led by an expert technician in greenhouse installations, provides the necessary knowledge to transfer to a reinforcement learning model. On the other hand, deploying the required set of sensors and access to external data sources increases the capacity of the learning model to be deployed to current installations. The proposed system was tested in automated greenhouse facilities under the supervision of a specialized technician, validating the usefulness of the proposed system.

Keywords: Reinforcement Learning · Q-Learning · Smart Greenhouse

1 Introduction

Automated industrial systems require the programming of a set of parameters so that the different actuators perform the appropriate tasks. This programming is based on choosing the appropriate set-points for the different subsystems to act (climate, lighting, irrigation control, etc.). An automated greenhouse is one of these systems requiring correct programming of these parameters and control rules, which are normally reactive. In recent years, several works have developed

J. Bravo and G. Urzáiz (Eds.): UCAmI 2023, LNNS 842, pp. 3–13, 2023.
https://doi.org/10.1007/978-3-031-48642-5_1

models based on set-point selection strategies using different artificial intelligence or heuristic paradigms with the deployment of predictive solutions. This work is based on the following premises:

- Automated greenhouses require technicians who know how to handle the control parameters, properly selecting the set-points and control rules.
- The technicians modify the set-points to comply with the control requirements.
- The control technicians have their own experience in adjusting the set-points in each installation.

With this background, a case of a greenhouse that complies with these premises has been analyzed. The control technician supervises the evolution of the variables that intervene in the correct operation of the installation and makes corrections that can be made daily.

The technician has acquired experience and knowledge through an apprenticeship in which the adjustment of the set-points depends on each installation. This type of learning is aligned with the reinforcement learning (RL) paradigm [1,2]. In reinforcement learning, an agent (algorithm) learn behavior through trial-and-error interactions in an environment. In general, the system is not told what action to take, rather, it must figure out which actions give the most benefit. The agent is connected to an environment through perception and action. Installations based on the IoT paradigm allow this interaction in a simple way. On the other hand, introducing the user-centered design method facilitates the development of the RL algorithms.

Therefore, the intelligent agent in an RL system performs actions and makes decisions in the same way as the control technician in the installation. If the algorithms of an RL-based development can execute tasks performed manually by the control technician, then we can optimize and increase the degree of automation of your control. In an automated greenhouse, these conditions concur, so the design and use of a system that increases the degree of automation through RL improves the operation of this type of facility.

Given that different subsystems can be controlled in a greenhouse, we developed the first prototype in environmental control, the source of 80% of energy consumption.

In this work, the objective is to develop an intelligent agent that decides to adjust the set-point values for optimal climate control regulation in a greenhouse. Currently, this task is carried out by the agronomic technician who knows the installation's behavior. The technician ensures the greenhouse temperature is between a maximum and a minimum threshold. With the integration of an intelligent RL agent, both temperature regulation and energy consumption optimization can be set as a combined objective. This RL agent will be able to select the set-point at any time in a predictive way, freeing the technician from this task (see Fig. 1).

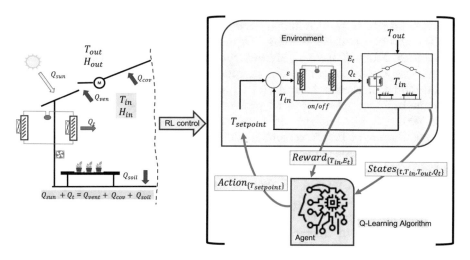

Fig. 1. Use Case. Reinforcement learning application in the greenhouse environmental control

2   Related Work

The area of building energy management systems and temperature control has gained a significant amount of interest in recent years. The advanced control strategies provide great potential to reduce building energy costs and improve grid energy efficiency and stability [8]. In the same way that buildings, greenhouses deploy similar facilities and require the same type of solutions. The number of consumption meters has increased considerably, making it a necessary item in any installation. With the growing number of smart meters, building operation data have recently been more accessible through the building automation systems (BAS) [9]. Empowered by large amounts of data, algorithms, and significant computing power, lots of studies looked at the various data-driven approaches employed in building energy management systems. A recent approach combines various machine learning (ML) [3,4] tools with model predictive control to design data-driven strategies that preserve reliability while reducing the time and computational complexities during online implementation. Another trend is to use data-driven methods to enhance control performance, combined with reinforcement learning (RL), which has a good decision-making capability as a branch of ML.

RL refers to a computational approach to learning whereby an agent tries to maximize the total amount of reward it receives while interacting with a complex and uncertain environment. Additionally. RL has been successfully applied in many areas and can also be used for advanced control strategies. Examples of RL applications are: adjusting weather and thermal loads, operating chillers at optimal conditions, and controlling blinding devices [10–12].

RL has powerful advantages for building energy management solutions. Some of these advantages are their possibility to naturally adapt to their environment, the capability to introduce human feedback into their logic, and their generalizability through transfer learning. Despite these achievements and promising capabilities, RL still faces significant challenges for its practical implementation [13], especially for continuous complex systems, as in the case of buildings. In [14] model predictive control (MPC) and RL are compared and merged for the same building optimal control problem. MPC effectively uses the controller model while pure RL violates the constraints.

Next, the use of Reinforcement Learning for climate control of a greenhouse is detailed in [15], this paper will 1) propose MPC and RL approaches for greenhouse climate control in an unified framework; 2) analyze connections and differences between MPC and RL from a mathematical perspective; 3) compare performance of MPC and RL in a simulation study and afterwards present and interpret comparative results into insights for the application of the different control approaches in different scenarios.

3   Reinforcement Learning Using Predictive Control in the Greenhouse

The Reinforcement Learning problem involves an agent exploring an unknown environment to achieve a goal. RL is based on the hypothesis that all goals can be described by maximizing the expected cumulative reward. The agent must learn to sense and perturb the state of the environment using its actions to derive maximal reward. In the use case presented in this work, the agent performs control actions on the set-points to keep the greenhouse's inside temperature between the maximum and minimum established limits.

Model Predictive Control (MPC) uses a control-oriented system model to generate output predictions for predefined N time steps in the future, known as the prediction horizon (PH), using the available input-output information at the current time step t. The controller optimizes the manipulated input variables for an interval K, known as the control horizon (CH), keeping the control inputs. The goal is to minimize a problem-specific objective function based on the open-loop predictions of controlled outputs.

In the case of use of this work, the predictive model is based on the use of the prediction of the outside temperature. This allows calculations in the RL algorithm to predict the best control action to perform. In reinforcement learning [5], the agent learns to maximize accumulated rewards from an environment that they can interact with by observing and taking actions. Usually, these environments satisfy a Markov property and are treated as Markov Decision Processes (MDPs)[6].

The RL control system has the following elements:

- Create the environment. This work uses OpenAI Gym [7] library, a standardized and open framework that provides many different environments to train agents against through a simple API.
  The states are: $\chi = \{T_{in}, T_{out}, T_{setpoint}, Air_{on\_off}\}$

The actions are: $\mathcal{A} = \{(T_{max} - 1), (T_{max} - 2), (T_{max} - 3), (T_{max} - 4)\}$

The reward function is: $\mathcal{R} = \omega \cdot (\mathcal{R}_1) + \theta \cdot (\mathcal{R}_2)$

$\mathcal{R}_1$ is the reward related with the *on_off* air conditioned connections.

$\mathcal{R}_2$ is the reward related with de T_{in} control

$$\mathcal{R}_1 \begin{Bmatrix} -1 \ if \ air: \ ON \\ 0 \ if \ air: \ OFF \end{Bmatrix} \quad \mathcal{R}_2 \begin{Bmatrix} -1 \ if \ T_{in} \geq T_{max} \\ 0 \ if \ T_{in} < T_{max} \end{Bmatrix}$$

– Create the $Q(s, a)$ table (see Fig. 2). Q-learning is a model-free, value-based, off-policy algorithm that will find the best series of actions based on the current state. The Q values represent how valuable the action is in maximizing future rewards.

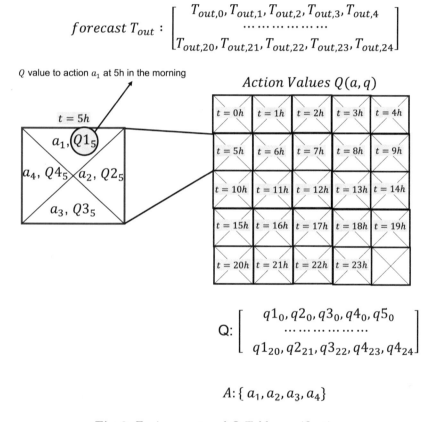

$$forecast \ T_{out} : \begin{bmatrix} T_{out,0}, T_{out,1}, T_{out,2}, T_{out,3}, T_{out,4} \\ \cdots \cdots \cdots \cdots \\ T_{out,20}, T_{out,21}, T_{out,22}, T_{out,23}, T_{out,24} \end{bmatrix}$$

$$Q : \begin{bmatrix} q1_0, q2_0, q3_0, q4_0, q5_0 \\ \cdots \cdots \cdots \cdots \\ q1_{20}, q2_{21}, q3_{22}, q4_{23}, q4_{24} \end{bmatrix}$$

$$A: \{a_1, a_2, a_3, a_4\}$$

Fig. 2. Environment and Q-Table specification

– Create the target policy $\pi(s)$. Target Policy is the policy that Agent uses to learn from the rewards received for its actions which are used to update the Q-value. Actions chosen by the target policy should be selected occasionally by the behavior policy if the target policy decides that the action should be executed.

- Create the exploratory policy $b(s)$. The agent explores the environment and learns from the outcomes of the actions directly, without constructing an internal model.
- Implement the algorithm. The goal is to learn the optimal Q-value function iteratively using the Bellman Optimality Equation. To do so, we store all the Q-values in a table that we will update at each time step using the Q-Learning iteration. The algorithm searches for Q values that optimize the control objectives. The steps are detailed in Algorithm 1.

Algorithm 1 Q-learning: Learn function $Q : \mathcal{X} \times \mathcal{A} \to \mathbb{R}$

Require:
  States $\mathcal{X} = \{1, \ldots, n_x\}$
  Actions $\mathcal{A} = \{1, \ldots, n_a\}$, $A : \mathcal{X} \Rightarrow \mathcal{A}$
  Reward function $R : \mathcal{X} \times \mathcal{A} \to \mathbb{R}$
  Learning rate $\alpha \in [0, 1]$, $\alpha = 1$
  Discounting factor $\gamma \in [0, 1]$
  procedure QLEARNING($\mathcal{X}$, A, R, T, α, γ)
      Initialize $Q : \mathcal{X} \times \mathcal{A} \to \mathbb{R}$ arbitrarily
      while Q is not converged **do**
         Start in state $s \in \mathcal{X}$
         while s is not terminal **do**
            Calculate π according to Q and exploration strategy
            $a \leftarrow \pi(s)$
            $r \leftarrow R(s, a)$ $\triangleright$ Receive the reward
            $s' \leftarrow T(s, a)$ $\triangleright$ Receive the new state
            $Q(s', a) \leftarrow Q(s, a) + \alpha \cdot (r + \gamma \cdot \max_{a'} Q(s', a'))$
            $s \leftarrow s'$
      return Q

In the following section, the use of the Internet of Things and user-centered design in the creation of the RL model is detailed.

4 Internet of Things and User-Centered Design for the RL Model

The Internet of Things (IoT) has productive ways to automate with the use of low-cost hardware (sensors/actuators) and communication (Internet) technologies. IoT protocols provide resources to capture and communique all data in control systems. Objects/things (sensors/actuators) can be connected and processed using nodes on wireless sensor networks with IoT protocols. In this work, for the RL agent to interact with the system, IoT protocols allow both capturing real data for pre-training and for application in production conditions.

To capture the expert's knowledge and apply it to the agent of the RL model, techniques based on user-oriented design are also used. User-centered or human-centered design principles are recognized as an important part of ML development across a range of sectors [16]. Figure 3 shows the relationship between the different actors in the proposed model: IoT, RL, and User-centered design. It is a layered architecture that starts from the capture and management of data captured from the IoT. The data feeds the RL agent layer, which is the main actor in decision-making for control actions. In order to interact with the platform, an interface layer is developed for the application and maintenance that users perform. It is a standard architecture that adapts to each type of installation and service required.

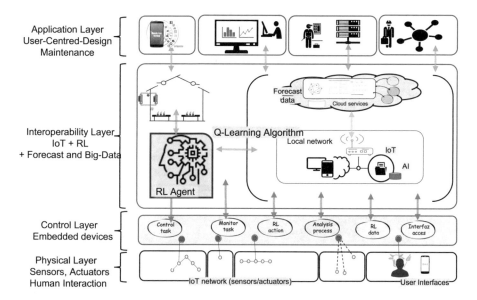

Fig. 3. Layered technological architecture. Relationship between IoT, user-centered-design and RL

For the RL agent to be able to reproduce the control technician's rules and optimize the selection of set-points, a wireless sensor network was deployed to capture the state of the system and transmit control orders efficiently. In the following section, we detail the use case in which we apply the proposed RL model for temperature control in an automated greenhouse.

5 Application in an Automated Greenhouse

In this work, we conduct a use case in which the greenhouse's inside temperature T_{in} (see Fig. 1) is controlled between maximum and minimum limits, considering the energy consumption produced in the control itself. The CH was 24 h (a day), and the time interval was (1 h).

To design and test the RL algorithm, initial training is used by simulating the temperature variation inside the greenhouse, considering the second law of thermodynamics. The use case was carried out in a greenhouse of the Science

Park at the University of Alicante. In this greenhouse, the control technician programs the set-points every 2 h; he must consider the external conditions to establish the values; therefore, it is a manual process.

The objective of the RL algorithm is to automate the task of the manual configuration. This way, the algorithm will decide the most suitable set-point (RL) in a predictive way (MPC).

The RL algorithm interacts with this simulated model by providing actions to change the set-point values. Then, it observes the thermal behavior of the building using minute intervals. Further, the information about the air-conditioned connections (on/off) is provided to the RL algorithm. The RL algorithm learns the use of energy to minimize consumption and keeps the inside temperature between the defined maximum and minimum thresholds.

During the training phase, the RL agent interacts with the greenhouse environment and takes sequences of actions (A_i) and rewards (R_i). The environment model used in this work is a cool model with an on/off switch capability.

To start the algorithm, the greenhouse's outside temperature is recorded ($forecast T_{out}$). The four actions (($T_{max} - 1$), ($T_{max} - 2$), ($T_{max} - 3$), ($T_{max} - 4$)) of the state space and the number of episodes that the algorithm must consider are specified. Every hour, the algorithm determines the Q values for every action considering the obtained rewards.

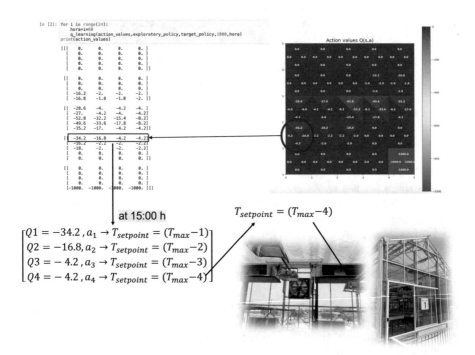

Fig. 4. The result is a table indicating, every hour, the rewards for each action. The graph highlights 15:00 as an example. At this time the Q3 and Q4 actions are proposed by RL agent

Figure 4 shows the results obtained after applying the RL model in the greenhouse. In the figure, the Q-values per hour are detailed. The table indicates 24 h a day. The rewards obtained by applying the four proposed control actions are shown in each box. One of the hours where the $(T_{max} - 4)$ action is optimal is highlighted.

In the next section, we present the conclusions obtained from this research.

6   Conclusions

In this paper, we present a novel approach to greenhouse control by leveraging a single-agent RL centralized controller. The strategy adopted by the controller is explicitly designed to outperform a commonly used rule-based control strategy in real greenhouse scenarios. The RL controller implements Q-Learning, a continuous control algorithm, which learns from experience through an off-policy strategy involving an exploratory policy and a target policy. The control performance analysis demonstrates the advantages of coordinated energy management and temperature control achieved by the RL agent.

The results of this study show that the RL-based approach can effectively automate and optimize set-point selection in greenhouse control systems. By employing Q-Learning, the controller can adapt and improve its decision-making process over time based on learned experiences provided by a greenhouse technician. This adaptability leads to better control outcomes, outperforming traditional rule-based methods in managing greenhouse environments.

Overall, the findings of this research highlight the potential of single-agent RL with Q-Learning in advancing greenhouse control systems. Combining the benefits of coordinated energy management and temperature control, we demonstrate the RL capability to optimize resource utilization and enhance overall greenhouse productivity. As the agricultural industry seeks more efficient and sustainable solutions, the insights from this study contribute valuable knowledge towards the development of intelligent and autonomous greenhouse control systems.

This work is associated with in-progress doctoral research on developing a technological architecture to guide smart farming application development and deployment. The RL model forms part of an integral smart farming platform that monitors and controls greenhouse conditions in a sustainable way, considering multiple subsystems and controllers. RL algorithms often require a learning period during which the system may not perform optimally. To reduce this limitation, it is proposed to capture a large number of data on the real behavior of the system without the use of RL. Based on these data, the contribution of the RL algorithm applied as pre-training is analyzed. Future work may focus on scaling up the approach for multi-agent systems and further exploring the RL algorithm's performance in various agricultural settings.

Acknowledgements. We are very grateful to Palmeera Farms (Palmeera) Biotechnology-based company, member of the Spanish Association of Biotechnology Companies (Asebio) and attached to the Alicante Science Park (PCA) for their collaboration in this work.

Funding. This study is part of the AGROALNEXT program (AGROAL-NEXT/2022/048) and has been supported by MCIN with funding from the European Union NextGenerationEU (PRTR-C17.I1) and the Generalitat Valenciana. This study was partially supported by the Research Center for Communication and Information Technologies (CITIC) and the School of Computer Science and Informatics (ECCI) at the University of Costa Rica, Research Project No. 834-B9-189.

References

1. Botvinick, M., Ritter, S., Wang, J.X., Kurth-Nelson, Z., Blundell, C., Hassabis, D.: Reinforcement learning, fast and slow. Trends Cogn. Sci. **23**(5), 408–422 (2019). https://doi.org/10.1016/j.tics.2019.02.006. ISSN 1364–6613
2. Kaelbling, L.P., Littman, M.L., Moore, A.W.: Reinforcement learning: a survey. J. Artif. Intell. Res. **4**, 237–285 (1996)
3. Yang, S., Wan, M.P., Chen, W., Ng, B.F., Dubey, S.: Model predictive control with adaptive machine-learning-based model for building energy efficiency and comfort optimization. Appl. Energy **271**, 115147 (2020). http://www.sciencedirect.com/science/article/pii/S0306261920306590
4. Drgona, J., Picard, D., Kvasnica, M., Helsen, L.: Approximate model predictive building control via machine learning, Appl. Energy **218**, 199–216 (2018). http://www.sciencedirect.com/science/article/pii/S0306261918302903
5. Sutton, R.S., Barto, A.G.: Reinforcement Learning: An Introduction. MIT press, Cambridge (1999)
6. Puterman, M.L.: Markov Decision Processes: Discrete Stochastic Dynamic Programming. Wiley, Hoboken (2005)
7. Brockman, G., et al.: Openai gym. ArXiv Preprint. ArXiv:1606.01540 (2016)
8. Missaoui, R., Joumaa, H., Ploix, S., Bacha, S.: Managing energy smart homes according to energy prices: analysis of a building energy management system. Energy Build. **71**, 155–167 (2014). https://doi.org/10.1016/j.enbuild.2013.12.018
9. Maddalena, E.T., Lian, Y., Jones, C.N.: Data-driven methods for building control—a review and promising future directions. Control Eng. Pract. **95**, 104211 (2020). https://doi.org/10.1016/j.conengprac.2019.104211
10. Chen, B., Cai, Z., Bergés, M.: Gnu-RL: a precocial reinforcement learning solution for building HVAC control using a differentiable MPC policy. In: Proceedings of 6th ACM International Conference on System Energy-Efficient Buildings, Cities, Transports, pp. 316–325 (2019)
11. Wen, Z., O'Neill, D., Maei, H.: Optimal demand response using device-based reinforcement learning. IEEE Trans. Smart Grid **6**(5), 2312–2324 (2015)
12. Wei, T., Wang, Y., Zhu, Q.: Deep reinforcement learning for building HVAC control. In: Proceedings of 54th Annual Design Automation Conference, pp. 1–6 (2017)
13. Mankowitz, D., Hester, T.: Challenges of real-world reinforcement learning. ArXiv arxiv:1904.12901 (2019)

14. Arroyo, J., Manna, C., Spiessens, F., Helsen, L.: Reinforced model predictive control (RL-MPC) for building energy management. Appl. Energy **309**, 1 (2022). https://doi.org/10.1016/j.apenergy.2021.118346
15. Morcego, B., Yin, W., Boersma, S., van Henten, E., Puig, V., Sun, C.: Reinforcement learning versus model predictive control on greenhouse climate control. arXiv preprint arXiv:2303.06110 (2023)
16. Gillies, M., Fiebrink, R., Tanaka, A.: Human-Centred machine learning. In: Proceedings of the 2016 CHI Conference Extended Abstracts on Human Factors in Computing Systems. Association for Computing Machinery, New York (2016)

Performance Evaluation of Face Detection Algorithms for an Emotion Recognition Application in a School in the Department of Nariño - Colombia

Andrés Díaz-Toro[1]([✉]), Álvaro Cervelión-Bastidas[1], Sixto Campaña-Bastidas[1],
Abel Méndez-Porras[2], Jorge Alfaro-Velasco[2], Efrén Jiménez-Delgado[2],
and Luis Calvo-Valverde[2]

[1] Universidad Nacional Abierta y a Distancia UNAD, Calle 14 # 28-45, Pasto, Colombia
`{andres.diaz,alvaro.cervelon,sixto.campana}@unad.edu.co`
[2] Instituto Tecnológico de Costa Rica, 17, 69121 Cartago, Costa Rica
`{amendez,joalfaro,efjimenez,lcalvo}@itcr.ac.cr`

Abstract. Emotion recognition in digital images, based on the facial expressions of people, can add value in different areas such as education, shopping centers, hotels, entertainment centers, restaurants, among others, since it allows a better understanding of the requirements of the people, improve services, and predict sales trends. In a classroom, this technology allows to identify in real time the reaction of students to the development of the class, and in this way, the teacher can make the necessary adjustments to improve the learning process. The first step for this application is to detect faces of multiple students present in the scene, with efficient algorithms that process good-quality images. In this paper, the performance of six face-detection algorithms is determined using images taken in a classroom, in the town of Túquerres, in the department of Nariño, Colombia. The results show that a good camera resolution of 5 megapixels or higher, and good lighting conditions are determinant for successful face detection in classrooms of approximately 46 m^2. In addition, the best performance was obtained with RetinaFace algorithm, which is more robust to different facial postures, achieving an accuracy of 96.5% with poor lighting conditions and 97.84% with good lighting conditions.

Keywords: Face Detection · Emotion Recognition · Group of Students · Classroom · Feedback · Performance Evaluation

1 Introduction

Emotions are critical in driving the teaching-learning process, influencing student interaction with educational content, decision-making, and information appropriation [1]. In this sense, technology allows the transition towards validations with greater objectivity based on emotions and perception in real-time. This translates into alternatives for educators to better understand the emotional needs of their students and adapt their teaching approaches accordingly.

J. Bravo and G. Urzáiz (Eds.): UCAmI 2023, LNNS 842, pp. 14–20, 2023.
https://doi.org/10.1007/978-3-031-48642-5_2

The first step for applications that involve emotion recognition of a group of people, like the proposed application in a classroom, is to detect faces of multiple students present in the scene. In this sense, this paper analyzes the performance of six face detection algorithms, considering the precision, recall and accuracy, and using our own dataset of images, taken in a classroom of a school in Túquerres, Nariño, Colombia.

The main contributions carried out in this paper are the following.

- A dataset of 256 images taken in a classroom, where a group of students have a class and interact with the teacher.
- A methodology to evaluate the performance of face detection algorithms applied to a group of students.
- The results of the performance evaluation of six face detection algorithms applied to a group of students.

This paper is organized as follows. In Sect. 2 some outstanding works in emotion recognition applied to education are presented. In Sect. 3, the configuration of the working space is described. In Sect. 4, the performance evaluation of six face-detection algorithms is presented. In this section the algorithm with the best performance for the described working space and under good and poor lighting conditions, is identified. Finally, the conclusions and references are presented.

2   Related Work

Technological advances such as facial expression analysis and sentiment analysis can provide educators with valuable information about students' emotional states. Regarding the monitoring and determination of student attention during the learning process, the approaches are varied. In [1] researchers propose an automated system that allows to capture and summarize student behaviors in the classroom as part of data acquisition for the decision-making process. They evaluated it in three main phases: the student's identification, their position, and the gaze. The students' gaze can provide educators with feedback on the teaching-learning process. For example, if a student is constantly looking at their classmates while the educator is giving a presentation or explanation, this could indicate a lack of attention on the part of the student or a possible lack of clarity in the instructor's explanation.

By analyzing student emotions, educators can adapt their teaching methods to foster positive emotions and optimize learning. The study by [2] proposes the creation of a system to recognize the student's emotions from their faces that consists of three phases: the detection of the face by Haar Cascades, the normalization and the recognition of the emotion through a CNN on FER 2013 database with seven types of expressions: surprise, fear, disgust, sad, happy, angry, and neutral. In [3], a Deep Learning model based on VGG (using the FER-2013, with one less convolutional block) and HOG features + SVMs is used, to improve engagement recognition from images. In [4], an approach is presented to automate these observation cues of students' facial expressions and identify their attention through a neural network model.

3 Configuration of the Working Space and Capture of Images

The images were captured in a classroom of the main headquarters of the Teresiano school, located in the municipality of Túquerres, which is 68 km from Pasto, in the department of Nariño. The room is 7.8 m long by 5.9 m wide, which corresponds to an area of 46.02 m^2. On this space are distributed 15 tables and 30 chairs. It has two windows that extend the entire length of one of its sides.

The photographs were taken every 4.5 s for 10 min, when the students were attending the class taught by a teacher. This process was carried out twice, once under good lighting conditions and the other under poor lighting conditions. The lighting conditions were quantified with a lux meter, obtaining 50 lx when the curtains were closed and 420 lx when the curtains were open. Finally, a dataset of 256 images was obtained, with 128 images for each lighting condition.

The image capture was programmed with Python, version 3.8.13, making use of the computer vision library OpenCV, version 4.6.0.66. The development environment was Spyder version 5.3.3. The camera is from the SV3C brand, with reference B06W-5MP-HX. It is a 5 Megapixel resolution IP camera, which uses RTSP protocol.

4 Performance Evaluation of Face-Detection Algorithms

Six face-detection algorithms were used, which were executed offline for the images of the group of students in the classroom, with good and poor lighting conditions. The objective is to compare and determine which presents the best performance. A description of each algorithm is presented below, followed by the performance comparison.

Haar Cascade. This object detection algorithm was designed by Paul Viola and Michel Jones in 2001 [5]. Although it is not based on deep learning, this algorithm is widely employed today. The model is trained using positive images (with faces) and negative images (without faces). The Haar features extracted from the images (edge or line detectors) are obtained by convolution. Adaboost is used to select the most important features and the technique called attentional cascade for a more efficient search.

Dlib is an open-source toolkit written in C++ by Davis King in 2009, which offers many machine learning features [6]. It can be used for object pose estimation, face detection and face recognition. For detecting faces, it is based on histogram of oriented gradients (HOGs) and linear SVM. It has low resource consumption but has limited capabilities with non-frontal faces, faces with occlusions, or faces that appear in the image with a size smaller than 80×80 pixels.

YOLO (You Only Look Once) is an object detector published in 2016 by Joseph Redmon [7]. It is a regression-based algorithm that with a neural network efficiently predicts bounding boxes and the probability of classes, achieving real-time performance. It has 24 convolutional layers and 2 fully connected layers. The net was pretrained on the ImageNet 1000-class competition dataset and trained for about 135 epochs on the dataset from PASCAL VOC 2007 and 2012, with a batch size of 64, a momentum of 0.9 and a decay of 0.0005.

MultiTask Cascaded Convolutional Neural Network (MTCNN) [8] was proposed by Kaipeng Zhang et al. in 2016. It implements a cascading structure with three states of deep convolutional networks. In the first state it uses a shallow CNN to generate candidate windows. In the second and third states, it uses complex CNNs to refine the results and determine the location of facial features. When it was published, it achieved precision above the state of the art using the FDDB and WIDER FACE datasets.

MediaPipe [9] is an open-source framework from Google, introduced in 2019, that allows the implementation of different applications such as face detection, hand tracking, hair segmentation, object detection and tracking. Regarding the first application, it allows multiple faces to be detected in real time and generates coordinates of 6 facial landmarks (center of the right and left eye, tip of the nose, center of the mouth and tragion of the right and left ear).

Retina Face. This deep learning-based algorithm was developed in 2019 by Jiankang Deng et al. [10]. The tasks it executes are face detection, 3D face reconstruction and 2D face alignment. This algorithm incorporated some ideas from MTCNN, achieving to beat the state of the art. Its average precision (AP) is 91.4% with the WIDER FACE dataset. It includes concepts such as feature pyramid, single stage, context modelling and cascade multi-task loss. SGD optimizer is used for training with a momentum of 0.9, a weight decay of 0.0005, and a batch size of 8x4.

Table 1 and 2 present the results, under good and poor lighting conditions, respectively. There are 18.5 faces on average in each image of the dataset with good lighting conditions and 20 in each image of the dataset with poor lighting conditions.

Table 1. Performance comparison for six face detection algorithms. Good lighting conditions

Algorithm\Indicator	Avg. True positives	Avg. False negatives	Avg. False positives	Precision (%)	Recall (%)	Accuracy (%)
Haar Cascade	8.2	10.3	0.08	99,03	44,32	44,13
Dlib	5	13.5	0.01	99,80	27,03	27,01
YOLO	0.9	17.6	0	100,00	4,86	4,86
MTCNN	11.4	7.1	2.05	84,76	61,62	55,47
MediaPipe	0.8	17.7	0.22	78,43	4,32	4,27
RetinaFace	18.1	0.4	0	**100,00**	**97,84**	**97,84**

Note that the algorithm with best performance is RetinaFace, achieving an accuracy over 96% for both lighting conditions, while the algorithm with worst performance is Mediapipe with an accuracy under 5% for both lighting conditions. Figure 1 shows the results obtained with these algorithms.

Table 2. Performance comparison for six face detection algorithms. Poor lighting conditions

Algorithm\Indicator	Avg. True positives	Avg. False negatives	Avg. False positives	Precision (%)	Recall (%)	Accuracy (%)
Haar Cascade	6.9	13.1	0.13	98,15	34,50	34,28
Dlib	4.7	15.3	0.004	99,91	23,50	23,50
YOLO	0.7	19.3	0	100,00	3,50	3,50
MTCNN	7.7	12.3	1.2	86,52	38,50	36,32
MediaPipe	0.7	19.3	0	100,00	3,50	3,50
RetinaFace	19.3	0.7	0	**100,00**	**96,50**	**96,50**

a)   b)

Fig. 1. Results of face detection with a) RetinaFace (algorithm with the best performance) and b) MediaPipe (algorithm with the worst performance)

RetinaFace does not present false positives while MTCNN is the algorithm with the highest number of false positives (1.2 in poor lighting conditions and 2.05 in good lighting conditions, on average).

All the algorithms present an increase in accuracy when the lighting conditions in the classroom improve. The algorithms that present a greater increase in accuracy due to better illumination are MTCNN (19.15%) and Haar Cascade (9.85%).

MediaPipe, YOLO and Dlib are the algorithms that are most affected by facial postures, since they only present successful detections if the students are looking straight ahead. The most robust algorithm for different face postures is Retinaface and this is reflected in its accuracy greater than 96% in both lighting conditions. Figure 2 shows the result when the students are looking to one side, using Retinaface and Dlib.

a)   b)

Fig. 2. Results of face detection when the students are looking to one side of the classroom using a) RetinaFace and b) Dlib.

5 Conclusions

Retinaface is the face detection algorithm with the best performance in the tests carried out at the school in the town of Túquerres, Nariño. An accuracy and a recall of 96.5% and a precision of 100% were obtained for poor lighting conditions while for good lighting conditions the accuracy and recall were 97.84% and the precision of 100%. This indicates that the algorithm is very robust to different lighting conditions and different facial postures. In addition, the tests did not present results with false positives. All these features make it the ideal face detection algorithm for emotion recognition applications in classrooms with students who are receiving classes.

MTCNN is the second-best performing algorithm, after RetinaFace, but it is the one with the most false positives. This makes its accuracy is only 36.32% for poor lighting conditions and 55.47% for good lighting conditions. MediaPipe, and YOLO are the worst performing algorithms due to their limitations in detecting non-frontal faces. This makes its accuracy less than 5% in both lighting conditions. Haar cascade, despite being one of the oldest, surpasses Dlib, YOLO and MediaPipe, achieving an accuracy of 34.28% for poor lighting conditions and 44.13% for good lighting conditions. With good lighting conditions, better indicators were produced in all the algorithms, but the greatest impact was evident in MTCNN, which achieved an increase in accuracy of 19.15%, followed by Haar Cascade, which achieved an increase of 9.85%.

References

1. Anh, B.N., et al.: A computer-vision based application for student behavior monitoring in classroom. Appl. Sci. **9**(22), 4729 (2019). https://doi.org/10.3390/app9224729
2. Lasri, I., Solh, A., Belkacemi, M.: Facial emotion recognition of students using convolutional neural network. In: 2019 3rd International Conference on Intelligent Computing in Data Sciences, pp. 1–6 (2019). https://doi.org/10.1109/ICDS47004.2019.8942386
3. Mohamad Nezami, O., Dras, M., Hamey, L., Richards, D., Wan, S., Paris, C.: Automatic recognition of student engagement using deep learning and facial expression. In: Brefeld, U., Fromont, E., Hotho, A., Knobbe, A., Maathuis, M., Robardet, C. (eds.) ECML PKDD 2019. LNCS (LNAI), vol. 11908, pp. 273–289. Springer, Cham (2020). https://doi.org/10.1007/978-3-030-46133-1_17

4. Tabassum, T.: Non-intrusive identification of student attentiveness and finding their correlation with detectable facial emotions. In: ACMSE 2020 - ACM Southeast Conference, pp. 127–134 (2020)

5. Viola, P., Jones, M.: Robust real-time face detection. Int. J. Comput. Vision **57**, 137–154 (2004). https://doi.org/10.1023/B:VISI.0000013087.49260.fb

6. King, D.: Dlib-ml: a machine learning toolkit. J. Mach. Learn. Res. **10**, 1755–1758 (2009)

7. Redmon, J., et al.: You only look once: unified, real-time object detection. In: Proceedings of the IEEE Conference on Computer Vision and Pattern Recognition, pp. 779–788 (2016), https://doi.org/10.48550/arXiv.1506.02640

8. Zhang, K., et al.: Joint face detection and alignment using multitask cascaded convolutional networks. IEEE Signal Process. Lett. **23**(10), 1499–1503 (2016). https://doi.org/10.48550/arXiv.1604.02878

9. Bazarevsky, V., et al.: Blazeface: sub-millisecond neural face detection on mobile GPUs (2019). arXiv preprint arXiv:1907.05047. https://doi.org/10.48550/arXiv.1907.05047

10. Deng, J., et al.: Retinaface: single-shot multi-level face localisation in the wild. In: Proceedings of the IEEE/CVF Conference on Computer Vision and Pattern Recognition, pp. 5203–5212 (2020). https://doi.org/10.1109/CVPR42600.2020.00525

An Open Framework for Nonverbal Communication in Human-Robot Interaction

Ernesto A. Lozano[1(✉)], Carlos E. Sánchez-Torres[2], Irvin H. López-Nava[1], and Jesús Favela[1]

[1] Computer Science, CICESE, Carretera Ensenada - Tijuana, 22860 Ensenada, Baja California, Mexico
ernesto@cicese.edu.mx, {hussein,favela}@cicese.mx
[2] Computer Science, UABC, Km. 106 Carretera Tijuana - Ensenada, 22860 Ensenada, Baja California, Mexico
hello@sanchezcarlosjr.com

Abstract. Nonverbal communication plays a vital role in human interaction. In the context of Human-Robot Interaction (HRI), social robots are designed primarily for verbal-based communication with humans, making nonverbal communication an open research area. We present a flexible, open framework designed to facilitate nonverbal interactions in HRI. Among its components is a P2P Browser-Based Computational Notebook, leveraged to code, run, and share reactive programs. Machine-learning models can be included for real-time recognition of gestures, poses, and moods, employing protocols such as MQTT. Another key component is a broker for distributing data among different physical devices like the robot, wearables, and environmental sensors. We demonstrate this framework's utility through two interaction scenarios: (i) the first one employing proxemics and gaze direction to initiate an impromptu encounter, and (ii) a second one incorporating object recognition and a Large-Language Model to suggest meals to be cooked based on available ingredients. These scenarios illustrate how the framework's components can be seamlessly integrated to address new scenarios, where robots need to infer nonverbal cues from users.

Keywords: Human-Robot Interaction · Nonverbal communication · Broker architectural pattern · Langchain · ChatGPT · Computational Notebook

1 Introduction

A Social Robot is an autonomous physical agent that communicates with humans. While verbal communication remains the main mode of interaction between social robots and humans, nonverbal communication has been explored to enhance human-robot interaction [1]. This involves enacting communication acts by the robot, such as gestures with the hands or face and/or the recognition by the robot of human nonverbal interactions such as gaze or body movement.

E. A. Lozano, C. E. Sánchez-Torres, I. H. López-Nava and J. Favela—These authors contributed equally to this work.

J. Bravo and G. Urzáiz (Eds.): UCAmI 2023, LNNS 842, pp. 21–32, 2023.
https://doi.org/10.1007/978-3-031-48642-5_3

The benefits of social robots being equipped with the capability to detect different types of non-verbal behaviors are vast. By using models to analyze what a person is doing at a given time, social robots can enhance their understanding of human intentions, emotional states, and preferences [2]. Thus they can tailor their responses and actions accordingly, creating a more personalized and meaningful interaction.

This paper presents a framework aimed at facilitating the recognition of human nonverbal communication to adapt the response of a social robot. This framework allows the integration of various sensor technologies, such as cameras, microphones, and depth sensors, to capture a wide range of non-verbal behaviors. By incorporating different models, such as computer vision algorithms and natural language processing techniques, social robots can analyze and interpret the collected data, gaining a comprehensive understanding of human actions and intentions.

The flexibility provided by such a framework enables researchers and developers to continuously enhance the capabilities of social robots. As new sensors and models are developed, they can be seamlessly integrated into the framework, expanding the robot's ability to recognize and respond to a broader array of non-verbal behaviors and tasks. This adaptability promotes innovation and paves the way for more sophisticated social robots that can cater to diverse human needs and contexts.

This work is organized as follows. Section 2 discusses the analysis of other existing social robots, showing the differences with the implemented architecture. Section 3 describes the architecture of the framework proposed and its main components. Section 4 presents how the framework is used to implement a variety of usage scenarios that involve nonverbal communication. Finally, Sect. 5 concludes this work and discusses possible future work.

2 Nonverbal Human-Robot Interaction

Nonverbal communication is generally defined as any transfer of messages that does not involve the use of words [3]. While verbal communication tends to dominate human interactions, nonverbal communication, such as gestures and eye gaze, augments and extends spoken communication. Nonverbal communication happens bi-directionally in an interaction, so social robots must be able to both recognize and generate nonverbal behaviors. To be effective in the real world, this nonverbal awareness must occur in real time. In recent years, there has been a notable surge in both interest and scholarly inquiry regarding robotic aides capable of executing tasks centered on social integration, the establishment of affective bonds, fostering companionship, and facilitating cognitive training [4].

Rocha, et al. proposed a 3D simulator [5] for the social robotics platform FRED that can be used for educational applications and health therapies. It has elements of verbal and non-verbal communication and can express emotions through the eyes and mouth. In fact, this work uses a part of our architecture to give it the ability to communicate with the simulator. Another work proposed by him was a based language and a simulator for the robot EVA [6], capable of executing its interactive sessions and assisting in the development of programs for the robot. The implemented EVA is an affective robot focused on social, communication integration and companionship tasks, giving the opportunity to work with many kinds of people, such as children with autism, old people with dementia and so on. It is important to remark that the previous works mentioned do not have this kind of architecture. State-of-art robots and assistants usually incorporate mechanisms to improve human-like natural behavior, but they keep missing the aspects of recognizing nonverbal interactions on humans.

Recent years have seen an increased interest and research in robotic assistants that are able to perform tasks focused on social integration, affective bonds and cognitive training [7]. These aspects are precisely what has motivated the development of an open architecture that grant us the possibility to enhance EVA with the integration of different IoT devices and artificial intelligence algorithms.

2.1 Types of Nonverbal Interactions in HRI

In human-robot interaction, various types of nonverbal interactions can take place. These interactions encompass a wide range of communication modalities and cues that go beyond spoken language [8]. Here are some key types of nonverbal interactions in human-robot interaction:

1. **Facial Expressions**: Analyze a person's facial expressions, such as smiles, frowns, raised eyebrows, or narrowed eyes, to infer emotions like happiness, sadness, surprise, or anger.
2. **Gestures**: Such as hand movements or arm motions, to communicate with humans. Gestures can include pointing to objects, waving, or making other meaningful movements to indicate actions, directions, or engage in collaborative tasks.
3. **Body Language and Posture**: By observing a person's posture and overall body movements, a robot can interpret cues like open or closed body position, leaning in or away, hand movements, or crossed arms, providing insights into the person's level of comfort, engagement, or defensiveness.
4. **Proxemics**: Understand the distance between herself and the person using sensors such as the camera, allowing her to assess the person's comfort level with proximity and adjust her behavior accordingly.
5. **Eye Contact**: Determine whether a person is making direct eye contact or avoiding it. This can indicate interest, attention, distraction, or discomfort.

These types of nonverbal interactions play a crucial role in making human-robot interactions more natural, intuitive, and socially engaging.

2.2 Design Scenarios

We describe a couple of scenarios of nonverbal communication in HRI developed to identify the design requirements for our nonverbal communication framework.

2.2.1 Scenario 1: Gaze and Proxemics to Initiate an Interaction

Eva is a social robot used to assist Roberto, a patient who lives alone at home and exhibits signs of dementia. He often feels disoriented and lonely. When he feels like talking to someone, he approaches Eva to get her attention and initiate an interaction. As he gets closer to her, the robot is able to analyze and calculate how far away is Roberto from her by means of an integrated camera. While this happens, if the distance between them is more than three meters, Eva will begin to move her neck to track him. Nevertheless, as Roberto becomes closer and keeps looking at the robot, it will change its behavior to one that is more interactive and affective. When Eva notices that Roberto keeps looking at her for more than two seconds, she will initiate an interaction: **"Hi Roberto, how have you been in this day?"**, the conversation will continue until Roberto decides to quit.

2.2.2 Scenario 2: Suggesting a Cooking Recipe

Luis, a young man fond of Mexican cuisine, finds himself in need of knowing more about dishes typical from his country. However, he has little experience preparing these dishes, and thus asks the social robot Eva about recipes that he can cook with the ingredients he has at hand. He points at a table where the ingredients are placed and asks: **"Eva, what can I prepare with this?"**. Eva, with the help of her camera, recognizes the ingredients, and recommends some recipes for him to prepare. Luis iterates with Eva to decide on a recipe to his liking and the robot will guide him with step-by-step instructions.

3 Architecture of the Nonverbal HRI Framework

We describe the nonverbal Human-Robot Interaction architecture proposed, abstracting implementations details to focus on the features it provides from the perspective of a discourse of **domain**. In this regard, the competence of our architecture refers to the domain (i.e. nonverbal scenarios, behaviors, and tasks) over which the system may operate. Meanwhile, the performance is contingent upon implementation specifics, such as technology choices, the algorithms used, and the adopted practices.

Our system compromises microservices that collaborate through streams. It is inspired by [9] but with an emphasis on distributed and data-driven strategies.

Ubiquitous computing involves distributed algorithms, which run on multiple inter-connected nodes without rigid centralized control, and online algorithms (as well as stream and dynamic algorithms), which receive their input incrementally over time rather than having the entire input available from the start. This makes them akin to non-monotonic logic. Additionally, reactive programming introduces operational semantics that react to external events and transparently manage time-related data flows, such as handling asynchronous signals from the environment.

The proposed software architecture for human-robot interaction is event-driven and employs a mix of publish/subscribe models through various communication protocols and layers. In this context, an **event** is defined as an immutable object, denoted as the evaluation of the signal, $s(t_0)$, or the change in the state of the system, originating from a subsystem. Over discrete time increments, this object yields a **stream**, the signal $s(t)$. A producer generates an event that multiple consumers can potentially process.

The messaging system can operate in one of three ways: (i) either the producers send direct messages without intermediary nodes, (ii) the producer sends a message to the consumer via a message broker, and (iii) through event logs. For the purpose of this work, we denote both producers and consumers as microservices. Contrary to monolithic software, microservices are independently deployable and loosely coupled. What can be done with the stream once it is in the hands of the consumer? and, how the consumer can process it and when we achieve a specific competence? Broadly, those are classified as data synchronizers, tasks, and pipeline stages. Data synchronizers are tasked with the responsibility of storing or presenting data on a storage system. Tasks push events that affect the environment. Pipeline stages receive events from multiple input streams to produce one or more output streams that go through different stages until they reach a task or data synchronizers applying pipe-and-filter operations. Therefore, a service is a graph of dependencies within microservices joined by different protocols, which is designed to achieve a specific competence.

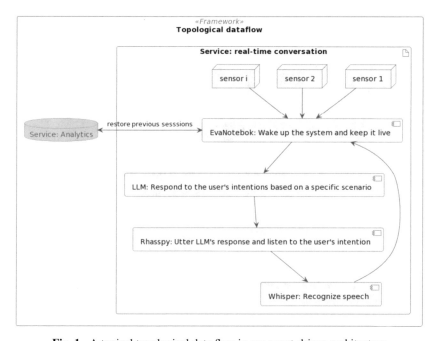

Fig. 1. A typical topological data flow in our event-driven architecture

Figure 1 illustrates the data flow within an example of a service, where nodes symbolize microservices and edges represent communication protocols. A real-time conversation, which is a specific competence, with a social robot involves understanding nonverbal communication to either wake up or put the system to sleep. Instead of hots words, the rest of the system remains largely inoperative until we have sufficient evidence to make the decision. Additionally, a conversation involves the user managing sessions, where the social robot can recall the voice, face, and other user characteristics in order to interact in the best way tailored for that user. When EvaNotebook, a component designed to build microservices on the fly, wakes up the system based on available information, it customizes the Large Language Model (LLM) to respond based on a specific scenario by injecting a system prompt and restoring previous sessions. Rhasspy then vocalizes the LLM's response and listens for the user's intention. Whisper transforms speech to text and sends it back to EvaNotebook. Therefore, we form a feedback loop where different consumers emit events that we need to inject into the LLM and react accordingly. In the background, the system records data for posterior analysis. The remainder of the discussion focuses on the logical components needed to build services such as Eva, wearable and environmental sensors, MQTT Broker, Rhasspy, and EvaNotebook.

3.1   The Eva Social Robot

EVA is an open-hardware robotic platform characterized by its modular design and its low-cost components devices. This platform was originally developed as a social assistant robot to interact and assist older adults who suffer from dementia [10]. EVA has been used as a platform for other related experiments, such as assisting children with autism in emotion regulation.

Fig. 2. EVA hardware components

3.2   MQTT Broker

We have adopted a pub-sub architecture that provides loosely coupled components, allowing us to scale and adapt more quickly than we could with a centralized architecture. This approach enables us to externalize the processing and view Eva, as well as the rest of the environment, as both subscribers and emitters of events. Consequently, we require a protocol to distribute these events. The current standard protocol for.

IoT is MQTT, a lightweight messaging protocol designed for efficient communication in constrained environments. We have deployed an instance of Eclipse Mosquitto, an open-source message broker that implements the MQTT protocol, on our local network to guarantee low latency and ensure responsiveness.

3.3   Wearable and Environmental Sensors

Wearable and environmental sensors integrated into social robots like Eva, in the user and/or the environment, are used to gather data. By capturing and analyzing data, these sensors provide valuable insights into human emotions, physical cues, and environmental context. Leveraging this information, Eva can deliver personalized and adaptive interactions, fostering a range of benefits for individuals. Some of the advantages that this offers to multiple interactions are:

- **Emotional understanding.** By analyzing physiological signals and facial expressions, Eva can perceive a person's emotional state and respond accordingly.
- **Contextual adaptation.** Environmental sensors, including microphones and cameras, allow the EVA robot the ability to sense and respond to the environment around it.
- **Supportive and assistive capabilities.** The framework process the data to offer support to individuals. For instance, if Eva detects an increase in heart rate or signs of stress through wearable sensors, it can provide relaxation techniques, breathing exercises, or soothing music to alleviate anxiety.
- **Data sharing and integration.** Eva's ability to deliver information captured by these sensors via MQTT to other devices or algorithms opens up opportunities for further analysis and integration.

3.4 Rhasspy

Rhasspy is an open-source, privacy-focused voice assistant toolkit that allows users to build their own voice-controlled applications. It provides a way to create custom voice interfaces that can be used to interact with various devices and services on MQTT. One of the key features of Rhasspy is its emphasis on privacy. By default, Rhasspy operates entirely offline, which means that all the voice data and processing stay within the user's local network. This local processing approach helps protect user privacy by avoiding the need to send voice data to external servers for processing [11].

3.5 EvaNotebook

EvaNotebook is a computational notebook specifically designed to operate solely within a browser environment, without the need for a client-server architecture. Its design lends itself to the development of prototypes, experimental processes, and system scenarios with a main emphasis on multilingual programming for event-driven architectures. It achieves this with the assistance of a decentralized database and incorporates various application protocols, including WebRTC, WebSockets, and MQTT, providing a higher level of abstraction.

3.6 Data Analysis Models

To infer nonverbal cues the framework can make use of pre-trained gesture, emotion of posture recognition modules. Several EvaNotebooks have been developed to recognize gaze direction, proximity, object-recognition, posture, etc. These can be adapted or customized to the problem at hand and run synchronously with other tasks.

4 Implementation of Scenarios

In this section we describe how the two scenarios presented in Sect. 2.2 can be integrated with the architecture, which are implemented using the framework components.

4.1   Scenario 1: Gaze and Proxemics to Initiate an Interaction

Figure 3 shows a sequence diagram to explain the interaction of Scenario 1 described in Sect. 2.2.1. The social robot Eva initiates an interaction when it detects the presence of an individual at a certain distance. She user her camera and a facial recognition algorithm to infer gaze direction and proximity.

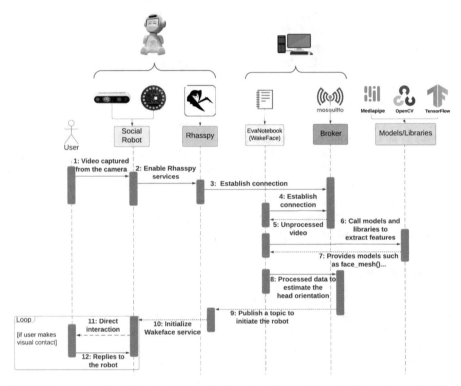

Fig. 3. Eva proactively initiates an interaction based on the user´s proximity and gaze direction.

This is implemented in an EvaNotebook that extends the"WakeFace" microservice proposed in [12] as an alternative to the wakeword method to initiate an interaction. The extension consists on using proximity besides gaze direction to adapt the interaction according to the following criteria:

1. **Distance greater than three meters**: When Eva detects a person walking at a distance of more than three meters, the robot activates its visual tracking function. Eva turns its neck and directs its gaze toward the person as long as the person keeps moving. The robot adjusts the speed and range of motion of its neck to follow smoothly and naturally the movement of the person. However, at this distance, Eva does not initiate verbal interaction or asks questions.
2. **Distance less than three meters**: When the person comes within approximately three meters of Eva, the robot emits a soft auditory signal to capture the person's attention

and then initiates a verbal interaction. For example, it might say,"Hello! How are you today?" or"Welcome! Can I help you with something?".

3. **The service of a real-time conversation**: Once Eva has initiated the conversation, it waits for the user's response. If the person responds positively or shows interest, Eva continues the dialogue smoothly and naturally.

The code below is an extract from the WakeFace microservice written in EvaNotebook, specifically in a WebWorker of JavaScript. It illustrates how the microservice subscribes to relevant MQTT events. The obtained camera data is then sent to the ML modules responsible for assessing gaze direction and proximity. To ensure clarity and express reactivity, the code follows the RxJS semantics, utilizing a pipe-and- filter architecture to handle asynchronous events, allowing for streaming processing via operators.

```
Begin
// Import necessary modules like mediapipe and MQTT

stream1 = Create Pipeline from MQTT 'camera stream' with this steps:
  1. Map each frame to detect face landmarks using mediapipe
  2. Filter out frames where no landmarks are detected
  3. Map remaining frames to test if the user is looking at the
     camera
stream2 = Create Pipeline from MQTT 'camera stream' with this steps:
  1. Map each frame to detect face landmarks using mediapipe
  2. Filter out frames where no landmarks are detected
  3. Map remaining frames to test if the user is within 3 meters
// Combine the two pipelines into a single pipeline
Combined_Pipeline = Zip(stream1, stream2) with the following steps:
  1. Filter pairs where both conditions are met
  2. For each remaining pair, publish 'wakeup' to MQTT
End
```

Listing 1: WakeFace microservice that runs on Evanotebook

Mediapipe is used to detect face landmarks. Afterwards, the code estimates whether the person is looking at Eva. There are different approaches to achieve this, such as solving PnP, Procrustes Analysis, SVD or defining a ROI. On the other hand, we calculate the user's distance with respect to the camera using, for instance, the pinhole camera model. Then we combine both streams to determine whether to wake up the system or not.

4.2 Scenario 2: Suggesting a Cooking Recipe

In this scenario, the user points to a table that contains several cooking ingredients. Eva will turn her head in the direction to which the user is pointing capture an image and call a classification algorithm to identify the ingredients (see the Fig. 4).

The following is a brief description of the interaction flow between the robot and the person:

1. The person places his plate with ingredients in front of Eva and asks for some recipes that can be cooked with them.

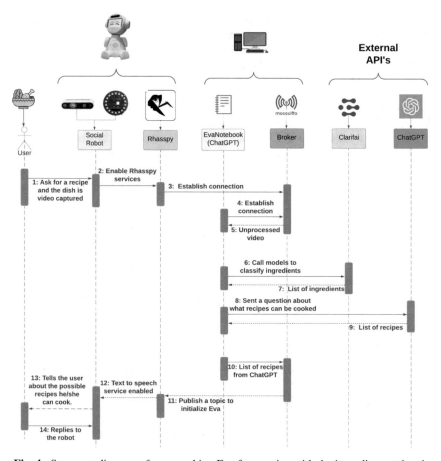

Fig. 4, Sequence diagram of a user asking Eva for a recipe with the ingredients at hand.

2. Eva uses her computer vision capabilities through the food-item-recognition algorithm from Clarifai, to identify the ingredients present on the plate. The algorithm classifies and labels the ingredients.
3. Once the ingredients have been identified, they will pass into list of grouped and classified ingredients.
4. EvaNotebook transmits the list of grouped ingredients to ChatGPT.
5. ChatGPT analyzes the list of ingredients and generates an answer with the names of the recipes and their preparation methods.
6. Eva reads the answer generated and transmits it to the person verbally.

The below code has the outline to achieve the competence.

The below code has the outline to achieve the competence.

```
// Previously, we wrote a system prompt to customize the Language
    Model.
...concatMap(_ => getOneFrameImage()),
  concatMap(frame => clarifai.recognizeFoodIn(frame)),
  concatMap(food => openai.reply(food)),
  tap(response => mqtt.publish('hermes/tts/say', response)))...
```
Listing 2: Suggesting a cooking recipe with RxJS syntax

5 Conclusion

Human-Robot interaction plays a central role in social robotics and while advances in speech recognition and synthesis have empowered verbal communication significantly, nonverbal communication helps make interaction more natural and efficient. To this end, we have developed a framework to facilitate the development of interaction scenarios that support the recognition of nonverbal cues from the user to adapt the behavior of a social robot. Two key components of the framework are broker server that distributes messages from sensors, actuators and analysis modules, and a P2P computational network used to quickly develop and test prototypes that can become components of more complex scenarios. We illustrate the use of the framework with two multimodal scenarios that make use of diverse sensors and inference models. While the framework was originally developed for the social robot Eva, we are currently integrating it with another robot that has different affordances, such as moving its legs and feet. The scenario involves the robot teaching dance moves to the user, and allowing us to test the capacity of EvaNotebook to work cooperatively, as the physical robot is in a different country from the user.

References

1. Saunderson, S., Nejat, G.: How robots influence humans: a survey of nonverbal communication in social human–robot interaction. Int J. Soc. Rob. **11**, 575–608 (2019). https://doi.org/10.1007/s12369-019-00523-0
2. Juan, E., Frum, C., Bianchi-Demicheli, F., Wang, Y.W., Lewis, J., Cacioppo, S.: Beyond human intentions and emotions. Front. Human Neurosci. **7**, 99 (2013) https://doi.org/10.3389/fnhum.2013.00099
3. Hall, J.A., Horgan, T.G., Murphy, N.A.: Nonverbal communication. Annu. Rev. Psychol. **70**, 271–294 (2019). https://doi.org/10.1146/annurev-psych-010418-10314
4. Robaczewski, A., Bouchard, J., Bouchard, K., Gaboury, S.: Socially assistive robots: the specific case of the Nao. Int. J. Social Rob. **13**, 795–831 (2021). https://doi.org/10.1007/s12369-020-00664-7
5. Rocha, E.: Débora: a 3D simulator proposal for the social robotics platform. FRED (2023)
6. Rocha, M.M., Muchaluat-Saade, D.C.: Evaml e evasim: Proposta de linguagem baseada em xml e simulador para o robô eva. In: Anais do XXXVI Concurso de Teses e Dissertações, pp. 98–107. SBC (2023). https://doi.org/10.5753/ctd.2023.23008
7. Robinson, H., MacDonald, B., Broadbent, E.: The role of healthcare robots for older people at home: a review. Int. J. Soc. Robot. **6**, 575–591 (2014). https://doi.org/10.1007/s12369-014-0242-

8. Urakami, J., Seaborn, K.: Nonverbal cues in human–robot interaction: a communication studies perspective. ACM Trans. Human-Robot Interact. **12**(2), 1–21 (2023). https://doi.org/10.1145/357016

9. Pineda, L.A., Rodríguez, A., Fuentes, G., Rascon, C., Meza, I.V.: Concept and functional structure of a service robot. Int. J. Adv. Rob. Syst. **12**(2), 6 (2015) https://doi.org/10.5772/6002

10. Cruz-Sandoval, D., Morales-Tellez, A., Sandoval, E.B., Favela, J.: A social robot as therapy facilitator in interventions to deal with dementia-related behavioral symptoms. In: Proceedings of the 2020 ACM/IEEE International Conference on Human-Robot Interaction. HRI 2020, pp. 161–169. Association for Computing Machinery, New York (2020). https://doi.org/10.1145/3319502.3374840

11. Filipe, L., Peres, R.S., Tavares, R.M.: Voice-activated smart home controller using machine learning. IEEE Access **9**, 66852–66863 (2021). https://doi.org/10.1109/ACCESS.2021.3076750

12. Villa, L., Hervás, R., Cruz-Sandoval, D., Favela, J.: Design and evaluation of proactive behavior in conversational assistants: approach with the eva companion robot. In: Bravo, J., Ochoa, S., Favela, J. (eds.) Proceedings of the International Conference on Ubiquitous Computing & Ambient Intelligence (UCAmI 2022), pp. 234–245. Springer, Cham (2023). https://doi.org/10.1007/978-3-031-21333-52

A First Approach to the Generation of Linguistic Summaries from Glucose Sensors Using GPT-4

Carmen Martinez-Cruz$^{(\boxtimes)}$ (ID), Juan F. Gaitan Guerrero (ID),
Jose L. Lopez Ruiz (ID), Antonio J. Rueda (ID), and M. Espinilla (ID)

Department of Computer Science, University of Jaen, Jaen, Spain
cmcruz@ujaen.es

Abstract. The use of activity monitoring sensors on users with some type of disease or dependence is very useful for health technicians, for family members or for the user himself. The knowledge of these values in real time allows alerting of a possible crisis or starting correcting actions to prevent a serious health problem. For this reason, many proposals have been made to summarize in words the huge amount of measures taken by these sensors in order to highlight only what is really important for the end user, family or medical staff. The emergence of new text generation tools based on Artificial Intelligence (AI), such as the latest GPT-4, is having a major impact in the healthcare field. In this article we analyze how the latest version of ChatGPT, allows the generation of linguistic summaries in natural language from glucose sensor measurements. We also learn how to ask the right questions to obtain the type of output adapted to the user, whether or not it is necessary to perform some kind of preprocessing on the data to be analyzed and what are the strengths and drawbacks of this technology.

Keywords: Linguistic summaries · Computing with words · Glucose level · Sensors · ChatGPT · Artificial Intelligence · GPT-4 · Large language models

1 Introduction

The Generative Pre-trained Transformer (GPT) family of Large Language Models, such ChatGPT o Bard, have revolutionized the current society and are going to set a new milestone in the technological race of the last decades [7]. These tools are able to produce themselves as humans providing answers to any question, in a fluent, agile, coherent, and natural way. In addition, they can respond

This work has been partially supported by the Government of Spain through the projects RTI2018-098979-A-I00 MCIN/ AEI/10.13039/501100011033/, ERDF "A way to make Europe", B-TIC-744-UGR20 ADIM: Accesibilidad de Datos para Investigación Médica of the Junta de Andalucía and the University of Jaén under Action 1 with reference EI_TIC1_2021.

J. Bravo and G. Urzáiz (Eds.): UCAmI 2023, LNNS 842, pp. 33–43, 2023.
https://doi.org/10.1007/978-3-031-48642-5_4

to different roles, in different styles, whether conversational, written, requests, scientific texts, etc. and even in different spoken languages or programming languages. Since the ChatGPT chatbot was released last 2022, it has been used in numerous applications with amazing results. In the healthcare field, its use is also being evaluated, especially with the new GPT-4 engine released in 2023, which outperforms ChatGPT-3.5 on all the tasks and is also capable of analyzing images [1].

In recent decades, there has been a dramatic increase in the number of devices for monitoring subjects to control and even prevent their degree of well-being. These sensors generate such a large amount of information that it is very difficult to process by healthcare personnel, end users or their relatives, either because of the large volume of data or because they lack the technical training needed to understand it correctly. The generation of linguistic summaries of Time Series (TS) has provided a solution to these problems since it represents in text form the relevant information adapted to the needs of the end user. The generation of these summaries requires the supervision and knowledge of an expert who specifies the quality measures that the generated summary must meet.

In this proposal, an analysis of the performance of ChatGPT with GPT-4 in the generation of linguistic summaries of TS of blood glucose measurements is carried out. These measurements are taken through a monitoring device that controls glucose every 5 min. The values obtained are provided to this Artificial Intelligence (AI) and its conclusions are analyzed and different alternatives in the processing of the series are proposed to try to obtain the best possible output.

2   Chatgpt

ChatGPT is an artificial intelligence chatbot developed by OpenAI [1,20,25]. ChatGPT, using GPT-4 model, excels at creating fluent and natural texts, even in academic scenarios, where the incorporation of citations and references is quite acceptable. However, the bibliographic references and other sources of data, depending on the version used may be outdated, incorrect or invented [3]. Another problem of this AI is the generation of hallucinations [16], understood as the generation of seemingly realistic sensory experiences that do not correspond to any real-world input. Moreover, it exhibits pervasive biases making the use of this technology controversial in different areas. It also can not analyze audio or video or generate arithmetic errors that a simple calculator would never have [12]. Despite these problems, it has been tested in many scenarios with impressive results. In academics, GPT-4 achieves top scores in several exams for accessing to universities or certifications. It is able to write automatically various types of documents, including reports, summaries and even book reviews. In medicine it is being proved that this AI is able to provide assistance to the practitioners in many areas [7].

3   Generation of Linguistic Summaries

The process of generating linguistic summaries from Time Series, formally called Generation of Linguistic Descriptions of Time Series (GLiDTS), has been largely studied in the literature (see compilations of [6,17,18,21]). Its main objective is to summarize raw data (mostly obtained from sensors) in a set of words, highlighting the most relevant phenomena identified in the TS, and adapting the language to the end user. Just as it would happen if it were a human interpreting these data. The purpose of the GLIDTS is very versatile, from the possibility of transmitting these data through text messages or to be audible, as the possibility of interpretation of technical data by users who have not expertness in the field of application of these data. There are many applications of these techniques in the literature, some examples are those related with summarizing weather reports using Natural Language Procesing (NLP) as we can see in the works of Reiter et al. [22] and Ramos-Soto et al. [21]. Trivino et al. in [4,10,23], [24] generate summaries of streaming data in different application areas such as the energy consumption, gait analysis, etc. Banaee et al. [5] propose a partial trend detection algorithm to describe particular changes of health parameters in physiological TS data. Marin et al. in [8,9] developed a mechanism that describes TS where time is expressed in different granularities. Finally, Martinez-Cruz et al. [19] presented a way of describing TS using the patterns detected in the TS, at different levels of abstraction. However, these techniques are complex as described in [18] since a knowledge model should be developed in order to obtain the most suitable summary according to the final user requirements. Moreover, the final text description must accomplish a set of quality standards specified by the domain experts. In this proposal, we try to avoid these steps by seeking the knowledge provided by the AI and, in particular, the one recently released, GPT-4.

4   Linguistic Summaries Using ChatGPT

The process of generating linguistic summaries from TS or any other data source through the use of chatbot is a subjective process. Not only the process of generating summaries is opaque because of the nature of the technology, but the handling of natural language itself is subjective as well. For this reason, a methodology is presented here to allows us to analyze the results as objectively as possible. The proposal also describes different improvement processes that can be applied to increase the quality of the final text.

4.1   Methodology

The methodology used in this proposal consists in six phases illustrated in Fig. 1 and described below:

Step 1. Data acquisition. A device embedded under the skin collects data that are transmitted to a database.

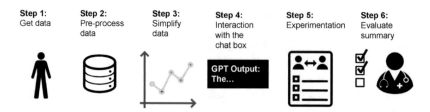

Fig. 1. System methodology.

Step 2. Preprocessing data. Data are processed to avoid inaccuracies generated
    by the sensors themselves, such as transmission errors or noise, empty periods
    of data, etc.
Step 3. Data simplification. This process allows highlighting the most relevant
    values of the TS while reducing the amount of data to be processed.
Step 4. Interaction with the chatbot. It consists of providing the chatbot with
    a textual description of the simplified TS and some prompting in the appro-
    priate manner to obtain the best possible output.
Step 5. Experimentation. Analysis of the output obtained from a set of TS.
Step 6. Evaluate output. Evaluation of the linguistic summaries obtained by the
    chatbot.

4.2   Description of the Problem

Data have to be acquired, processed and in some cases, preprocessed (steps 1
and 2). In this proposal we have used data obtained through an invasive glucose
sensor connected via NFC (Near-Field Communication) and BLE (Bluetooth
Low Energy). This sensor gets the interstitial fluid data, processes it and sends
the samples continuously (one sample every five minutes) to a server running
a MongoDB database. This data is accessible through a web application. In
this proposal, glucose data are collected and preprocessed to fill spaces with no
data below a threshold set at 30 min, to avoid any error caused by temporary
connection problems between the sensor and the server. In this case, a moving
average operation is performed on the data around the empty spaces, with a
window of size 5.

4.3   Simplification Process

When a Time Series contains many points, it is advisable to reduce this number
of points so that its processing is simplified, in terms of efficiency and complexity. For example, the time series used in this experimentation are 288 points
(with some exceptions e.g. empty periods), resulting from monitoring a user's
glucose every 5 min over the course of a day. This number may be too large to
be processed by an artificial intelligence with satisfactory results. To simplify
this number, a TS segmentation algorithm has been chosen using a technique
similar to that performed by humans when observing a graph, where the most

relevant points (peaks or valleys) are highlighted, and the rest are omitted. For this purpose, the Douglas-Peucker method has been used, whose algorithm can be seen in [11]. There are other segmentation methods, such as the Breakpoint Detection Algorithms [2], the Minimum Description Length [13], Moving Average Algorithm or those indicated in [14,15]. However, it has been considered that geometric techniques are very useful in these cases, as discussed in [19].

This method filters those points whose distance is less than a given threshold (ϵ), so that only the most representative points of the series remain in a natural way. For this experiment, three thresholds have been used: $\epsilon = 0$ i.e. no threshold, $\epsilon = 0.2$ that generates a mild segmentation, and $\epsilon = 0.4$, which develops a more drastic segmentation. An example of this process is illustrated in Fig. 2 from a TS of glucose data recorded on a person on December 31st, 2022.

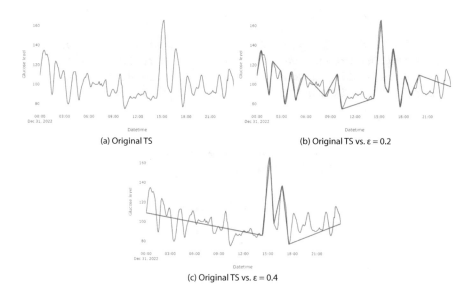

Fig. 2. TS Simplification (December 31st, 2022).

5   Experimentation

The goal of this experimentation is to obtain a textual description from a TS of sufficient quality to help the end user to understand the evolution of the measure, either with or without access to the graph. Moreover, the output must be short and highlight the most relevant events, which in the context of glucose, are related to relevant sugar highs or lows. To do so, in this experimentation we analyze:

– The choice of words as the prompt of the chatbot.
– The number of points included in the TS for obtaining the best description.

The interaction with ChatGPT (using GPT-4) consists in the creation of new conversations, where the AI uses a contextual memory to store the relevant details it needs to improve its answers. For this reason, we can perform multiple data insertions in the same conversation to obtain the desired result. Also, any response generated by this technology can be rephrased, so that the AI generates another output with the same purpose but using different words. However, the optimization of resources to obtain an answer must be taken into account, and performing a large number of interactions with the AI or generating different phrases for the same query is time and effort consuming.

We have try the following prompts once provided the TS[1] to the chatbot:

```
- Make a short linguistic summary of the following TS of
glucose.
- Make a brief summary of the previous TS of glucose levels
stressing the trends, variability and most relevant features.
- Generate a brief summary of the previous TS of glucose
measures, highlighting the most relevant features.
```

The results obtained using any of these prompts are similar, highlighting the most representative elements of the TS, but varying the length and quality of the summary depending on the number of points entered. The prompt chosen for this experimentation is as follows:

```
Make a summary of this Time Series in less than 100 words with the
most relevant information.
```

The output obtained from the AI has been analyzed for three sets of data: i) The original time series (described in Sect. 4.3) that is illustrated with an example in Fig. 2(a), ii) A *"medium"* simplification used with an $\epsilon = 0.2$, where the number of points obtained keeps the general trend of the graph, as can be seen in Illustration 2(b), and iii) A *"strong"* simplification generated with $\epsilon = 0.4$, which gives us a very segmented TS where only the most significant points are maintained, as shown in Illustration 2(c). The GPT-4 responses to those inputs are shown in Table 1.

---

[1] Although GPT-4 allows the analysis of figures, this is still a beta feature and therefore has not been used in our experimentation.

Table 1. Output of GPT-4 for a TS recorded on December 31st, 2022

Original TS ($\epsilon = 0$). Number of points: 282

The glucose level time series for one person throughout the day shows a gradual decrease from 135.0 to 84.0 in the first few hours, reaching a minimum of 75.0. From there, the glucose levels gradually increase, peaking at 166.0 before stabilizing around 90.0–110.0. The most notable patterns include a decrease during the early morning hours, a rise during the afternoon, and a subsequent decrease during the evening. Overall, the glucose levels exhibit fluctuations within a range, with occasional spikes and dips throughout the day.

"Medium" simplification ($\epsilon = 0.2$). Number of points: 22.

The provided time series represents the glucose levels of an individual over a day. The range of glucose levels varies from 75.0 to 166.0 units. The average glucose level appears to be around 105.0 units. There are fluctuations throughout the day, with some notable spikes and drops. The lowest glucose reading was 75.0 units, while the highest was 166.0 units. It's important to note that this summary is based on the limited information provided and does not account for potential patterns or underlying factors that may influence glucose levels. Consulting a medical professional for a thorough analysis is advised.

"Strong" simplification ($\epsilon = 0.4$). Number of points: 7.

The provided time series data represents a person's glucose levels throughout the day. The initial measurement at midnight is 109.0, and it gradually decreases to 86.0 by 14:20. However, there is a significant spike at 15:15, with the glucose level reaching 166.0. Subsequently, it drops to 98.0 at 15:45 and then increases to 137.0 at 16:45. A sudden decrease is observed at 17:35 to 77.0. Towards the end of the day, at 23:55, the glucose level is recorded as 98.0 again. The data showcases fluctuations in glucose levels, with notable variations at specific time points.

---

6   Evaluation

To assess the performance of this technology based on its outcomes, various criteria have been established to compare how effectively the AI system aligns with human-expected statements.

C1 Highlight of relevant values (peaks, valleys, sharp increases or decreases, etc.).

C2 Highlight of more complex structures involving sets of peaks or valleys, or significant fluctuations.

C3 Summarize the value of the majority of registered instances through statistical values.

C4 Summarize the value of the majority of relevant registered instances.

C5 Omit not significant measures (i.e.: normal glucose levels).

C6 Indicate the day time a relevant situation is registered (hour or moment of the day)

C7 Describe TS progressively in time.

C8 Detect no-data-collection-intervals along the TS.

C9 Describe the TS values in a semantic manner (i.e.: high glucose levels).
C10 Include vague recommendations or superfluous information.
C11 Include incorrect statements.
C12 Omit relevant information.
C13 Describe TS briefly.
C14 Describe TS redundantly.
C15 Warn about the need of more input data for the description.

For this experimentation, 5 days of data have been tested, i.e. a total of 15 Time Series with different amount of data. The results obtained and their analysis are discussed below. The scale defined in Fig. 3 refers to the percentage of the TS for which the AI satisfies a certain criterion (Table 2).

Never	Rarely	Sometimes	Almost always	Always
0	0.2	0.4	0.8	1

Fig. 3. GPT-4 performance scale.

Table 2. Checklist to validate the ChatGPT output.

Criterion	$\epsilon = 0$	$\epsilon = 0.2$	$\epsilon = 0.4$
C1	Always	Always	Always
C2	Never	Never	Never
C3	Rarely	Almost always	Rarely
C4	Never	Never	Never
C5	Rarely	Rarely	Never
C6	Always	Almost always	Always
C7	Always	Sometimes	Almost always
C8	Never	Never	Never
C9	Rarely	Rarely	Never
C10	Almost always	Almost always	Almost always
C11	Almost always	Rarely	Rarely
C12	Always	Almost always	Never
C13	Almost always	Always	Always
C14	Rarely	Almost always	Rarely
C15	Never	Sometimes	Never

The generated results allow the following conclusions to be drawn:

– Using the total number of glucose measures collected as input data leads the AI to generate incorrect sentences, as it tries to describe a huge amount of data in a few words.
– The generated output from a segmented TS with a *"strong"* simplification includes a more progressive linguistic description over time, due to the low number of points involved.
– All the linguistic descriptions generated by GPT-4 include values and times of the day which are quite accurate. It must be noticed that non-significant glucose measures are not filtered, i.e. normal glucose levels or irrelevant fluctuations which do not contribute to a transcendental knowledge.
– GPT-4 is unable to detect intervals of no-data, assuming a complete collection of registered instances in the input data set.
– The different outputs provided by GPT-4 contain trivial sentences or recommendations which move away from the main purpose of describing a graphical representation of a TS.
– GPT-4 randomly generates linguistic summaries following a correct time sequence but tends to describe the TS in a "discrete way" by jumping between relevant events. Therefore it does not provide a proper description of the trend and evolution of the TS over time.

7 Conclusion

The generation of linguistic summaries from time series facilitates the task of interpreting data obtained, for example from IoT devices, for family members or non-experts. Specifically, in the context of glucose, given the nature of the data, it is advisable to obtain only the relevant information, peaks or valleys with high or low values, and when they have taken place, avoiding redundancies or data that only result in extending the size of the final text. The use of ChatGPT to perform this type of summaries has shown us that the data output is totally dependent on the number of TS measures. If a large number of data is provided, ChatGPT will not return a correct summary, since it omits relevant information and performs a description that does not follow a temporal sequence. On the other hand, a TS with few points does not generate a good result either, since instead of an actual summary, it generates a straightforward text transcription of the TS values. A medium number of points is the most advisable option. However, it should be noted that in general, ChatGPT returns superfluous and not very relevant information, such as tips or statements that only lengthens a description that is expected to be short.

Also, the lack of open source code for GPT, in any of its versions, makes it difficult to know how to introduce data in this platform in order to obtain good results. However, it is possible to train the tool in certain areas of knowledge, so that a large data bank would be needed to obtain the desired results, in the form that the end user needs.

In any case, the results provided by this tool are impressive, and we believe that it is only a matter of time before ChatGPT or any LLM-based chatbot is used as a standard method for the generation of quality user-tailored summaries to the user's needs in a variety of different contexts.

References

1. Gpt-4 technical report (2023). https://arxiv.org/pdf/2303.08774.pdf
2. Ali, A., Aggarwal, J.: Segmentation and recognition of continuous human activity. In: Proceedings IEEE Workshop on Detection and Recognition of Events in Video, pp. 28–35 (2001). https://doi.org/10.1109/EVENT.2001.938863
3. Alkaissi, H., Mcfarlane, S.: Artificial hallucinations in chatGPT: implications in scientific writing. Cureus **15**, e35179 (2023). https://doi.org/10.7759/cureus.3517
4. Alvarez-Alvarez, A., Triviño, G.: Linguistic description of the human gait quality. Eng. Appl. AI **26**(1), 13–23 (2013). https://doi.org/10.1016/j.engappai.2012.01.022
5. Banaee, H., Ahmed, M.U., Loutfi, A.: A framework for automatic text generation of trends in physiological time series data. In: 2013 IEEE International Conference on Systems, Man, and Cybernetics, pp. 3876–3881 (2013). https://doi.org/10.1109/SMC.2013.661
6. Boran, F.E., Akay, D., Yager, R.R.: An overview of methods for linguistic summarization with fuzzy sets. Expert Syst. Appl. **61**, 356–377 (2016)
7. Bubeck, S., et al.: Sparks of artificial general intelligence: early experiments with gpt-4 (2023)
8. Castillo-Ortega, R., Marín, N., Sánchez, D.: A fuzzy approach to the linguistic summarization of time series. J. Multiple-Valued Logic Soft Comput. **17**, 157–182 (2011)
9. Castillo-Ortega, R., Marín, N., Sánchez, D.: Linguistic query answering on data cubes with time dimension. Int. J. Intell. Syst. (IJIS) **26**(10), 1002–1021 (2011)
10. Conde-Clemente, P., Alonso, J.M., Trivino, G.: Toward automatic generation of linguistic advice for saving energy at home. Soft. Comput. **22**(2), 345–359 (2016). https://doi.org/10.1007/s00500-016-2430-5
11. Douglas, D.H., Peucker, T.K.: Algorithms for the reduction of the number of points required to represent a digitized line or its caricature. Cartographica Int. J. Geogr. Inf. Geovisual. **10**, 112–122 (1973). https://doi.org/10.3138/FM57-6770-U75U-7727
12. Etzioni, O.: Commentary: Openai's gpt-4 has some limitations that are fixable - and some that are not (2023). https://www.geekwire.com/2023/commentary-openais-gpt-4-has-some-limitations-that-are-fixable-and-some-that-are-not/. Accessed 14 Mar 2023
13. Farsi, N., Mahjouri, N., Ghasemi, H.: Breakpoint detection in non-stationary runoff time series under uncertainty. J. Hydrol. **590**, 125458 (2020). https://doi.org/10.1016/j.jhydrol.2020.125458
14. Fu, T.C.: A review on time series data mining. Eng. Appl. Artif. Intell. **24**(1), 164–181 (2011)
15. Höppner, F.: Time series abstraction methods - a survey, pp. 777–786 (2002)
16. Ji, Z., et al.: Survey of hallucination in natural language generation. ACM Comput. Surv. **55**(12), 1–38 (2023). https://doi.org/10.1145/3571730

17. Kacprzyk, J., Zadrozny, S.: Fuzzy logic based linguistic summaries of time series: a powerful tool for discovering knowledge on time varying processes and systems under imprecision. Wiley Interdisc. Rev. Data Mining Knowl. Disc. **6**(1), 37–46 (2016). https://doi.org/10.1002/widm.1175
18. Marín, N., Sánchez, D.: On generating linguistic descriptions of time series. Fuzzy Sets Syst. **285**, 6–30 (2016). Special Issue on Linguistic Description of Time Series
19. Martinez-Cruz, C., Rueda, A.J., Popescu, M., Keller, J.M.: New linguistic description approach for time series and its application to bed restlessness monitoring for eldercare. IEEE Trans. Fuzzy Syst. **30**(4), 1048–1059 (2022). https://doi.org/10.1109/tfuzz.2021.3052107
20. OpenAI: Chatgpt - release notes (2023). https://help.openai.com/en/articles/6825453-chatgpt-release-notes
21. Ramos-Soto, A., Bugarín, A., Barro, S.: On the role of linguistic descriptions of data in the building of natural language generation systems. Fuzzy Sets Syst. **28**, 31–51 (2016). https://doi.org/10.1016/j.fss.2015.06.019
22. Reiter, E., Sripada, S., Hunter, J., Davy, I.: Choosing words in computer-generated weather forecasts. Artif. Intell. **167**, 137–169 (2005)
23. Sanchez-Valdes, D., Eciolaza, L., Triviño, G.: Linguistic description of human activity based on mobile phone's accelerometers. In: Ambient Assisted Living and Home Care - 4th International Workshop, IWAAL 2012, Vitoria-Gasteiz, Spain, 3–5 December 2012, Proceedings, pp. 346–353 (2012)
24. Trivino, G., Sugeno, M.: Towards linguistic descriptions of phenomena. Int. J. Approx. Reas. **54**(1), 22–34 (2013). https://doi.org/10.1016/j.ijar.2012.07.004. http://www.sciencedirect.com/science/article/pii/S0888613X12001375
25. Wang, F.Y., Miao, Q., Li, X., Wang, X., Lin, Y.: What does chatGPT say: the dao from algorithmic intelligence to linguistic intelligence. IEEE/CAA J. Automatica Sinica **10**(3), 575–579 (2023). https://doi.org/10.1109/JAS.2023.123486

Efficient and Accountable Industry 5.0 Production Scheduling Mechanism for Mass Customization Scenarios

Borja Bordel[(✉)], Ramón Alcarria, Guillermo de la Cal Hacar, and Tomás Robles Valladares

Universidad Politécnica de Madrid, Madrid, Spain
{borja.bordel,ramon.alcarria,tomas.robles}@upm.es,
guillermo.delacal@alumnos.upm.es

Abstract. Industry 5.0 refers to a new industrial revolution characterized by the humanization of industrial systems, applications, and services: from cognitive robots to customizable products. But at the same time, Industry 5.0 must preserve the efficiency and profitability of Industry 4.0 mass production scenarios. Although, personalization and product adaptation to the individuals are traditionally understood as obstacles to take the maximum advantage of exponential business models, economies of scale and mass markets; some authors envision technologies such as cognitive Cyber-Physical Systems or swarm intelligence to overcome this challenge and enable new profitable markets for mass customizable products and services. However, Industry 5.0 is one of the most recent technological paradigms, even Industry 4.0 paradigm is not fully developed yet, and no tangible or specific proposal is still reported to achieve this efficient mass customization. This paper addresses this challenge. We propose an efficient and accountable Industry 5.0 production scheduling mechanism based on a transparent Blockchain-enabled marketplace and particle swarm optimization algorithms. Customers can request a customized product using a prosumer environment and Smart Contracts. Later, Industry 5.0 producers will combine three functions to define an optimization problem and find the most efficient production schedule. Particle Swarm Optimization algorithms are employed to calculate the most profitable production schedule. The proposed mechanism is experimentally validated using simulation tools. Results show the economy of scale is preserved, contrary to traditional customized product markets, and efficiency is just 8% lower than common mass Industry 4.0 production systems.

Keywords: Industry 5.0 · production scheduling · mass customization · Blockchain marketplace · particle swarm optimization · simulation

1 Introduction

Industry 5.0 [1] refers to the next industrial revolution, where humanization [2] is the main objective. This new paradigm includes a wide catalog of innovative applications: from cognitive robots [3] making interactions between workers and production systems

© The Author(s), under exclusive license to Springer Nature Switzerland AG 2023
J. Bravo and G. Urzáiz (Eds.): UCAmI 2023, LNNS 842, pp. 44–56, 2023.
https://doi.org/10.1007/978-3-031-48642-5_5

and environments more natural and user-friendly, to fully customized products adapted to customers' needs and preferences [4]. In general terms, Industry 5.0 services and applications must replace mass production strategies with individualized designs and on-demand manufacturing [5]. However, this new paradigm should be an improvement of the previous Industry 4.0 revolution [6], and then new industrial business models must preserve the efficiency and profitability of previous mass production schemes. But traditionally, customization and product adaptation are understood as an obstacle to achieve an exponential cost reduction and an economy of scale.

In fact, reported works on Industry 4.0 solutions generally distinguish between two different scenarios [7]. On the one hand, purely industrial applications where total efficiency and productivity must be achieved through intelligent mechanisms and next generation communication technologies. On the other hand, handcraft production centers where humanized Cyber-Physical Systems [8] and intrinsic interfaces must be deployed to facilitate the integration of workers in digital control systems. But strategies to combine the added value of both models are still unexplored. Actually, enabling technologies for efficient mass customization scenarios is one of the major open challenges in Industry 5.0.

However, even Industry 4.0 has not yet been fully developed, and problems such as production scheduling or distributed manufacturing are still unsolved. Due to this fact, Industry 5.0 solutions are still scarce. Specifically, three basic technologies have been proposed to achieve an efficient mass customization: artificial intelligence mechanisms [9], cognitive Cyber-Physical Systems [10] and swarm intelligence [11]. However, all reported works are exploratory and initial ideas about possible research lines, with no tangible or particular technological contributions. Therefore, innovative engineered systems and proposals are needed to achieve efficient mass customization in Industry 5.0 scenarios.

In this paper, we address this challenge. We propose an innovative production scheduling mechanism, so the customized product manufacturing is organized in such a way that final industrial tasks and costs are globally as efficient as in Industry 4.0 mass production applications. In the proposed approach, customers can design their own products through a prosumer environment, where all existing production options (and their restrictions) are represented. Customized products are later uploaded to a Blockchain marketplace, where users can buy them using cryptocurrencies and Smart Contracts. The marketplace determines a minimum price to be paid, but users can pay more in order to increase the priority of their products in the production scheduling. All transactions in the marketplace are transparent and accountable to all customers. An automatic production scheduling systems acquired information from the marketplace and decomposes products in their elemental components. Three functions are employed to represent the production process of these elemental components and their final assembling process: the revenue function (depends on monetary phenomena), the expense function (depends on the production costs), and the user satisfaction function (the production delay is the main factor to be considered). These three functions are combined to define an optimization problem to be solved using Particle Swarm Optimization algorithms, whose output defines the most efficient production scheduling.

The rest of the paper is organized as follows. Section 2 analyzes the state of the art on Industry 5.0 and mass customization. Section 3 presents the proposed production scheduling technology. Section 4 describes the experimental validation and its results. Finally, Sect. 5 concludes the paper.

2 Industry 5.0 Technologies and Mass Customization Solutions

Industry 5.0 is one of the most recent paradigms in information technologies, so works on this topic are still exploratory. Most authors discussing Industry 5.0 introduce enabling technologies [12] or future research opportunities [13]. Some works have also been reported that analyze the meaning and impact of this new paradigm [14]. Furthermore, articles studying the evolution from Industry 4.0 to Industry 5.0 [15] or even a future sixth industrial revolution [16] can be found. Furthermore, different analyzes have also been reported that describe how Industry 5.0 could improve humanization and sustainability [17], or the viability of such future technologies [18]. But, in general, all these works are abstract descriptions about future technologies to be designed and implemented, which have not been described or experimentally evaluated in a relevant scenario yet.

However, mass customization is a much older paradigm, originally reported in the late 1980s and early 1990s [19]. Recent work has clearly identified product modularity [20] and additive manufacturing [21] as key industrial strategies to enable mass customization. In this approach, basically, products are compositions of different elemental modules, which are assembled (or not) in an additive chain. Technologies such as biomaterials [22], human-robot collaborative scenarios [23], or 3D printing [24] have been studied as enabling solutions for this additive manufacturing. Besides, cost analyses [25] and comparative studies [26] to identify the advantages and disadvantages of this new vision compared to traditional production schemes may be found. But all these previous works are partial and only describe how mass customization could be enabled. No full end-to-end description is provided, showing how users could order or buy customized products and efficiently make them available to exploit the potential of modular product and additive manufacturing. Some authors proposed digital twins as the technology to allow an efficient production scheduling [27, 28], but this approach (again) is just an enabling solution. The actual mechanism to define an efficient mass customization using digital twins (or any other possible technology) is not described.

This paper covers this gap. In the proposed solution, products are modular, and manufacturing follows an additive scheme. But customers can create their own products by combining these elemental modules through a prosumer environment. Besides, later, they can buy those customized products using a Blockchain-enabled marketplace. This transparent and automatic commercialization mechanism is connected to an optimization algorithm that uses information from the market to define the most profitable production schedule, so that mass customization turns out to be economically viable.

3 Efficient and Accountable Industry 5.0 Production Scheduling

In Industry 5.0 mass customization scenarios, products are modular. Then a product P is the composition of N_P different elemental modules p_i (1). Each one of these modules has a unitary production cost c_i. Additionally, there is an assembly cost c_P to manufacture product P from the elemental modules p_i.

$$P = \{p_i i = 1, \ldots N_P\} \tag{1}$$

Mass customization solutions must allow customers to design their own products by combining the existing elemental modules as they desire, but the manufacturing of these modules and the assembly of the global product must be scheduled in such a way that there is no difference with common mass production industrial applications. Figure 1 represents the proposed Industry 5.0 architecture to address this challenge.

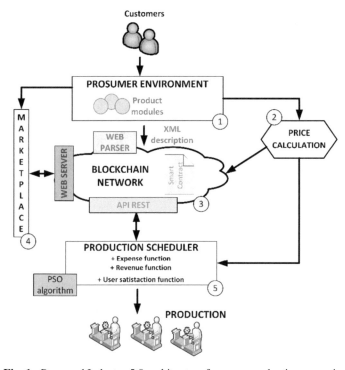

Fig. 1. Proposed Industry 5.0 architecture for mass production scenarios

Customers (non-industrial experts) can create products by combining the exiting elemental modules through a prosumer environment ①. When the product design is finalized, it is analyzed by a price calculation engine ②, which calculates the minimum price the customer needs to pay for this new customized product (considering unitary costs c_i and the assembly cost c_P). This cost, together with the product design, is employed to create a Non-Fungible Token (NFT) in a transparent and accountable Blockchain-enabled marketplace ③. Through a web marketplace and monetization technologies ④, customers can buy customized products and order their production. Before starting the diary production, pending orders are collected and a production scheduler ⑤ determines in which order modules are manufactured and products assembled, so revenue and user satisfaction are maximized, and expense minimized. The Particle Swarm Optimization algorithm is employed to calculate the most efficient production scheduling.

3.1 Prosumer Environment and Price Calculation

Customers can design their own customized products through a prosumer environment. This environment is a web interface where customers can design any product P by combining elemental modules p_i from a catalogue (see Fig. 2-left-). Graphical tools are employed to facilitate the interaction with non-industrial experts. However, some restrictions could be applied to this customization. Then, this prosumer environment describes restrictions associated to each elemental module p_i using Linear Temporal Logic (LTL) Rules. The output of this prosumer environments is an XML document describing the customized product (basically, a list of elemental modules), and some required metadata such as a unique identifier, the customer's name, etc. Figure 2(right) shows an example of this XML product description.

Fig. 2. **(left)** Prosumer environment **(right)** XML product description

This XML description, on the one hand, is sent to the Blockchain network to be transformed into a buyable NFT. On the other hand, the XML description is sent to a price calculation engine. In this engine, the minimum price π_{min} to be paid by customers, so the product manufacturing is profitable is calculated (2). This price includes the unitary cost c_i of each elemental module p_i making up the product P; the assembly cost c_P and the profit margin m (percentage) to ensure the business viability. To calculate the assembly cost c_P, we employ an exponential cost function $f_{cp}(\cdot)$ depending on the number of elemental modules N_P to be ensembled, and a configuration parameter τ_{cp} (positive number) controlling the cost growing rate and τ_0 is the assembly cost for a product that includes only one module (3).

$$\pi_{min} = \left(\sum_{i=1}^{N_P} c_i + c_P \right) \cdot (1 + m) \tag{2}$$

$$c_P = \tau_0 \cdot exp\left(\frac{N_P}{\tau_{cp}} \right) \tag{3}$$

3.2 Transparent and Accountable Blockchain-Enabled Marketplace

The XML description is sent to an Ethereum Blockchain network. This Blockchain is provided with a web server which can receive and parse the XML document. After the parsing, the product design is transformed into a digital object: a Non-Fungible Token (NFT). This NFT is maintained inside the Blockchain network and can be exchanged in a transparent and accountable manner. To create this NFT, the web parser must invoke the appropriate function (constructor) contained in the Smart Contract describing the token structure. The web server is based on JavaScript technologies, and the Blockchain network is supported by technologies from the Truffle project. Both infrastructures are connected using the web3 library. The Smart Contract describing the NFT structure follows the standard ERC1155, see Fig. 3(left). The ERC1155 standard is a "multitoken" model, so it can be used to describe, create, and manage fungible tokens (monetization units) and NFT using some common functions and logic. It replaces the ERC20 and ERC721 standards, usually employed to create fungible tokens and NFT independently. This approach reduces the complexity of the software and the Ethereum gas consumed (installation costs) consumed during the deployment.

On the other hand, a second Smart Contract (SC) is deployed. This SC is employed to support a web marketplace. This marketplace shows all created NFT and the minimum price π_{min} (in fungible tokens or monetization units) the customer must pay to order to product manufacturing. Figure 3(right) shows the web interface. But this marketplace allows users to pay any price π_{custom} above the minimum π_{min} (4). This strategy can be used by customers in order to improve the priority of their products' production, as industrial agents always try to maximize both: the revenue and the user satisfaction.

$$\pi_{custom}(P) \geq \pi_{min}(P) \tag{4}$$

The Blockchain-enabled marketplace is based on the web3 library and JavaScript technologies too. Customers can pay products using of the digital wallets currently available, such as Metamask. This new interface is hosted on a different webserver, which can communicate with the production scheduler as well. Then, it includes a REST (Representational State Transfer) API (Application Programming Interface) so the industrial production scheduler can retrieve information about pending products automatically at any moment (and specifically before starting the production day). In addition, this SC includes a data field to maintain the current product state. This state can be updated through the REST API and the web interface by the production scheduler. This information, besides, is public, accountable, and transparent. And all customers can consume and check the production policy and schedule.

3.3 Efficient Production Scheduling

At this point, the production scheduler (5) acquired a collection of M pending products P_j to be manufactured. These products are modular and Q different industrial production systems must manufacture all the elemental modules p_i^j (6). Thus, the list of pending products can be decomposed. As a result, Q different sub-lists ℓ_q may be created (one

```
/* is ERC165 */
interface ERC1155 {
event TransferSingle(address,
address, address, uint256,
uint256);
event ApprovalForAll(address,
address, bool);
event URI(string, uint256);
function balanceOf(address,
uint256) external view
returns (uint256);
function
isApprovedForAll(address,
address) external view
returns (bool);
}
```

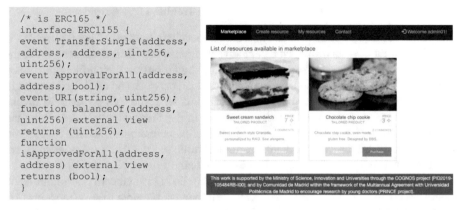

Fig. 3. (**left**) ERC1155 standard -extract- (**right**) Blockchain-enabled marketplace

for each industrial production systems) representing all the elemental modules every production system must manufacture (7).

$$\{P_j \quad j = 1, \dots, M\} \tag{5}$$

$$\left\{p_i^j \quad j = 1, \dots, M \quad i = 1, \dots, N_{P_j}\right\} \tag{6}$$

$$\{\ell_q q = 1, .., Q\} = \left\{\left\{p_i^j\right\}_q q = 1, \dots, Q\right\} \tag{7}$$

The q-th industrial production systems requires T_q seconds to produce an elemental module p_i^j. And then, a maximum of K_{max}^q elemental modules can be produced in a working day of T_{wd} seconds (8). In general, this number K_{max}^q will be lower than the number of pending modules to be manufactured by that industrial production system. On the other hand, all these modules need to be assembled. The assembling process requires T_{asb} seconds and a maximum of K_{max}^{asb} products can be assembled in a working day (9). Again, this number K_{max}^{asb} will be much lower than the number of pending products, M. Therefore, a scheduling algorithm is required to select which elemental modules p_i are actually manufactured and assembled, and in which order.

$$K_{max}^q = \frac{T_{wd}}{T_q} \tag{8}$$

$$K_{max}^{asb} = \frac{T_{wd}}{T_{asb}} \tag{9}$$

The i-th elemental module in the j-th pending product (hereinafter, a particle), p_i^j, must be manufactured by the q-th production system in the $\theta_{i,j,q}$ slot (known as the particle's position). This particle's position may take any value in the interval $[0, K_{max}^q]$, where the null value indicates the p_i^j is not scheduled to be manufactured (10). To calculate the most efficient and optimum position $\theta_{i,j,q}$ for each particle p_i^j, a Particle Swarm Optimization (PSO) algorithm is employed (11). R_q different iterations are executed, where r represents the current iteration, $\theta_{i,j,q}^{best}$ the best particle's position according to function $\mathcal{F}(\cdot)$ and $\theta_{global,q}^{best}$ particle's position ever created (according to function $\mathcal{F}$ as well). Additionally, parameters $\gamma_{(0,1)}^{1,2,3}$ are random numbers whose probability distribution is uniform in the interval $[0, 1]$. And parameters $\alpha_{1,2}$, w_{max} and w_{min} are configuration values.

$$\theta_{i,j,q} \in \left[0, K_{max}^q\right] \tag{10}$$

$$\theta_{i,j,q}^r = \theta_{i,j,q}^{r-1} + v_{i,j,q}^r$$

$$v_{i,j,q}^r = \begin{cases} \left\lfloor \varphi_{i,j,q}^r \right\rfloor & if \ \gamma_{(0,1)}^3 \\ \left\lceil \varphi_{i,j,q}^r \right\rceil & otherwise \end{cases} \tag{11}$$

$$\varphi_{i,j,q}^r = \alpha_1 \cdot \gamma_{(0,1)}^1 \cdot \left(\theta_{i,j,q}^{best} - \theta_{i,j,q}^{r-1}\right) + \alpha_2 \cdot \gamma_{(0,1)}^2 \cdot \left(\theta_{i,j,q}^{best} - \theta_{i,j,q}^{best}\right) + w_r \cdot v_{i,j,q}^{r-1}$$

$$w_r = w_{max} - \frac{(w_{max} - w_{min}) \cdot r}{R_q}$$

This PSO algorithm is executed in parallel for every industrial production system. As well as for the assembly process (12). But, in that case, some additional restrictions must be considered. A product P_j can only be assembled if all its elemental modules p_i^j are manufactured before the proposed assembly slot. If this condition is not met, the particle's position $\theta_{j,a}$ is set to the null value (meaning, "not assembled").

$$\theta_{j,a}^r = \begin{cases} \hat{\theta}_{j,a}^r \ if \ \hat{\theta}_{j,a}^r > \hat{\theta}_{j,a}^r \ \forall i, q \\ 0 \ otherwise \end{cases}$$

$$\hat{\theta}_{j,a}^r = \theta_{j,a}^{r-1} + v_{j,a}^r$$

$$v_{j,a}^r = \begin{cases} \left\lfloor \varphi_{j,a}^r \right\rfloor & if \ \gamma_{(0,1)}^3 \\ \left\lceil \varphi_{j,a}^r \right\rceil & otherwise \end{cases} \tag{12}$$

$$\varphi_{j,a}^r = \alpha_1 \cdot \gamma_{(0,1)}^1 \cdot \left(\theta_{j,a}^{best} - \theta_{j,a}^{r-1}\right) + \alpha_2 \cdot \gamma_{(0,1)}^2 \cdot \left(\theta_{j,a}^{best} - \theta_{j,a}^{best}\right) + w_r \cdot v_{j,a}^{r-1}$$

$$w_r = w_{max} - \frac{(w_{max} - w_{min}) \cdot r}{R_a}$$

All these parallel PSO algorithms are interconnected through the cost function $\mathcal{F}(\cdot)$ (13). This function must be minimized. This function is the combination of three partial sub-functions: the revenue function $\rho(\cdot)$, the expense function $\xi(\cdot)$ and the user satisfaction function $\mu(\cdot)$. Revenue and user satisfaction functions must be maximized, while expense function needs to be minimized. Parameters $\beta_{1,2,3}$ are employed to strengthen the relevance of some sub-functions against the others.

$$\mathcal{F}\left(\vec{\theta}\right) = \frac{\left(\xi\left(\cdot\right)\right)^{\beta_3}}{\left(\rho\left(\vec{\theta}\right)\right)^{\beta_1} \cdot \left(\mu\left(\vec{\theta}\right)\right)^{\beta_2}} \tag{13}$$

The revenue function $\rho(\cdot)$ is quite easy to calculate. The incomes associated to a production schedule are the prices π_{custom} paid by customers for all customized products to be (partially or totally) manufactured according to that schedule (14). Values are normalized considering the potential revenue for all pending products in the marketplace. The expense function $\xi(\cdot)$ can be obtained considering the unitary costs c_i of all elemental modules p_i making up products totally or particularly manufactured (15). Together with a storage cost c_{st} (16) for all those products P_j which are not totally manufactured or assembled and must be handled while finalized in a future time (increases linearly at a speed of c_{st}^0 monetary units). Values are normalized, considering the nominal production costs of all manufactured products. Finally, the user satisfaction function $\mu(\cdot)$ is a mean value but composed of decreasing exponential functions (17). As production delays go up, user satisfaction goes down. Function is normalized and τ_{user} is a configuration parameter.

$$\rho\left(\vec{\theta}\right) = \sum_{\forall j:\exists\theta_{i,j,q}\neq 0} \pi_{custom}\left(P_j\right) \tag{14}$$

$$\xi\left(\vec{\theta}\right) = \sum_{\forall j:\exists\theta_{i,j,q}\neq 0} \frac{\left(\sum_{i=1}^{N_{P_j}} c_i + c_P\right) + c_{st}^j}{\sum_{i=1}^{N_{P_j}} c_i + c_P} \tag{15}$$

$$c_{st}^j = c_{st}^0 \cdot \delta_{pd}^j \tag{16}$$

$$\mu\left(\vec{\theta}\right) = \frac{1}{M} \sum_{j=1}^{M} exp\left(-\frac{\left(\max_i\{\delta_{pd}^j\}-1\right)}{\tau_{user}}\right) \tag{17}$$

$$\delta_{pd}^j = \begin{cases} T_{wd} & if\ \exists\theta_{i,j,q}^r = 0 \vee \theta_{j,a}^r = 0 \\ \max_i\{\theta_{i,j,q}^r, \theta_{j,a}^r\} \cdot T_q & otherwise \end{cases}$$

4 Experimental Validation: Simulation and Results

In order to evaluate the performance of the proposed Industry 5.0 solution for mass customization scenarios, an experimental validation was designed and carried out. The experiment was based on numerical simulations, supported by MATLAB 2022a software suite. The proposed simulation scenario represented a bakery industry, where a large catalog of different products is available, including many ingredients and sub-products customers could combine freely, to design their customized products.

One hundred and twenty (120) sub-products and ingredients were considered and integrated in the Industry 5.0 solution, including the prosumer environment and the price calculation engine. Customized products were automatically and randomly designed and purchased. The Blockchain-enabled marketplace was deployed using the Ganache tool to recreate an entire Ethereum network using only local resources. Tokens and prices were also randomly invoked from the MATLAB suite, where the simulation scenario was managed. Products were designed and bought at random time instants, but with an average speed of seventy-five (75) products per hour. The total number of products to be manufactured was variable and was employed as a control parameter. The simulation scenario represented a real bakery industry [29], including eight different production lines, where all sub-products must be manufactured, and the final bakery product "assembled". The working day in this industrial scenario has a duration of twenty hours and products can be manufactured at a maximum rate of one hundred sub-products per hour (including the final assembly).

In this context, the proposed PSO production scheduling algorithm was implemented using native numerical programming in the MATLAB suite. The global efficiency (or cost-profit ratio) was monitored and calculated. For every configuration of the scenario, simulation results were generated twelve times to mitigate the impact of exogenous variables and effects. Additionally, results are compared to the real efficiency reported by the bakery industry used as a reference for an equivalent number of products (mass production), and to the efficiency in a traditional business for custom-made bakery products. Simulations were supported by a Linux server (Ubuntu 22.04 LTS) with the following hardware characteristics: Dell R540 Rack 2U, 96 GB RAM, two processors Intel Xeon Silver 4114 2.2G, HD 2TB SATA 7,2K rpm.

Figure 4 shows the experimental results. In this benchmark we compare the performance of the proposed solution against other existing strategies and technologies. As can be seen, the proposed Industry 5.0 preserves the economy of scale, typical of Industry 4.0 mass production scenarios. Contrary to traditional custom-made product

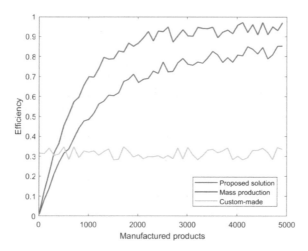

Fig. 4. Experimental results

businesses, where exponential models are not applied, and efficiency is constant. However, the proposed solution for mass customization applications shows a lower efficiency than previous Industry 4.0 mass production strategies. Although efficiency may be up to 25% lower for some situations, in average this distance is just around 8%. It is important to note here that transaction delays in real Blockchain network could also affect these results, although in general they will be negligible compared to production delays.

In general, as difference in efficiency is below 10% (typical error level in this kind of experiments), we can conclude that the proposed Industry 5.0 solution points in the right direction towards an efficient solution.

5 Conclusions and Future Work

In this paper, we propose an efficient and accountable Industry 5.0 production scheduling mechanism based on a transparent Blockchain-enabled marketplace and particle swarm optimization algorithms. Customers can create and buy customized products thanks to a prosumer environment and Smart Contracts which generate a transparent agreement between users and industrial agents. Later, Industry 5.0 producers combine three functions to define an optimization problem and find the most efficient production schedule: the revenue function, the expense function, and the user satisfaction function. Particle Swarm Optimization algorithms are employed to calculate the most profitable production schedule. Results show efficiency is still 8% (in average) lower than in previous Industry 4.0 mass production applications, but points towards a future efficient solution.

In future works, new types of industrial applications and different scenarios will be considered, in order to guarantee the general validity of these first results.

Acknowledgments. This work is supported by the Ministry of Science, Innovation and Universities through the COGNOS project (PID2019-105484RB-I00); and by Comunidad de Madrid within the framework of the Multiannual Agreement with Universidad Politécnica de Madrid to encourage research by young doctors (PRINCE project).

References

1. Leng, J., et al.: Industry 5.0: prospect and retrospect. J. Manuf. Syst. **65**, 279–295 (2022)
2. Bordel, B., Alcarria, R., Hernández, M., Robles, T.: People-as-a-Service dilemma: humanizing computing solutions in high-efficiency applications. Multidisc. Digital Publ. Inst. Proc. **31**(1), 39 (2019)
3. Shimoda, S., et al.: What is the role of the next generation of cognitive robotics? Adv. Robot. **36**(1–2), 3–16 (2022)
4. Bordel, B., Iturrioz, T., Alcarria, R., Sanchez-Picot, A.: Provision of next-generation personalized cyber-physical services. In 2018 13th Iberian Conference on Information Systems and Technologies (CISTI), pp. 1–6. IEEE (2018)
5. Lu, Y., Xu, X.: Cloud-based manufacturing equipment and big data analytics to enable on-demand manufacturing services. Rob. Comput.-Integr. Manuf. **57**, 92–102 (2019)
6. Bordel, B., Alcarria, R., Robles, T.: Recognizing human activities in Industry 4.0 scenarios through an analysis-modeling-recognition algorithm and context labels. Integr. Comput.-Aided Eng. **29**(1), 83–103 (2022)

7. Bordel, B., Alcarria, R., de Rivera, D.S., Robles, T.: Process execution in cyber-physical systems using cloud and cyber-physical internet services. J. Supercomput. **74**, 4127–4169 (2018)
8. Bordel, B., Alcarria, R., Robles, T., Martín, D.: Cyber–physical systems: extending pervasive sensing from control theory to the Internet of Things. Perv. Mob. Comput. **40**, 156–184 (2017)
9. Rožanec, J.M., et al.: Human-centric artificial intelligence architecture for industry 5.0 applications. Int. J. Prod. Res. **61**, 1–26 (2022)
10. Alohali, M.A., Al-Wesabi, F.N., Hilal, A.M., Goel, S., Gupta, D., Khanna, A.: Artificial intelligence enabled intrusion detection systems for cognitive cyber-physical systems in industry 4.0 environment. Cogn. Neurodyn. **16**(5), 1045–1057 (2022)
11. Shami, T.M., El-Saleh, A.A., Alswaitti, M., Al-Tashi, Q., Summakieh, M.A., Mirjalili, S.: Particle swarm optimization: a comprehensive survey. IEEE Access **10**, 10031–10061 (2022)
12. Maddikunta, P.K.R., et al.: Industry 5.0: a survey on enabling technologies and potential applications. J. Ind. Inf. Integr. **26**, 10257 (2022)
13. Akundi, A., Euresti, D., Luna, S., Ankobiah, W., Lopes, A., Edinbarough, I.: State of Industry 5.0—analysis and identification of current research trends. Appl. Syst. Innov. **5**(1), 27 (2022)
14. Xu, X., Lu, Y., Vogel-Heuser, B., Wang, L.: Industry 4.0 and Industry 5.0—inception, conception and perception. J. Manuf. Syst. **61**, 530–535 (2021)
15. Skobelev, P.O., Borovik, S.Y.: On the way from Industry 4.0 to Industry 5.0: from digital manufacturing to digital society. Industry 4.0 **2**(6), 307–311 (2017)
16. Yavari, F., Pilevari, N.: Industry revolutions development from Industry 1.0 to Industry 5.0 in manufacturing. J. Ind. Strat. Manag. **5**(2), 44–63 (2020)
17. Grabowska, S., Saniuk, S., Gajdzik, B.: Industry 5.0: improving humanization and sustainability of Industry 4.0. Scientometrics **127**(6), 3117–3144 (2022)
18. Ivanov, D.: The Industry 5.0 framework: viability-based integration of the resilience, sustainability, and human-centricity perspectives. Int. J. Prod. Res. **61**(5), 1683–1695 (2023)
19. Kotler, P.: From mass marketing to mass customization. Plan. Rev. **17**(5), 10–47 (1989)
20. Zhang, M., Guo, H., Huo, B., Zhao, X., Huang, J.: Linking supply chain quality integration with mass customization and product modularity. Int. J. Prod. Econ. **207**, 227–235 (2019)
21. Ashima, R., Haleem, A., Bahl, S., Javaid, M., Mahla, S.K., Singh, S.: Automation and manufacturing of smart materials in Additive Manufacturing technologies using Internet of Things towards the adoption of Industry 4.0. Mater. Today: Proc. **45**, 5081–5088 (2021)
22. Guzzi, E.A., Tibbitt, M.W.: Additive manufacturing of precision biomaterials. Adv. Mater. **32**(13), 1901994 (2020)
23. Inkulu, A.K., Bahubalendruni, M.R., Dara, A., SankaranarayanaSamy, K.J.I.R.: Challenges and opportunities in human robot collaboration context of Industry 4.0-a state of the art review. Ind. Robot: Int. J. Rob. Res. Appl. **49**(2), 226–239 (2021)
24. Shahrubudin, N., Lee, T.C., Ramlan, R.J.P.M.: An overview on 3D printing technology: technological, materials, and applications. Procedia Manuf. **35**, 1286–1296 (2019)
25. Kuhl, J., Rennpferdt, C., Krause, D.: Characteristic-oriented complexity cost analysis for evaluating individual product attributes. In: Andersen, A.-L., Andersen, Rasmus, Brunoe, Thomas Ditlev, Larsen, Maria Stoettrup Schioenning., Nielsen, Kjeld, Napoleone, Alessia, Kjeldgaard, Stefan (eds.) CARV/MCPC -2021. LNME, pp. 686–693. Springer, Cham (2022). https://doi.org/10.1007/978-3-030-90700-6_78
26. Pereira, T., Kennedy, J.V., Potgieter, J.: A comparison of traditional manufacturing vs additive manufacturing, the best method for the job. Procedia Manuf. **30**, 11–18 (2019)

27. Leng, J., et al.: Digital twin-driven rapid reconfiguration of the automated manufacturing system via an open architecture model. Rob. Comput.-Integr. Manuf. **63**, 101895 (2020)
28. Aheleroff, S., Xu, X., Zhong, R.Y., Lu, Y.: Digital twin as a service (DTaaS) in industry 4.0: an architecture reference model. Adv. Eng. Inf. **47**, 101225 (2021)
29. Bordel, B., Alcarria, R., Robles, T., de la Torre, G., Carretero, I.: Digital user-industry inter-actions and Industry 4.0 services to improve customers' experience and satisfaction in the European bakery sector. In: 2021 16th Iberian Conference on Information Systems and Technologies (CISTI), pp. 1–10. IEEE (2021)

Non-invasive Synthesis from Vision Sensors for the Generation of 3D Body Landmarks, Locations and Identification in Smart Environments

Aurora Polo-Rodriguez[1,4], Mathew Burns[2], Chris Nugent[2], Francisco Florez-Revuelta[3], and Javier Medina-Quero[4(✉)]

[1] Department of Computer Science, University of Jaén, 23071 Jaén, Spain
apolo@ujaen.es
[2] School of Computing, Ulster University, Belfast, Northern Ireland, UK
[3] Department of Information Technology and Computing, University of Alicante, Alicante, Spain
[4] Department of Computer Engineering, Automation and Robotics, University of Granada, 18071 Granada, Spain
javiermq@ugr.es

Abstract. This work proposes 3D body landmarks, location, and identification in multioccupancy contexts from vision sensors. Methods include high-performance vision tools, such as Yolo, DeepFace, and MediaPipe, to estimate 3D body landmarks and identification. First, we sense a smart space where a vision sensor is deployed to collect the activities of inhabitants. Our proposed model computes, identifies, tracks and obtains 3D body landmarks in multi-occupancy contexts. Third, 2D location over the floor is estimated based on homography projection, enabling fusing multiple vision sensors' information. Third, tracking and face recognition are integrated with non-supervised tracking to identify the inhabitants in the smart environment and relate the landmarks to them. A case study is presented to illustrate the proposal with an encouraging performance (f1-score: 0.98) in tracking multi-occupancy of two inhabitants with five scenes in two rooms.

Keywords: Computer Vision · Smart Environments · Machine Learning

1 Introduction

In recent times, the way we interact with our daily spaces has been revolutionised by smart environments [2]. In particular, the integration of 3D body landmarks, precise location estimation, and robust user identification has become a major focus of research [12]. This paper introduces a cutting-edge solution to facilitate the computation of 3D body landmarks in multi-occupancy contexts, which allows for the noninvasive synthesis of real environments from vision sensors.

© The Author(s), under exclusive license to Springer Nature Switzerland AG 2023
J. Bravo and G. Urzáiz (Eds.): UCAmI 2023, LNNS 842, pp. 57–68, 2023.
https://doi.org/10.1007/978-3-031-48642-5_6

Our approach combines the capabilities of vision sensors deployed in a smart space with deep learning-based applications to create virtual representations.

By taking advantage of the computing power of nearby vision sensors, we can compute 3D body landmarks [24], which are essential for comprehending human motion and interactions in the smart environment [20].

To achieve accurate 2D location estimation within the smart space, we introduce a homography projection based on foot estimation to locate the user in the real-world space [1]. In addition, an interesting method for non-supervised tracking based on features of deep learning [4,9] is also included. Finally, a face identification [13] enables the relation of body shapes with inhabitants and enriches the user experience of the virtual world where we project the synthesis of their poses.

Synthesis of vision is a promising research topic [16], where the shape of the human body generates human-like avatars that preserve intrinsic characteristics and anatomic coherence [21]. It is also partially related to the creation of digital twins, which provides rich scenarios of industry 4.0 [14], as well as in recognition of human activity, where virtual representations of individuals are created to mirror their behaviours, movements, and activities in the real world [6].

On the methods and applications proposed, the core of our methodology lies a set of high-performance vision tools, including Yolo [8], YoloFace [3], Resnet, DeepFace [18] and MediaPipe [19], which play a crucial role in estimating 3D body landmarks accurately and efficiently.

The remainder of this paper is organised as follows. Section 2 describes the materials and method approached in this work. Section 3 shows the results of a real-life case study. In the end, Sect. 4 concludes the paper and discusses ongoing works.

2 Materials and Methods

This section describes materials and methods for generating 3D body landmarks, locations, and identification in smart environments using vision sensors. First, the sensor, deployment, and setup are detailed. Next, the methods for segmentation and 3D landmark body estimation, computing 2D location, identification and tracking in multi-occupancy are described. Ultimately, we describe the tool to represent data in a virtual environment.

2.1 Sensor, Deployment and Setup

This section describes the vision sensors and deployment in smart environments. The target of this approach is spectrum visible sensors, such as IP cameras or embedded vision sensors, to collect data in real-time. The location of the devices is strongly recommended to be deployed in the corners of the room to provide a wide range of vision. This camera location, regarding other approaches located in zenithal or roof view, is key to provide face identification and trace body and foot of inhabitants at the same time.

Once the deployment of vision sensor is included, the methodology proposed is designed for one single vision sensor, so the next short-term setup is necessary:

– Real world floor coordinates. The location of marks or measurements to provide the 2D location of the user in real-world coordinates. Several points
between source images and related coordinates are necessary to model a
homography automatically.
– Facial database. Each inhabitant needs to walk for a short time across the
smart environment to build a facial database to identify in the scenes. The
approach automatically collects frontal and lateral faces.

After the camera setup is deployed with this minimal configuration, the vision
processing methods of this work are described in the next sections.

2.2   Body Segmentation and 3D Landmark Body Estimation

This section describes the body segmentation and 3D landmark body integrated
into this approach. The first stage consists of detecting objects from the scene
using the well-known tool You Only Look Once (YoLo) (v7), which provides
state-of-art performance in heterogeneous domains [22,23]. Using YoLo-v7, the
objects in the image and the people represented can be detected using bounding
boxes. This work will focus on inhabitants, but future works aim to integrate
other key objects in the virtual environment.

Once the bounding box for each person is detected, the second stage consists
of integrating Mediapipe for human pose estimation [10,11]. In concrete, we
compute the 3D body landmarks for each bounding box selected. The 3D body
landmarks and pixel segmentation of Mediapipe are key in this approach:

– To provide a non-invasive descriptive synthesis of the body of inhabitants in
a virtual environment which guarantees their privacy.
– To compute the 2D location on the floor of the inhabitants using a homography from image to floor coordinates in monocular vision.
– To remove background pixel from body segmented image, which reduces noise
in person tracking based on clothes and body features.

In Fig. 1, we describe an example of an image with one inhabitant, bounding
boxes, and 3D body landmarks.

A) original image   B) Object segmentation   C) 3D body landmarks
                                                           +pixel segmentation

Fig. 1. Example of an image with one inhabitant (A) with bounding boxes from YoLo
(B) and 3D body landmarks and pixel segmentation from Mediapipe (C).

The output data generated by computing the body segmentation and 3D landmark body estimation provides the main inputs for the next components of 2D location and tracking multi-occupancy from vision sensors, which are described in next sections.

2.3 Computing 2D Location of Inhabitants

In this section, we describe how computing the 2D location of the inhabitants in the smart environment. Here, we propose a monocular solution for each vision sensor. The reason behind this is integrating an approach where vision sensors are integrated autonomously in the first step and next share information to provide global information of the scene with real-world coordinates. We assume the hypothesis that the inhabitants will always have their feet on the floor. So, straightforwardly, the floor's surface from the cenital view in metres is related to points in pixels in the image, which correspond to the feet of the inhabitant. For calibration of cameras and translation of points from image to real world coordiantes, we use RANSAC (Random Sample Consensus) algorithm [5] to estimate intrinsic and extrinsic parameters of a camera from a data set containing corresponding points in the real world and in the captured image. For that, we relate a few points from the floor in the source image (x_1, y_1) to the real-world coordinates (x_2, y_2). The homography $(x_2, y_2) = H(x_1, y_1)$ is computed by RANSAC. We note the precision of translating the location of feet from image pixels to real-world coordinates of this method is affected by the location and distance of the vision device, whose granularity is expressed by pixel by meters and potentially affected by estimation errors.

In Fig. 2, we detail an example of 2D real-world location of inhabitants based on foot estimation.

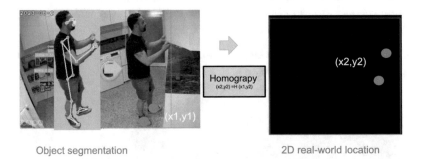

Object segmentation   2D real-world location

Fig. 2. 2D real-world location of inhabitants based on feet estimation and homography from the monocular vision sensor.

2.4 Non-supervised Deep Learning Tracking in Multi-occupancy

In this section, we describe the tracking of inhabitants in a multioccupancy context. The main challenge in this step is to relate the sequence of images for each person for each time step. A note to the reader is that the frequency of image collection is not (relatively) high due to the computational cost of the approach, so 1 FPS has been evaluated. It is due because the deep learning algorithms require high processing power or a big memory to store the deep models. As the FPS has not a high frequency, during this time, the movements and relations of the inhabitants change quickly in orientation, shape and colours. The SHIFT methods [25] have been discarded for that.

In this approach, we provide innovative non-supervised tracking based on deep learning. We use the pre-trained ResNet50 model [7] to extract features [17] for evaluating candidate similarity. The flattened layer of ResNet model contains the relevant features F_k from lower and high abstraction in rotation, translation for each inhabitant k. Using Cosine Similarity, we compute a similarity matrix between previous inhabitants j and candidates of the current image i: $S[i][j] = |F_i - F_j|$ where $S[i][j]$ determines a similarity score.

Next, we apply a Greedy search where minimal value and positions of the matrix $S^* = min(S), i^*, j^*$. If the value S^* overcomes a threshold, we consider it a new inhabitant due to significance differences; otherwise, we relate the potential candidate i^* with the previous inhabitant j^*. In the next step: i) we void the fill and row i^*, j^*, and ii) update the characteristics of inhabitants F_k, repeating the process. We note that an inhabitant can generate several candidates when the shape changes abruptly (if it sits after walking, for example). So, the final candidates are related to clusters of inhabitant shapes, which are tracked to one single person based on face recognition, which is described in the next section.

In Fig. 3 and Algorithm 1, we describe a multi-occupancy tracking example of images based on this approach.

Algorithm 1. Tracking clustering method

$features_{candidate} \leftarrow \emptyset$
while $frame \in camera$ **do**
    $inhabitants \leftarrow YoLo(frame)$
    $features \leftarrow ResNet(inhabitants)$
    $S \leftarrow cosine(features_{candidate}, features)$
    while $inhabitant \in inhabitants$ **do**
        $S^* \leftarrow min(similarity[inhabitant])$
        if $S^* < \alpha$ **then**
            $features_{candidate}[inhabitant] = features[inhabitant]$
        else if $S^* > \alpha$ **then**
            $features_{candidate} \cup features[inhabitant]$
        end if
    end while
end while

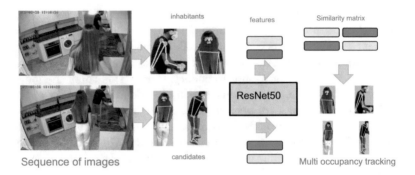

inhabitants   features   Similarity matrix

ResNet50

Sequence of images   candidates   Multi occupancy tracking

Fig. 3. Multi-occupancy tracking example of images based on this approach

2.5 Face-Based Identification

In this section, we describe the face identification of inhabitants for multi-occupancy developed in this work. To face this challenge, we have integrated DeepFace using VGG-Face [15,18] and YoloFace [3].

For that three stages are included, which are described in Fig. 4:

- We note, as we introduced in Sect. 2.1, that the setup of this work requires a short-term configuration which includes face recognition. For that, some minutes of face identification to generate a facial database for each inhabitant was developed previously to evaluate multi-occupancy context.
- YoloFace is applied to the bounding boxes for each inhabitant computed by YoLo-v7. We note YoloFace is properly trained over frontal and lateral faces, which is encouraging for our proposal to reduce the false positives in face-based identification. Thanks to the multi-occupancy tracking, we can trace the inhabitants at the scene and label them with the identification only in some points of the track.
- When YoloFace detects a frontal or lateral face in the image's bounding box, DeepFace searches in the facial database of the users and provides cosine similarity for each image. A k-nearest neighbour (K = 3) is developed, and average cosine similarity is threshold-ed to obtain hardy performance.

We note the face recognition related to images in tracking users with a single a camera performs a low rate of recognition. For that, the facial detection is related to a clusters of inhabitant shapes, which is described in previous section, in order to identify the track of before and after frames which configure the cluster.

2.6 Representing Data in a Virtual Environment

In this section, we describe how to fuse the information extracted from the vision sensor source and represent it in a virtual environment, preserving privacy but preserving the synthesis from collected images.

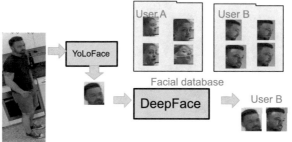

Fig. 4. Components for face-based identification on this approach

For that, a Unity application has been developed. Unity is a popular game engine and development platform allowing creators to design, animate, and program these 3D avatars relatively easily.

The application summarizes the information in a non-invasive synthesis using an avatar. The data can be uploaded from file format and MQTT in real-time, showing the information from the different components described in this work in avatars:

- 2D locations of inhabitants) which translates the avatar to a given real-world coordinates over the floor.
- Identification) which provides a different colour and head to the avatar based on the face recognition.
- Tracked 3D body landmarks) which describes the pose of the identified and tracked inhabitants to create a new avatar or replace their poses in real time.

The avatar of each inhabitant developed by Unity is related to a digital twin, presenting a virtual representation of the real-world person in the smart environment. Digital twins, on the other hand, are primarily used for analysis, monitoring, and optimization of real-world systems and monitoring. In future works, the movements of 3D avatars developed by Unity can be learned from real-world contexts to provide synthetic data to replicate the behaviour of physical assets or processes related to activity recognition.

In Fig. 5, an image is presented that displays the Unity representation compared to the real image.

3   Case Study

In this section, we describe the case study developed in this work to evaluate the performance of the approach for the generation of non-invasive synthesis (3D body landmarks, locations and identification) from vision sensors in a smart environment.

For that, real-life scenes with one camera have been deployed. An IP camera (Internet Protocol Camera) has been located in a room of the corner, and two

Office

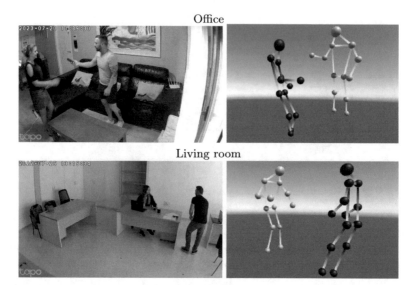

Living room

Fig. 5. Example of inhabitants in two different contexts (office and living room). Example of Unity representation (right) from spectrum visible images (left).

inhabitants developed daily tracks in 5 scenes. A man and a woman have participated in 5 scenes in two contexts (4 in the living room and one in the office room). Scene 1 refers to a woman reading on the sofa and a man coming into the room to watch TV next to her. Scene 2 describes a man coming to make a phone call and then giving the phone to the other person to end the conversation. In scene 3, a man is working on his laptop when the woman comes to consult and talk to him. In scene 4, the two have a coffee and talk while sitting on the sofa. Finally, scene 5 begins with the woman and the man working on two laptops, and next, he prepares her a drink.

In Fig. 5, we show an example of images collected in the two rooms.

The data collected, and the code of the models are available in: https://github.com/AuroraPR/Real2Virtual3D-UCAmI.

3.1   Results

In this section, we describe the results through the proposed methods of this work.

First, we evaluate the Non-supervised Deep Learning Tracking in multi-occupancy and Yolo performance in body detection. In Tables 1 and 2, we detail the number of frames where YoLo detects an inhabitant (detected body shapes), the number of clusters of body shapes computed by the multi-occupancy facial-body tracker (FBT) and the accuracy and error in separating the inhabitants by FBT (error = number of body shapes of another person not related to this user).

We highlight the good performance in detecting faces and body shapes in the frames by Deep Face and YoLo, as well as the relevance of FBT for discriminating clusters of candidates and relating the facial identification to them, which notably improves the performance of these tools in an isolated way. On the metric of false positive or error:

– Error FBT. We note that they represent the number of seconds (FPS = 1) where an inhabitant is traced as another.
– None detection of FBT. They represent the number of clusters of body shapes unrelated to facial identification, so in this case, we detect the presence of another person but not who is.

Table 1. Faces detected by YoLo and recovered faces to body shapes using FBT by scene.

Face and tracking/scenes	Man	Woman	None	Total
Face identified(1)	46	74	14	134
FBT(1)	51	79	4	134
Face identified(2)	18	14	34	66
FBT(2)	33	32	1	66
Face identified(3)	67	16	28	111
FBT(3)	75	34	2	111
Face identified(4)	69	33	36	138
FBT(4)	90	48	0	138
Face identified(5)	23	88	65	176
FBT(5)	40	136	0	176

Table 2. Detected body shapes, number of cluster, error and accuracy for discriminating clusters of candidates of FBT.

	Detected body shapes	N. of cluster	Error FBT	Accuracy FBT
Scene-1	134	5	7	0,95
Scene-2	66	3	3	0,95
Scene-3	111	3	3	0,97
Scene-4	138	3	5	0,96
Scene-5	176	5	0	1,00

Second, we have evaluated the Face-based identification provided by YoLo Face and Deep Face along the five scenes and two contexts. Three faces were unrelated to real faces (other objects), and only three false positives were related to user identification (recall: 1.0, precision: 0.97, f1-score: 0.98). We detail the confusion matrix in Table 3.

Table 3. Confusion matrix of face recognition by Deep Face in the five scenes.

predic\real	woman	man	None
woman	180	0	0
man	3	225	3
None	0	0	0

3.2 Limitations of the Study

The results of this work are mainly limited to the next points:

- Performance on real-life scenes: The evaluation is conducted on a short-term simple case study with two rooms and two inhabitants. In ongoing works, assessing how the proposed system performs in more complex real-world scenarios with varying lighting conditions, occlusions, and multiple occupants engaged in different activities is essential. So, the unreliability of our approach to diverse environments, demographics, and body type needs will be addressed in the next works to evaluate different body shapes, clothing, and cultural variations that impact the effectiveness of body landmark estimation and identification.
- Real-time Processing: The computational complexity of the employed vision tools (Yolo, DeepFace, MediaPipe) might limit real-time performance, especially when dealing with multiple occupants in the environment. The case study provides insights into the processing speed and hardware requirements for practical implementation with relatively low FPS=1.
- Single camera. In this work, a single camera is deployed, and the location and partial visibility of inhabitants is critical for the performance of the approach. In future works, multi-view cameras will be approached to fuse information from different locations and points of view in order to provide more robust and stable results regardless of the user location in the room with respect to a single camera location.

4 Conclusions and Ongoing Works

In this work, the authors have proposed a comprehensive system for 3D body landmark location and identification in a smart space using vision sensors and deep learning-based methods for describing inhabitants in a non-invasive way. A wide range of deep learning applications (YoLo, MediaPipe, DeepFace) is evaluated and processed together by our approach, including a facial body tracker. Results from 5 scenes of the case study show an encouraging performance. The presented work constitutes a significant step towards achieving accurate and real-time 3D body landmark location and identification in smart environments.

Despite the significant achievements in this work, there are several avenues for future research and improvement.

- Scalability: The proposed system's performance and accuracy should be evaluated in larger-scale smart spaces with more vision sensors and occupants which were able to improve the approach to provide robustness to different contexts in real-life applications.
- Real-Time Optimization: While Fog Computing reduces latency, further optimization techniques can be explored to enhance the real-time performance of the system. This may involve hardware acceleration, parallel processing, or more efficient algorithms.
- Integration with Smart Applications: The proposed system lays the foundation for multi-sensing approaches, such as personalized healthcare, smart home automation, and security systems. Future research will explore integrating the proposed model into these applications to demonstrate its practical value and effectiveness.
- Comparative Studies: Comparing the proposed model with other existing methods for 3D body landmark location and identification would provide insights into its strengths and limitations. A thorough comparison can help identify areas for further improvement.

By addressing the future research directions outlined above, researchers will contribute to advancing and practically implementing this technology in various domains.

Acknowledgements. This contribution has been supported by the Spanish Institute of Health ISCIII through the project DTS21-00047. Moreover, this research has received funding by EU Horizon 2020 Pharaon Project 'Pilots for Healthy and Active Ageing', Grant agreement no. 857188.

References

1. Castro, J., Delgado, M., Medina, J., Ruiz-Lozano, M.: An expert fuzzy system for predicting object collisions: its application for avoiding pedestrian accidents. Expert Syst. Appl. **38**(1), 486–494 (2011)
2. Castro, L.A., Bravo, J.: Modeling interactions in ambient intelligence. Pers. Ubiq. Comput. **26**(6), 1333–1335 (2022)
3. Chen, W., Huang, H., Peng, S., Zhou, C., Zhang, C.: Yolo-face: a real-time face detector. Vis. Comput. **37**, 805–813 (2021)
4. Cheng, S., Sun, J.X., Cao, Y.G., Zhao, L.R., et al.: Target tracking based on incremental deep learning (2015)
5. Chum, O., Matas, J., Kittler, J.: Locally optimized RANSAC. In: Michaelis, B., Krell, G. (eds.) DAGM 2003. LNCS, vol. 2781, pp. 236–243. Springer, Heidelberg (2003). https://doi.org/10.1007/978-3-540-45243-0_31
6. Gopinath, V., Srija, A., Sravanthi, C.N.: Re-design of smart homes with digital twins. In: Journal of Physics: Conference Series, vol. 1228, p. 012031. IOP Publishing (2019)
7. Jian, S., Kaiming, H., Shaoqing, R., Xiangyu, Z.: Deep residual learning for image recognition. In: IEEE Conference on Computer Vision & Pattern Recognition, pp. 770–778 (2016)

8. Jiang, P., Ergu, D., Liu, F., Cai, Y., Ma, B.: A review of yolo algorithm developments. Procedia Comput. Sci. **199**, 1066–1073 (2022)
9. Jiang, T., Zhang, Q., Yuan, J., Wang, C., Li, C.: Multi-type object tracking based on residual neural network model. Symmetry **14**(8), 1689 (2022)
10. Kim, J.W., Choi, J.Y., Ha, E.J., Choi, J.H.: Human pose estimation using mediapipe pose and optimization method based on a humanoid model. Appl. Sci. **13**(4), 2700 (2023)
11. Lin, Y., Jiao, X., Zhao, L.: Detection of 3d human posture based on improved mediapipe. J. Comput. Commun. **11**(2), 102–121 (2023)
12. Liu, P.L., Chang, C.C.: Simple method integrating openpose and rgb-d camera for identifying 3d body landmark locations in various postures. Int. J. Ind. Ergon. **91**, 103354 (2022)
13. Masi, I., Wu, Y., Hassner, T., Natarajan, P.: Deep face recognition: a survey. In: 2018 31st SIBGRAPI Conference on Graphics, Patterns and Images (SIBGRAPI), pp. 471–478. IEEE (2018)
14. Nikolakis, N., Alexopoulos, K., Xanthakis, E., Chryssolouris, G.: The digital twin implementation for linking the virtual representation of human-based production tasks to their physical counterpart in the factory-floor. Int. J. Comput. Integr. Manuf. **32**(1), 1–12 (2019)
15. Parkhi, O., Vedaldi, A., Zisserman, A.: Deep face recognition. In: BMVC 2015-Proceedings of the British Machine Vision Conference 2015. British Machine Vision Association (2015)
16. Pereira, A., Carvalho, P., Pereira, N., Viana, P., Côrte-Real, L.: From a visual scene to a virtual representation: a cross-domain review. IEEE Access **11**, 57916–57933 (2023)
17. Saraee, E., Jalal, M., Betke, M.: Visual complexity analysis using deep intermediate-layer features. Comput. Vis. Image Underst. **195**, 102949 (2020)
18. Serengil, S.: Deepface (2020). https://github.com/serengil/deepface
19. Singh, A.K., Kumbhare, V.A., Arthi, K.: Real-time human pose detection and recognition using mediapipe. In: eddy, V.S., Prasad, V.K., Wang, J., Reddy, K. (eds.) International Conference on Soft Computing and Signal Processing, vol. 1413, pp. 145–154. Springer, Heidelberg (2021). https://doi.org/10.1007/978-981-16-7088-6_12
20. Song, L., Yu, G., Yuan, J., Liu, Z.: Human pose estimation and its application to action recognition: a survey. J. Vis. Commun. Image Represent. **76**, 103055 (2021)
21. Tian, Y., Zhang, H., Liu, Y., Wang, L.: Recovering 3D human mesh from monocular images: a survey. arXiv preprint arXiv:2203.01923 (2022)
22. Wang, C.Y., Bochkovskiy, A., Liao, H.Y.M.: Yolov7: trainable bag-of-freebies sets new state-of-the-art for real-time object detectors. In: Proceedings of the IEEE/CVF Conference on Computer Vision and Pattern Recognition, pp. 7464–7475 (2023)
23. Yang, F., Zhang, X., Liu, B.: Video object tracking based on yolov7 and deepsort. arXiv preprint arXiv:2207.12202 (2022)
24. Zago, M., Luzzago, M., Marangoni, T., De Cecco, M., Tarabini, M., Galli, M.: 3d tracking of human motion using visual skeletonization and stereoscopic vision. Front. Bioeng. Biotechnol. **8**, 181 (2020)
25. Zhu, C.: Video object tracking using sift and mean shift (2011)

A Model to Support the Prediction of Indicators in the Diagnosis and Intervention of Autism Spectrum Disorder

Vanessa Ramos[1], Tania Mondéjar[2], Antonio Ferrández[3], Jesús Peral[4],
David Gil[1(✉)], and Higinio Mora[1]

[1] Department of Computer Technology and Computation, University of Alicante,
03690 Alicante, Spain
david.gil@ua.es
[2] Department of Psychology, University of Castilla la Mancha, Ciudad Real, Spain
[3] GPLSI Research Group, Department of Software and Computing Systems,
University of Alicante, 03690 Alicante, Spain
[4] LUCENTIA Research Group, Department of Software and Computing Systems,
University of Alicante, 03690 Alicante, Spain

Abstract. Autism Spectrum Disorder (ASD) is a developmental disability primarily characterized by challenges in social interaction and communication. Due to the unknown etiology of ASD, numerous computational psychiatry research studies have been carried out to identify pertinent features and uncover hidden correlations to detect this type of disability at an early stage. The aim of this ongoing project is to present the initial tests carried out on autistic children by analysing their conversations or writings to assess their social skills in order to find indicators for the most personalised intervention possible. This model would consist of the most advanced machine learning algorithms and Natural Language Processing techniques (e.g. Transformers or ChatGPT). The paper concludes by presenting a case study that utilized autism data to verify the efficacy of our proposed model, demonstrating remarkably promising findings.

Keywords: Autism Spectrum Disorder · Natural Language Processing · Transformers

1 Introduction

ASD is defined as a developmental disability characterized by deficits in social communication and interaction, as well as restricted and repetitive patterns of behavior, as outlined in the Diagnostic and Statistical Manual of Mental Disorders ("DSM-5"). In recent years, there has been extensive research attention directed toward Autism Spectrum Disorder (ASD) within the field of computational psychiatry. Consequently, ASD represents a heterogeneous disability with

J. Bravo and G. Urzáiz (Eds.): UCAmI 2023, LNNS 842, pp. 69–75, 2023.
https://doi.org/10.1007/978-3-031-48642-5_7

variations in symptom manifestation, severity, risk factors, research focus, and response to treatment.

Language difficulties are becoming increasingly common in children and are one of the main problems today with a prevalence of approximately 7% of the population [1]. Verbal ability is essential to understand the environment around us and to enable good communication and social interaction. There are some groups with language difficulties, either in their verbal or non-verbal communication, who can benefit from technology to achieve a better general understanding. If we follow the guidelines of the Diagnostic and Statistical Manual of Mental Disorders DSM5 we would focus on groups with language difficulties such as Specific Language Disorders (SLD), other disorders such as ASD or General Developmental Disorders (GDD) with communication difficulties. Such groups may include difficulties in discourse, pragmatics, phonology, word finding, semantics, etc. [2] fundamental to their development in early stages of life.

In the area of psychology, conventional practices have relied heavily on the utilization of numerous forms to establish diagnoses and facilitate treatment progress. Typically, these forms necessitated completion by family members or individuals within the immediate social circle of the person under assessment. However, this approach presented inherent complexities, including the time-consuming nature of form completion and the potential compromise in the quality of data obtained as a result [3]. Within the realm of computational psychiatry and ASD research frameworks, various Machine Learning (ML) approaches have been implemented [4,5], demonstrating their superiority over knowledge-based methods. Consequently, in the context of pediatric psychiatry, where optimal clinical decisions greatly impact the well-being of children with ASD, the application of ML approaches in the computational investigation of ASD has emerged as a prevailing trend for detection and management tasks.

Among the ML algorithms used for the prediction of autism diagnosis are Logistic Regression (LR) [6,7], Support Vector Machine (SVM) [6,8,9], Naive Bayes (NB) [6,7], ADTree [4,7,10], Convolutional Neural Network (CNN) [6], K-Nearest Neighbor (KNN) [6], and Artificial Neural Network [6], which have yielded good results [6]. Peral et al. [11] carried out a comparison with diverse algorithms, such as Decision Tree, Random Forest, Bayes, Adaboost, Part, Artificial Neural Networks, Support Vector Machines, and AttributeSelectedClassifier. Another paper by the same authors emphasised that despite the continuous advancements in machine learning and deep learning models, the issue of insufficient data remains a significant limitation [12].

There are also other studies that propose a deep learning approach combined with the feature selection method for the diagnosis of ASD using a functional magnetic resonance imaging (fMRI) dataset. The proposed method is evaluated on the worldwide fMRI dataset, known as ABIDE (Autism Brain Imaging Data Exchange). However, despite the potential of this dataset for extracting functional biomarkers for ASD classification and its widespread attention, the classification of ASD in this dataset is in its early stages, mainly due to the insufficient number of samples and its heterogeneity [13].

We can summarize that many of these scientific studies are based on data obtained through structured questionnaires, where data collection mostly occurs in specialized clinical or research settings and when there is already a suspicion of ASD in the individual [10]. This poses a problem not only due to the complexity of obtaining information but also because of the bias that can arise from not having representative information from individuals without ASD [10].

Nowadays, advances in ML, especially in the field of natural language processing [14], have opened up new opportunities for autism diagnosis [15–17]. One promising approach is the use of Transformer-based models, which are neural network architectures designed to capture and understand complex patterns in sequential data [18]. In the context of pediatrics psychiatry, where tics have a great impact on the wellbeing of children, tools for language detection and management tasks have emerged as a predominant trend, just as in the area of psychology. For these reasons, we propose a Transformer-based model where conversations of children with communication difficulties are analysed in order to establish early indicators to detect these difficulties in comparison with children who do not have such difficulties.

The aim of this study is to investigate the potential of Transformer-based models for autism diagnosis. A Transformer-based classification model has been developed to analyze and process the collected textual information. It is expected that the results of this study will provide a better understanding of the use of Transformer-based models in autism diagnosis. Additionally, the proposed model is expected to improve the accuracy and efficiency of diagnosis, thereby assisting healthcare professionals and clinicians in clinical decision-making and personalized intervention planning for individuals with autism.

The remainder of this paper is structured as follows: in Sect. 2, we present the experiments carried out with our computational model. Section 3 summarizes the conclusions of this ongoing project.

2   Case Study

This section describes the initial experiments that have been carried out in this project. A total of 12 participants were included in the study, divided into two groups: 6 with a neurotypical profile and 6 with language difficulties, all ranging in age from 4 to 8 years. The audio recordings collected for the research captured the participants' responses to various tests framed within the Weschler Test (WISC), designed to assess the overall verbal area as well as specific abilities in comprehension, similarity, information, and vocabulary. This methodology allowed us to compare typical responses within their respective age groups with the aim of identifying potential differences. It is important to note that the evaluation did not consider underlying emotional content or prosody.

The collected data consisted of verbal expressions in response to various prompts. The primary purpose of analyzing this audio data was to examine responses to specific questions and evaluate the description of concepts provided by the children, in order to study the influence of context and literalness of speech, as well as their ability to store and use verbal data, their competence in semantic grouping, and their comprehension and abstraction of knowledge. These skills are essential for communication and interaction with the environment, and the early detection of difficulties in these areas can significantly contribute to improving the quality of life for children. It is worth mentioning that this study followed the guidelines for this type of research in accordance with the Declaration of Helsinki and the university's code of ethics. To achieve this, after informing the parents of the study participants and obtaining their consent, the data was anonymized and kept solely in the possession of the person responsible for collecting them, ensuring that their identity could not be linked to the responses obtained in the study. The purpose of data collection is purely for research purposes.

The Transformer model will be trained and evaluated using a supervised learning approach, where labeled datasets of individuals with and without autism will be used. A pilot sample of 12 children between 4 and 8 years of chronological age has been selected in order to have a first test of the functioning of this computational model.

The conversations of each child have been stored as .ogg audio files. Then, these files have been transcribed to text by OpenAI whisper application (https://huggingface.co/spaces/openai/whisper), specifically the "small" model with Spanish language setting. The textual version of the conversation of each child has been tagged with the classes "withProblem" and "withoutProblem". Therefore, the whole corpus is formed by 12 instances (6 withProblem):

0. El pie de la dedo.. La promesa es real.. ¡Ac... withProblem
1. ¡Enciagel!. Macalones. para canciones. ¿Cómo... withProblem
2. El pie es una parte del cuerpo, tiene cinco ... withProblem
3. es que tiene dedos y uñas. que se tiene que ... withProblem
4. El Parandar. La comida es una cosa que sirve... withProblem
5. y esto. ¡Soy el ponerme en el bote!. ¡Nos de... withProblem
6. Es una parte del cuerpo.. es lo que comemos.... withoutProblem
7. Pedra mustica. Con lo que andas.. Lo que se ... withoutProblem
8. es una parte de cuerpo hasta una finalidad. ... withoutProblem
9. una parte del cuerpo para andar.. pa' despar... withoutProblem
10. Algo colucciando. Algo que se calló mi. La S... withoutProblem
11. Para que andemos. Para que comemos. ¡Soy Ada... withoutProblem

In the excerpt shown, some spelling errors are detected and assumed in the subsequent classification process, which has been performed by the Transformer model "tf_roberta_for_sequence_classification" (https://huggingface.co/PlanTL-GOB-ES) with 124,644,866 parameters. The training has consisted on 5 epochs:

Epoch 1/5. ETA: 26 s - loss: 0.6944 - categorical_accuracy: 0.5000
Epoch 2/5. ETA: 7 s - loss: 0.6909 - categorical_accuracy: 0.5000
Epoch 3/5. ETA: 7 s - loss: 0.6839 - categorical_accuracy: 0.6250
Epoch 4/5. ETA: 7 s - loss: 0.6764 - categorical_accuracy: 0.8750
Epoch 5/5. ETA: 8 s - loss: 0.6456 - categorical_accuracy: 0.8750

The results, obtained through a ten-fold cross-validation using 90% of the corpus for training, are presented in Table 1. Whereas the evaluation on the 10% of the test corpus has presented a 50% of accuracy. Despite the small corpus, and the transcription errors, these results encourage us to continue this research, specially by compiling a larger corpus.

Table 1. The 10-fold cross-validation results with 90% of the training corpus.

	precision	recall	f1-score	support
withProblem	0.89	0.93	0.89	5
withoutProblem	0.79	0.80	0.79	5
accuracy			0.88	10
macro_avg	0.84	0.87	0.84	10
weighted_avg	0.84	0.87	0.84	10

3   Conclusions

The aim of this study is to investigate the potential of Transformer-based models to detect significant discrepancies in language, specifically in oral communication in children as a primary indicator of language difficulties and possible associated disorders. Being aware that this work is a preliminary study, we can infer that these models, with adequate experience and exposure to both oral and written discourse, can be good detectors to aid in the diagnosis and treatment of autism spectrum disorders or other disorders with language and communication impairment. It's important to note that this work represents an innovative approach aimed at facilitating the early detection of ASD using Transformer models.

Some of the limitations encountered during the experiments included the limited number of participants and, regarding the audio recordings, their short length due to being responses to specific questions. Therefore, it would be advisable, for future research, to increase both the number of participants and the length and spontaneity of the speeches in order to observe a more natural language.

Acknowledgements. This research has been partially funded by the BALLADEER project (PROMETEO/2021/088) and the project NL4DISMIS (CIPROM/2021/21) from the Consellería Valenciana (Generalitat Valenciana). Furthermore, it has been partially funded by the AETHER-UA (PID2020-112540RB-C43) project from the MCIN and the R&D projects "CORTEX" (PID2021-123956OB-I00), funded by MCIN/ AEI/10.13039/501100011033/. This result has been supported through the Spanish Government by the FEDER project PID2021-127275OB-I00.

References

1. Vargas, A.L., Ahufinger, N., Igualada, A.J., Torrent, M.S.: Descripción del cambio del tel al tdl en contexto angloparlante. Revista de investigación en Logopedia **11**(1), 9–20 (2021)
2. RCSLT: Royal College of Speech & Language Therapists. Briefing paper on language disorder with a specific focus on developmental language disorder (2017)
3. Cai, L., Zhu, Y.: The challenges of data quality and data quality assessment in the big data era. Data Sci. J. **14**, 1–10 (2015)
4. Hyde, K.K., et al.: Applications of supervised machine learning in autism spectrum disorder research: a review. Rev. J. Autism Dev. Disord. **6**, 128–146 (2019)
5. Mak, K.K., Lee, K., Park, C.: Applications of machine learning in addiction studies: a systematic review. Psychiatry Res. **275**, 53–60 (2019)
6. Raj, S., Masood, S.: Analysis and detection of autism spectrum disorder using machine learning techniques. Procedia Comput. Sci. **167**, 994–1004 (2020)
7. Kosmicki, J., Sochat, V., Duda, M., Wall, D.: Searching for a minimal set of behaviors for autism detection through feature selection-based machine learning. Transl. Psychiatry **5**(2), e514 (2015)
8. Bozhilova, N., et al.: Profiles of autism characteristics in thirteen genetic syndromes: a machine learning approach. Mol. Autism **14**(1), 3 (2023)
9. Li, B., Sharma, A., Meng, J., Purushwalkam, S., Gowen, E.: Applying machine learning to identify autistic adults using imitation: an exploratory study. PLoS ONE **12**(8), e0182652 (2017)
10. Cavus, N., et al.: A systematic literature review on the application of machine-learning models in behavioral assessment of autism spectrum disorder. J. Personalized Med. **11**(4), 299 (2021)
11. Peral, J., Gil, D., Rotbei, S., Amador, S., Guerrero, M., Moradi, H.: A machine learning and integration based architecture for cognitive disorder detection used for early autism screening. Electronics **9**(3), 516 (2020)
12. del Mar Guillén, M., Amador, S., Peral, J., Gil, D., Elouali, A.: Overcoming the lack of data to improve prediction and treatment of individuals with autistic spectrum disorder and attention deficit hyperactivity disorder. In: Bravo, J., Ochoa, S., Favela, J. (eds.) Proceedings of the International Conference on Ubiquitous Computing & Ambient Intelligence, UCAmI 2022. LNNS, vol. 594, pp. 760–771. Springer, Cham (2023). https://doi.org/10.1007/978-3-031-21333-5_75
13. Zhang, J., Feng, F., Han, T., et al.: Detection of autism spectrum disorder using fMRI functional connectivity with feature selection and deep learning. Cogn. Comput. **15**, 1106–1117 (2023). https://doi.org/10.1007/s12559-021-09981-z
14. Liu, P., Yuan, W., Fu, J., Jiang, Z., Hayashi, H., Neubig, G.: Pre-train, prompt, and predict: A systematic survey of prompting methods in natural language processing. ACM Comput. Surv. **55**(9), 1–35 (2023)

15. Chi, N.A., et al.: Classifying autism from crowdsourced semistructured speech recordings: machine learning model comparison study. JMIR Pediatr. Parenting **5**(2), e35406 (2022)
16. Lee, J.H., Lee, G.W., Bong, G., Yoo, H.J., Kim, H.K.: Deep-learning-based detection of infants with autism spectrum disorder using auto-encoder feature representation. Sensors **20**(23), 6762 (2020)
17. Cho, S., Liberman, M., Ryant, N., Cola, M., Schultz, R.T., Parish-Morris, J.: Automatic detection of autism spectrum disorder in children using acoustic and text features from brief natural conversations. In: Interspeech, pp. 2513–2517 (2019)
18. Wolf, T., et al.: Transformers: state-of-the-art natural language processing. In: Proceedings of the 2020 Conference on Empirical Methods in Natural Language Processing: System Demonstrations, pp. 38–45 (2020)

ADLnet: A 1d-CNN for Activity of Daily Living Recognition in Smart Homes

Fabio Salice[1] , Andrea Masciadri[1] , Giuseppe Di Blasio[2], Matteo Venturelli[2], and Sara Comai[1]([⊠])

[1] Department of Electronics Information and Bioengineering, Politecnico di Milano, Milan, Italy
{fabio.salice,andrea.masciadri,sara.comai}@polimi.it
[2] Department of Physics, Politenico di Milano, Milan, Italy
{giuseppe.blasio,matteo.venturelli}@mail.polimi.it

Abstract. Human activity recognition (HAR) systems enable continuous monitoring of human behaviours in several areas, including activity of daily living (ADL) detection in ambient intelligent environments. The extraction of relevant features is the most challenging part of sensor-based HAR. Feature extraction influences algorithm performance and reduces computation time and complexity. However, the majority of current HAR systems rely on handcrafted features that are incapable of handling complex activities, especially with the influx of multimodal and high-dimensional sensor data. Over the last few decades, Deep Learning has been considered to be one of the most powerful tools to handle huge amounts of data. Thus, we developed ADLnet, a One-Dimensional Convolutional Neural Network (1d-CNN) for recognizing ADLs in Smart Homes, as part of an ambient assisted living framework to provide assistance to elderly inhabitants. We propose an innovative method to scan and classify time-series sensor data such as the CASAS dataset, which has been used for the training/validation/testing process. Testing results show very high performance.

Keywords: Deep Learning · Neural Networks · Human Activity Recognition · Ambient Intelligent environments · Feature extraction · ADL detection · Ambient Assisted Living · Time-series sensor data · CASAS dataset

1 Introduction

Based on data from the World Population Prospects [1], the global age structure is shifting, with a projected increase in the number of older individuals in the coming decades. By 2050, it is estimated that one in six people worldwide (16%) will be over the age of 65, compared to one in eleven (9%) in 2019. Furthermore, by 2050, approximately one in four individuals (25%) in Europe and Northern America will be aged 65 or older. In light of these demographic changes, it becomes imperative to enhance the quality of independent living for the elderly and their families.

Over time, various capabilities of human beings, including visual, auditory, physical, and cognitive functions, tend to decline, leading to two significant issues. The former is an increased likelihood of home accidents, such as falls and sudden illnesses. Falls alone account for an estimated 684,000 fatal incidents annually, ranking as the second leading cause of unintentional injury death worldwide, after road traffic injuries [2]. The latter is a gradual deterioration of self-sufficiency, which directly impacts the level of assistance required by older individuals. Possible declines are visible only in the long term. Since changes in the regular daily mobility routine of an elderly person can be an indicator or early symptom of developing health problems [3], sensor technology can be implemented in order to gain a detailed view of the daily mobility of a person at home when performing everyday activities.

Monitoring human behaviour poses significant challenges. The problem of detecting unusual changes in the daily behaviour of an older person who lives independently at home has been widely investigated in the literature [4]. Automated recognition of human activities has become crucial in ambient intelligence applications, and the integration of artificial intelligence (AI) into smart homes presents a valuable opportunity. Existing solutions for human activity recognition (HAR) in smart homes often rely on sensor-based systems that utilize both wearable and non-wearable sensors to track daily behaviour and provide alerts when deviations occur. However, previous studies indicate that wearable and camera-based sensors present computational complexities, and privacy concerns, and are not well-received by older individuals [5]. The elderly may feel discomfort wearing sensors continuously, might forget to wear them at times, or express concerns about their privacy when being monitored by home cameras.

For this reason, our focus is on Active and Assisted Living (AAL) solutions that utilize only non-invasive sensors strategically placed within the home environment, like environmental, movement, and contact sensors. DL models have shown the potential to outperform the state of the art in HAR [6]. In general, DL involves neural networks made of several layers that process nonlinear information organized hierarchically, where the output of each layer is the input for the next one. Well-known DL architectures for time series classification include One-Dimensional Convolutional Neural Networks (1d-CNNs) [7].

In this work, we built a 1d-CNN to accurately perform HAR since, in general, CNNs are good at reducing frequency variations and capturing local dependency [8]. The neural network has been trained on the CASAS dataset [9], which is one of the most used in this field.

2   Related Works

Several smart home projects have been implemented in the past decade, which use sensors for activity recognition inside the home environment [10]. ADL detection can be achieved by means of different Machine Learning (ML) approaches. Although classical ML techniques, such as Support Vector Machine (SVM), are simple to use, they require manual data feature extraction, which is time-consuming and heavily dependent on human knowledge of the domain [11]. To solve this problem, studies have delved into techniques that involve automatic feature extraction with less human effort using DL, which provides the ability to learn features from raw sensors data with minimal pre-processing [8]. Using multiple layers of abstraction, DL methods learn intricate features representation from raw sensors data and discover the best pattern to improve recognition performance [12]. Many researchers have demonstrated the potential of DL to push the state of the art in HAR: in [13,14] DL is applied to movement data captured with wearable sensors, regarding the whole body's movements, the gait or hand's movement, outperforming state-of-the-art algorithms. Some frequently used DL methods for human activity recognition are Convolutional Neural Networks (CNN), Vanilla Recurrent Neural Networks (RNN) and Long Short-Term Memory (LSTM), Deep Belief Networks (DBN), Autoencoders and Gated Recurrent Units (GRU).

Motivated by the recent success of CNNs architectures in different fields (e.g., image classification [15,16] and speech recognition [17]), researchers have started adopting them for time series analysis, showing great results [18]. In general, the input of the neural network is the original raw data and extracting features from this data tends to maximize the overall performance. CNNs are one of the most used DL approaches in the HAR field not only for their ability to automatically extract features from data but also for their activity classification capability [19].

Recent CNNs applications in time series analysis and ADL recognition are represented by [20,21]: here, the authors propose two different CNNs to detect human activities from accelerometer raw data. In [20] Yang et al. achieved very good accuracy (up to 93,8%) modifying the convolution kernels to adapt the characteristics of tri-axial acceleration signals, whereas in [21] Zeng et al. proposed a CNN which applies weight sharing to accelerometer data showing good accuracy too.

Several other works show that convolutional networks give excellent results compared to other algorithms. Wang et al. chose a different approach and encoded time series data as different types of images: this enabled the use of techniques from computer vision for classification. They used Tiled Convolutional Neural Networks on 12 standard datasets to learn high-level features. The classification results of their approach are competitive with five state-of-the-art approaches [22].

Cui et al. propose a Multi-scale Convolutional Neural Network (MCNN), which incorporates feature extraction and classification in a single framework. Leveraging a novel multi-branch layer and learnable convolutional layers, MCNN automatically extracts features at different scales and frequencies, leading to excellent feature representation comparable with other state-of-the-art methods [23]. Another attempt to enhance time series classification was proposed in [24], which employs the same idea of multiple branches within the CNN architecture, except that the input is not a transformed version of the time series signal fed into each branch, but rather a duplicate of the same time series signal fed into all the branches (three branches), showing similar results. Wang et al. proposed baseline models which are pure end-to-end without any heavy preprocessing on the raw data or feature crafting [25]. The proposed Fully Convolutional Network (FCN) achieves premium performance to other state-of-the-art approaches and their exploration of the very deep neural networks with the ResNet structure is also competitive.

In [26] the authors proposed a joint temporal model to conduct a parallel combination of LSTM and 1d-CNN architectures to improve the accuracy of HAR. The proposed method was extensively evaluated for ADL recognition using binary sensors dataset showing great results in terms of F1 score.

In [27] Alghamdi et al. used a 1d-CNN for ADL recognition in a smart home, and results showed that the model achieved really good performances in terms of recall, precision and F_1 score. The main goal of the paper was to study the potential energy saving through activity recognition. For this reason, the authors trained their model using ContextAct@A4H dataset, in which most of the sensors are bound to energy monitoring.

The authors in [10] implemented a CNN in the framework of activity recognition, comparing its results to other models such as LSTM, RNN and other ML algorithms, including Naive Bayes (NB), Hidden Markov Models (HMM), Hidden Semi-Markov Models (HSMM) and Conditional Random Fields (CRF). The experimental results on publicly available smart home datasets demonstrate that the performance of 1D-CNN is similar to LSTM and better than the other probabilistic models.

One more comparison was conducted in [6] to evaluate state-of-the-art ML techniques and the DL models for the classification of ADLs in smart homes, including CNNs, LSTM, SVM and HMM. They used a real-world dataset (ARAS) to evaluate and compare the performance of the models. The results of the comparison indicated that the CNN model had the best performance compared to the other evaluated techniques across both datasets in term of F_1 score.

3   Proposed Solution

Dataset Selection. For our study, we chose a dataset that met specific criteria: the dataset comprises a time series of sensor activations; to address the scenario of human activity recognition (HAR) for lonely elderly individuals, we focused on single-resident apartments; we also needed that each sensor activation in the time series corresponded to a single activity of daily living (ADL), such as cooking, rather than multiple ADLs like cooking and watching TV in order to provide a one-hot encoded ground truth to the neural network for ADLs classification; considering the types of sensors, we were not interested in datasets related to wearable sensors because of privacy issues and computational complexity. We evaluated various datasets with similar structures but varying dimensions and sensor types. The datasets we considered for selection included CASAS [9], van Kasteren [28], Ordonez [29], ContextAct [30], and the HAR Dataset of the University of Mannheim [31]. Ultimately, we opted for the CASAS dataset, which provides single-resident, one-hot labeled datasets associated with different houses and residents. This dataset incorporates sensor types such as PIR (passive infrared) sensors, door sensors, and light sensors. An excerpt of the CASAS dataset is shown in Fig. 1. It contains a timestamp, the sensor that was activated, the room and the location where it was activated, the sensor type, and the detected activity.

```
<DATE>      <TIME>              <SENSOR>  <ROOM>    <LOCATION>  <VALUE>  <SENSOR_TYPE>           <ACTIVITY>

[...]

2012-07-21  14:02:17.396243  MA016   Kitchen   Kitchen   ON    Control4-MotionArea   Wash_Lunch_Dishes
2012-07-21  14:02:19.053527  MA016   Kitchen   Kitchen   OFF   Control4-MotionArea   Wash_Lunch_Dishes
2012-07-21  14:02:19.304187  M002    Kitchen   Kitchen   OFF   Control4-Motion       Wash_Lunch_Dishes
2012-07-21  14:02:20.172207  MA016   Kitchen   Kitchen   ON    Control4-MotionArea   Wash_Lunch_Dishes
2012-07-21  14:02:21.306634  MA016   Kitchen   Kitchen   OFF   Control4-MotionArea   Wash_Lunch_Dishes
2012-07-21  14:02:24.148691  MA016   Kitchen   Kitchen   ON    Control4-MotionArea   Wash_Lunch_Dishes
2012-07-21  14:02:24.742626  M002    Kitchen   Kitchen   ON    Control4-Motion       Wash_Lunch_Dishes
2012-07-21  14:02:25.876328  M002    Kitchen   Kitchen   OFF   Control4-Motion       Wash_Lunch_Dishes
2012-07-21  14:02:26.940210  MA016   Kitchen   Kitchen   OFF   Control4-MotionArea   Wash_Lunch_Dishes
2012-07-21  14:02:28.145896  M002    Kitchen   Kitchen   ON    Control4-Motion       Wash_Lunch_Dishes
2012-07-21  14:02:28.253628  MA016   Kitchen   Kitchen   ON    Control4-MotionArea   Wash_Lunch_Dishes
2012-07-21  14:02:31.673533  M002    Kitchen   Kitchen   OFF   Control4-Motion       Wash_Lunch_Dishes

[...]
```

Fig. 1. An example of sensor events in CASAS dataset.

Data Preprocessing. We preprocessed the raw dataset by transforming it into a list of training pairs, which consisted of an instance and its corresponding ground truth. Each instance was represented by a vector, with each element uniquely associated with a specific sensor installed in the apartment. Each instance consists of a vector where each element is uniquely associated with a specific sensor placed in the apartment.

The vector elements were ordered in a way that sensors belonging to the same rooms were grouped together. For example, in Fig. 2, sensors 6 to 14 were placed in the kitchen, while sensors 15 to 21 were positioned in the living room.

In our experiments, we standardized the vector dimension to $N = 60$. If the number of sensors in a dataset was fewer than N, we evenly distributed empty values within the vector to maintain a clear separation between sensors located in different rooms. The choice of $N = 60$ was made to ensure an adequate sensor density across all the analyzed apartments.

To generate input instances for our model, we employed the sliding window technique, a well-documented approach in the literature [19]. This technique involves splitting the dataset into windows, with each window comprising W consecutive lines from the original dataset. Whenever a sensor value is 'ON' (for PIR sensors), 'OPEN' (for door sensors) or a number different from zero (for light sensors) the corresponding vector value associated with that sensor increments by one. For example, in the upper diagram of Fig. 2, the 12th sensor is activated three times in the input instance.

To ensure effective training on different activity lengths, we conducted a statistical analysis of the datasets and adjusted the window dimension, W, based on the smallest mean length of an activity in each dataset. The length of an

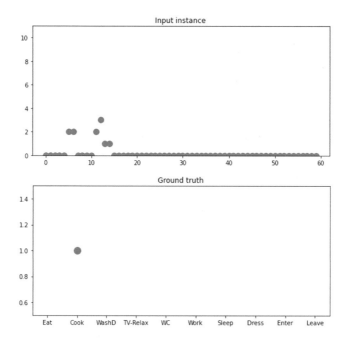

Fig. 2. Example of a training pair.

activity is measured by the number of lines in the dataset that correspond to that activity. In this way, our model is trained on both short activities and segments of longer ones and the network is able to self-adapt to different ADL lengths. For all the datasets considered in our study, we set W to 20.

Another important parameter to define in our model is the padding, denoted as P. Padding consists in adding additional elements to the input sequence to ensure that the convolutional filters have sufficient context and can capture information at the boundaries. We chose to set P to 1 the network is capable of classifying the activity of daily living (ADL) performed by the resident whenever a sensor signal is received. This enables real-time operation and makes it practical for real-world applications where timely assistance for residents is needed.

For each input instance in our model, there is a corresponding ground truth vector that serves as the target for classification. The ground truth vector has a dimension of n, where n represents the number of possible activity of daily living (ADL) classes that the network can classify. In our experimental dataset, we have ten distinct ADLs to be classified ($n = 10$, as indicated in the bottom diagram of Fig. 2. The ground truth vector is encoded in a one-hot format, where all values are set to zero except for the position associated with the dominant ADL, which is represented by a value of one. The dominant ADL is determined as the longest activity within each window, using the previously defined length calculation.

In this way, our approach to ADL recognition is based on detecting the most probable ADL starting from a sequence of sensors activations. This approach solves the problem of recognizing transitions between different ADLs. To help the network classify instances where the dominant ADL length is shorter than the window size W, we have implemented a systematic data augmentation (DA) technique which will be explained in the next section. Since several parameters depend on the input, this type of data preprocessing allows great adaptability to a large variety of datasets.

Data Augmentation. Data augmentation techniques are well-established in the literature and widely recognized for their effectiveness in improving the generalization, performance, and mitigating overfitting of machine learning models [32–34]. Given their proven benefits, along with the considerations discussed in the previous section, we have incorporated data augmentation into our approach.

Also in this case the method adapts to different datasets and requires a statistical analysis to ensure its effectiveness. Knowing the ADL mean length D of a dataset, the augmented one consists of X copies of it, where $X = \left\lceil \frac{D}{W} \right\rceil$ is the round-up integer of the ratio between the mean length of all ADLs and the mean length of the shortest ADL. Each copy i has a sliding window dimension $W_i = i \cdot W$, where i goes from 1 to X. For all $i \neq 1$ (all copies of the dataset whose window dimension W_i is more than W), training pairs in which the length of the dominant ADL is less than $0.75 \cdot W_i$ are deleted. If we define "noise" as the activations of the sensors associated with a non-dominant ADL in a specific window, this method of data augmentation adds instances with a noise percentage up to 25%. This should help the network to identify features related to low-noise windows because we are providing "cleaner" instances. As a consequence, this knowledge should also improve the classification of noisy data.

Data augmentation can provide bigger datasets, as we can see in Table 1. The table refers to five CASAS datasets: it contains the number of sensors used in the dataset, the quantity of training pairs generated from the original dataset before applying any data augmentation techniques, and the number of training pairs obtained after applying data augmentation.

Table 1. Number of sensors and number of generated training pairs (before and after Data Augmentation (DA)) for each of the analyzed datasets.

Dataset	Number of sensors	Training pairs before DA	Training pairs after DA
csh101	34	168 647	662 234
csh102	57	147 551	538 204
csh103	30	86 034	197 547
csh111	55	125 775	399 307
csh113	59	1 269 801	4 932 964

Time Management. The instances constituting the training datasets do not take into account information about the activation time of sensors. This is because we intended to prevent this information from being processed by the network. However, we do consider this information afterwards. In fact, after the classification network provides its output, we associate each instance with the time of its last sensor activation. For example, the output may indicate that the person is sleeping at 3:00 based on the sensor activations and the associated time information.

Neural Network. ADLnet is implemented in Python language using Tensor-flow and Keras libraries, which provide an interface to express machine learning algorithms and the implementation to execute these programs [35,36]. The model architecture was designed and developed as a CNN operating on the sensor activations instances described in the previous sections. Convolutional layers process the input as feature extractors, whereas fully connected layers receive convolutional feature maps to compute the final prediction. The structure of the CNN is shown in Fig. 3.

The first layer of the CNN consists of a batch normalization (Batch Norm) layer, which helps the network to increase the accuracy and to converge faster through normalization of the layers' inputs by re-centring and re-scaling [37] since Batch Norm ensures that the mean and standard deviation of the level inputs always remain the same. The convolutional stage of the CNN includes the following layers: 256 (8,1)-shaped kernels, two layers of 128 (6,1)-shaped kernels, two layers of 64 (4,1)-shaped kernels, and two layers of 32 (2,1)-shaped kernels. The following fully connected stage features three dense layers: a 32-neurons layer, a 16-neurons layer and an 8-neuron layer. Weight regularization of fully connected layers, with penalties L2 $=$ 0.1, was introduced in order to reduce overfitting. All layers mentioned before sharing the same activation function $\sigma(\mathbf{z})$, that is the Rectified Linear Unit (ReLU) $\sigma(\mathbf{z}) = \max(0, \mathbf{z})$, where $\mathbf{z}$ is the linear output of a single kernel in convolutional layers or of a single neuron in fully connected layers [38]. After the fully connected stage, a dropout layer with $rate = 0.5$ that helps to reduce overfitting is used [39]. Lastly, a dense layer of dimension 10 contains the output of the network. The activation

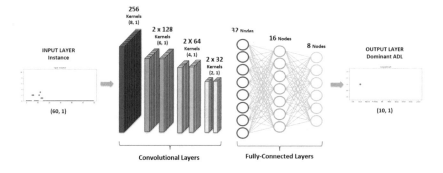

Fig. 3. ADLnet model architecture.

function of this layer is the *softmax* function, which converts a vector of values to a probability distribution [38]. *Softmax* is computed as $\sigma(\mathbf{z})_j = e^{z_j} / \sum_{i=1}^{K} e^{z_i}$, where $\mathbf{z}$ is the input vector to the *softmax* function and z_j are its elements, with j that goes from 1 to K, where K is equal to the number of classes. Through the training process, the CNN algorithm predicts dominant ADLs $\hat{\mathbf{y}}$ by applying the current parametric transfer function to the input, and, thanks to the ideal output $\mathbf{y}$ provided in the training dataset, it computes the distance between prediction and the ground truth.

Such a distance is quantified by means of a loss function L, which was here chosen to be the Categorical Crossentropy, which is the standard loss function used for multi-class classification [40]. This loss function is defined as $L = -\sum_{i=1}^{\text{outputsize}} y_i \cdot \log \hat{y}_i$, where $\hat{y}_i$ is the i-th scalar prediction of the model and y_i is the corresponding ground truth element. ADLnet trainable parameters are 357 492. The learning rate was tuned according to the adaptive moment estimation optimization technique (ADAM) [41]. For further details, the code is available online at [42].

The model was trained for 100 epochs with a batch size of 512. Each dataset was split into 3 parts: 70% for training, 15% for validation and 15% for testing. The metrics used for the evaluation of the model are accuracy ($a = (tp+tn)/(tp+tn+fp+fn)$), with tp true positives, tn true negatives, fp false positives and fn false negatives), precision ($p = tp/(tp+fp)$), recall ($r = tp/(tp+fn)$), and $F_1 = 2/(p^{-1} + r^{-1})$ score.

4   Experimental Results

Table 2 presents the results for 5 different datasets. These datasets, listed in Table 1, show great variety in dimensions and sensors number. Test cases were also different in terms of residents, rooms and sensors arrangement. The network featured excellent figures of merit on the testing datasets (Table 2).

The averaged metrics obtained are respectively: $a = 0.896$, $p = 0.896$, $r = 0.888$ and $F_1 = 0.892$. The computational time for data preprocessing (order of magnitude of a few ms) is not relevant, since it is reasonably lower than the average time interval between two subsequent sensor activations (order of magnitude of a few s). The training process took from 8 s to 15 s per epoch and is strictly dependent on the richness of the dataset and the GPU used. To train the network on the test cases, we used Google Colab Pro GPU.

Table 2. Experimental Results

	csh101		csh102		csh103		csh111		csh113	
	Before DA	After DA	Before DA	After DA	Before DA	After DA	Before DA	After DA	Before DA	After DA
Accuracy	0.900	0.926	0.856	0.886	0.892	0.920	0.833	0.887	0.836	0.859
Precision	0.899	0.926	0.859	0.888	0.900	0.918	0.855	0.887	0.839	0.861
Recall	0.900	0.925	0.854	0.885	0.883	0.916	0.810	0.857	0.833	0.856
F1 score	0.898	0.925	0.856	0.886	0.892	0.921	0.831	0.868	0.836	0.858

It is evident that the data augmentation process improved substantially the network performances in all test cases, and the mean increase of the metrics is 3%. In addition, the value of the Categorical Crossentropy loss function decreases from 0.35, before data augmentation, to 0.24, after data augmentation.

The implemented method proved to be consistent, showing excellent metrics, quite simple and general, as it benefits high adaptability capabilities.

Compared to the work in [27], performance results are very similar. However, the two approaches cannot be directly compared because of different goals and used sensors. Their work was mainly based on smart home energy saving, thus they used a dataset containing data related to energy consumption sensors. Furthermore, their particular feature selection reduces the number of training pairs, while our data preprocessing and augmentation system increases the size of the training dataset. For this reason, ADLnet can operate very well on different datasets with a reasonably lower number of epochs with respect to [27].

5 Conclusions

This paper presented ADLnet, a real-time activity recognition system based on a 1d-CNN architecture. ADLnet demonstrates excellent performance on multiple test cases, having great generalization capabilities. The whole data preprocessing of sensors time series is fast, adaptable to each dataset and it does not depend on human bias, thus guaranteeing the best classification performances. Future work includes evaluating both different datasets and the impact of variations in smart home environments.

Implementing this approach within an Ambient Assisted Living framework can enable effective real-time activity recognition for residents. Moreover, further enhancements to the network could facilitate the detection of sudden illnesses or behavioural changes, which are crucial factors for supporting the independent living of elderly individuals.

References

1. United Nations. Global issues: Ageing. https://www.un.org/en/global-issues/ageing
2. World Health Organization, fact sheets: Falls. https://www.who.int/news-room/fact sheets/detail/falls
3. Azefack, C., et al.: An approach for behavioral drift detection in a smart home. In: IEEE 15th International Conference on Automation Science and Engineering (CASE), vol. 2019, pp. 727–732 (2019)
4. Lowe, S.A., ÓLaighin, G.: Monitoring human health behaviour in one's living environment: a technological review. Med. Eng. Phys. 36(2), 147–168 (2014)
5. Steele, R., Lo, A., Secombe, C., Wong, Y.K.: Elderly persons' perception and acceptance of using wireless sensor networks to assist healthcare. Int. J. Med. Inf. 78(12), 788–801 (2009)
6. Alshammari, T., Alshammari, N., Sedky, M., Howard, C.: Evaluating machine learning techniques for activity classification in smart home environments. Int. J. Inf. Syst. Comput. Sci. 12, 48–54 (2018)

7. Albawi, S., Mohammed, T.A., Al-Zawi, S.: Understanding of a convolutional neural network. In: International Conference on engineering and Technology (ICeT), vol. 2017, pp. 1–6 (2017)
8. Bouchabou, D., Nguyen, S.M., Lohr, C., LeDuc, B., Kanellos, I.: A survey of human activity recognition in smart homes based on IoT sensors algorithms: taxonomies, challenges, and opportunities with deep learning. Sensors **21**(18), 6037 (2021)
9. Casas dataset. http://casas.wsu.edu/datasets/
10. Singh, D., Merdivan, E., Hanke, S., Kropf, J., Geist, M., Holzinger, A.: Convolutional and recurrent neural networks for activity recognition in smart environment. In: Holzinger, A., Goebel, R., Ferri, M., Palade, V. (eds.) Towards Integrative Machine Learning and Knowledge Extraction. LNCS (LNAI), vol. 10344, pp. 194–205. Springer, Cham (2017). https://doi.org/10.1007/978-3-319-69775-8_12
11. Ramasamy Ramamurthy, S., Roy, N.: Recent trends in machine learning for human activity recognition-a survey. Wiley Interdisc. Rev. Data Min. Knowl. Disc. **8**, e1254 (2018)
12. Nweke, H.F., Teh, Y.W., Al-garadi, M.A., Alo, U.R.: Deep learning algorithms for human activity recognition using mobile and wearable sensor networks: state of the art and research challenges. Exp. Syst. Appl. **105**, 233–261 (2018)
13. Hammerla, N.Y., Halloran, S., Ploetz, T.: Deep, convolutional, and recurrent models for human activity recognition using wearables (2016)
14. Yang, J.-B., Nhut, N., San, P., Li, X., Shonali, P.: Deep convolutional neural networks on multichannel time series for human activity recognition. In: IJCAI, July 2015
15. Chan, T.-H., Jia, K., Gao, S., Lu, J., Zeng, Z., Ma, Y.: PCANet: a simple deep learning baseline for image classification? IEEE Trans. Image Process. **24**(12), 5017–5032 (2015)
16. Szegedy, C., et al.: Going deeper with convolutions. In: Proceedings of the IEEE Conference on Computer Vision and Pattern Recognition (CVPR), June 2015
17. Richardson, F., Reynolds, D., Dehak, N.: Deep neural network approaches to speaker and language recognition. IEEE Sig. Process. Lett. **22**(10), 1671–1675 (2015)
18. Gamboa, J.C.B.: Deep learning for time-series analysis. CoRR, vol. abs/1701.01887 (2017)
19. Sadouk, L.: CNN approaches for time series classification, chap. 4. In: Ngan, C.-K. (ed.) Time Series Analysis, IntechOpen 2019, Rijeka (2019)
20. Chen, Y., Xue, Y.: A deep learning approach to human activity recognition based on single accelerometer. In: 2015 IEEE International Conference on Systems, Man, and Cybernetics, pp. 1488–1492 (2015)
21. Zeng, M., et al.: Convolutional neural networks for human activity recognition using mobile sensors. In: 6th International Conference on Mobile Computing, Applications and Services, pp. 197–205 (2014)
22. Wang, Z., Oates, T.: Encoding time series as images for visual inspection and classification using tiled convolutional neural networks. In: Twenty-Ninth AAAI Conference on Artificial Intelligence, January 2015
23. Cui, Z., Chen, W., Chen, Y.: Multi-scale convolutional neural networks for time series classification (2016)
24. Wang, W., Chen, C., Wang, W., Rai, P., Carin, L.: Earliness-aware deep convolutional networks for early time series classification (2016)
25. Wang, Z., Yan, W., Oates, T.: Time series classification from scratch with deep neural networks: a strong baseline (2016)

26. Hamad, R.A., Yang, L., Woo, W.L., Wei, B.: Joint learning of temporal models to handle imbalanced data for human activity recognition. Appl. Sci. **10**(15), 5293 (2020)
27. Alghamdi, S., Fadel, E., Alowidi, N.: Recognizing activities of daily living using 1D convolutional neural networks for efficient smart homes. Int. J. Adv. Comput. Sci. Appl. **12**(1), 1–11 (2021)
28. van Kasteren, T.L.M., Englebienne, G., Kröse, B.J.A.: Human activity recognition from wireless sensor network data: benchmark and software. In: Chen, L., Nugent, C., Biswas, J., Hoey, J. (eds.) Activity Recognition in Pervasive Intelligent Environments. Atlantis Ambient and Pervasive Intelligence, vol. 4, pp. 165–186. Atlantis Press (2011). https://doi.org/10.2991/978-94-91216-05-3_8
29. Morales, F., De Toledo, P., de Miguel, A.S.: Activity recognition using hybrid generative/discriminative models on home environments using binary sensors. Sensors (Basel, Switzerland) **13**, 5460–5477 (2013)
30. Lago, P., Lang, F., Roncancio, C., Jiménez-Guarín, C., Mateescu, R., Bonnefond, N.: ContextAct at A4H dataset, June 2017
31. University of Mannheim - HAR Dataset. https://sensor.informatik.uni-mannheim.de/
32. Taylor, L., Nitschke, G.: Improving deep learning with generic data augmentation. In: IEEE Symposium Series on Computational Intelligence (SSCI), vol. 2018, pp. 1542–1547 (2018)
33. Wen, Q., et al.: Time series data augmentation for deep learning: a survey. In: Proceedings of the Thirtieth International Joint Conference on Artificial Intelligence, August 2021
34. Iwana, B.K., Uchida, S.: An empirical survey of data augmentation for time series classification with neural networks. PLoS ONE **16**(7), 1–32 (2021)
35. Abadi, M., Agarwal, A., et al.: TensorFlow: large-scale machine learning on heterogeneous systems. Software available from tensorflow.org (2015). https://www.tensorflow.org/
36. Chollet, F., et al.: Keras (2015). https://github.com/fchollet/keras
37. Bjorck, J., Gomes, C., Selman, B., Weinberger, K.Q.: Understanding batch normalization (2018)
38. Keras activation functions. https://keras.io/api/layers/activations/
39. Srivastava, N., Hinton, G., Krizhevsky, A., Sutskever, I., Salakhutdinov, R.: Dropout: a simple way to prevent neural networks from overfitting. J. Mach. Learn. Res. **15**(56), 1929–1958 (2014)
40. Keras loss functions. https://keras.io/api/losses/probabilistic_losses/categoricalcrossentropy-class
41. Kingma, D.P., Ba, J.: Adam: a method for stochastic optimization (2017)
42. di Blasio, G., Venturelli, M.: ADLnet (2021). https://github.com/GDB-MV/ADLnet

NeeMAS: A Need-Based Multi-agent Simulator of Human Behavior for Long-Term Drifts in Smart Environments

Sara Comai[(✉)][ID], Andrea Masciadri[ID], Davide Zuccarello, and Fabio Salice[ID]

Politecnico di Milano, P.zza L. Da Vinci 32, Milan, Italy
{sara.comai,andrea.masciadri,davide.zuccarello,fabio.salice}@polimi.it
http://www.polimi.it

Abstract. Early identification of long-term changes in the behaviour of people monitored with smart environment solutions is essential to prevent health decline. However, data collection and analysis of human behaviour are challenging and time-consuming.

A potential solution consists in creating digital twins of the individuals to replicate the typical behaviours for advanced data analytics. The Assistive Technology Group (ATG) at Politecnico di Milano has developed NeeMAS (NEEd-based Multi-Agent Simulator), a novel simulator that effectively simulates human behaviour with physiological and social needs, cognitive decay, and behavioural drifts due to ageing or disease onset such as apathy or depression.

NeeMAS simulates a senior care facility with several individuals, spending part of their time in their rooms and in part sharing common indoor and outdoor spaces interacting with other people. Experimental results show the feasibility and flexibility of the proposed approach for the generation of typical human behaviours and their drifts.

Keywords: Simulated vitual environment · Health decline · Human behavior · Ageing · Senior care facility · Elderly people · Data generation · Monitoring activities of daily living · Behavioral drift

1 Introduction

With the population ageing [1], healthcare assistance for elderly people becomes increasingly necessary. Information Technology can support health care, in particular by monitoring possible changes in activities of daily living (such as bathing, sleeping, eating, etc.), which may be caused by the onset of diseases like for example, Alzheimer's or dementia. However, such changes can be slow and take a long time before they can be clearly observed.

In our solution, we turn to the concept of digital twins-replicas of individuals that faithfully mimic their typical behaviours. Such an approach holds immense potential for advanced data analytics, offering early insights into long-term 'behavioural drifts' that might otherwise go unnoticed.

The result is a simulator developed within the Assistive Technology Group (ATG) at Politecnico di Milano that effectively simulates human behaviour, cognitive decay, and behavioural drifts due to ageing or disease onset.

This work focuses on the modelling of individuals and their parameters to simulate not only physiological needs such as eating, drinking, sleeping, etc. but also social characteristics, spirituality and other self-actualization needs based on Maslow's hierarchy of needs [2]; moreover, behavioural characteristics of typical elderly diseases are considered, such as apathy, depression, and potential behavioural drifts in daily activities.

The simulated virtual environment is a senior care facility with several individuals, spending part of their time in their own rooms and in part sharing common indoor and outdoor spaces where they interact with other people. Experimental results show the feasibility and flexibility of the proposed approach for the generation of typical human behaviours.

2   Related Works

Generation of human activity data can be mainly achieved with two different approaches: instrumenting a smart home/environment and collecting real data or by means of agent-based simulators. For the generation of large amounts of data, the former approach has prohibitive costs to equip physical spaces and recruit human subjects for performing activities to generate and collect the needed data. Examples of datasets obtained with smart home test beds include: CASAS [3], TigerPlace [4], SmartLab [5], UbiHome [6], PlaceLab [7], and HomeLab [8].

Agent-based simulators have emerged as a cheaper alternative approach: they imitate real-life aspects by simulating the behaviour of the ecosystem with autonomous and interacting agents [9]. With the rise of the Internet of Things and the development of new machine-learning techniques, research in this area is becoming increasingly popular. In this context, two main approaches exist. *Model-based approaches* use pre-defined statistical models to generate datasets (e.g., specifying the order of the events, the probability of their occurrence, and the duration of each activity); this category includes tools like PerSim [10], SESim [11], SIMACT [12], DiaSim [13], CASS [14]. They generate large datasets, but cannot capture complex interactions. On the contrary, *interactive approaches* rely on capturing real-time activities using an avatar controlled by a human/simulated participant, like in UbiReal [15], V-PlaceSims [16], SimCon [17], OpenSHS [18], IE Sim [19]. Like in [20], we use a needs-driven approach but focus on a senior care facility with multiple residents to produce movement patterns; given the multiplicity of agents we consider also social aspects and include also long-term drift behaviours.

3   Proposed Approach

NeeMAS is a needs-driven multi-agent-based, discrete-time, model-driven simulator. It emulates the behaviour of one or more individuals and their interactions with each other. It takes into account both physiological needs (such as sleeping, eating, etc.) and social needs (people need to interact with others) and generates human behaviours in terms of movement patterns in various areas of the virtual environment.

Virtual Environment. The virtual environment can be characterized by outdoor places (e.g., gardens) and indoor places (e.g., bedroom, kitchen but also bar and activity rooms); the spaces can be personal, such as the bedroom, or common. The places are also characterized by the coordinates X, Y (identifying their barycenters), a name, and a unique identifier. An example is represented by the following JSON fragment:

```
{"id": 2,
 "name": "Bedroom",
 "outdoors": false,
 "social": false,
 "x": 1196, "y": 982 }
```

The environments are also characterized by access times. For example, in a residence, it may not be possible to leave the building where the rooms are located after 10 PM, or the dining room might only be accessible during designated lunch or dinner breaks. The practicability of places is represented by a binary value, with the default being 0 (non-accessible) and is established on a need-by-need basis; for example, the dining room, primarily designated for meals, may also serve as a space for socializing during other times of the day. For example, the following fragment specifies that place 5 is always accessible while place 1 is accessible from 8 AM to 7 PM.

```
"places" : [
        {"id" : 5,
         "practicability" : 1}
        {"id" : 1,
         "practicability" : [[0, 8, 0],[8, 19, 1],[19, 0, 0]]},
        ... ]
```

Invidual's Needs. Once the places have been defined, each individual is characterized through parameters that define their needs. Needs are characterized by "growth" and "satisfaction". The concept of "growth" refers to the gradual increase in intensity leading up to the point where the need must be satisfied. On the other hand, "satisfaction" (or "degrowth") represents the gradual fulfilment of the need. It is worth noting that the duration of the growth phase is typically much longer than that of the satisfaction phase. For example, the need for food gradually increases over the span of hours until the individual eventually consumes a meal, which typically takes only a few tens of minutes. The transition between the "growth" and "satisfaction" states, and vice versa, is determined by two thresholds: th_{max} and th_{min}, where $th_{max} \geq th_{min}$. These threshold values range between 0 and 1. Each need is further characterized by its priority level, by the possibility of being interrupted, and by an initial value. Consider the following example of the need of eating:

```
{"minexec-time" : 600,
 "initialval" : 0.35,
 "priority" : 0.8,
 "preemptible" : true,
 "thmax" : [ [ 0, 7, 1 ], //from 0 to 7 the maximum threshold is 1
             [ 7, 8, 0.8 ], //from 7 to 8 the maximum threshold is 0.8
             ... ]
 "thmin" : [ [ 0, 7, 0.7 ], //from 0 to 7 the minimum threshold is 1
             [ 7, 9, 0.2 ], //from 7 to 9 the minimum threshold is 0.2
             ... ] }
```

The minimum execution time is 600 s (note: this prevents an excessive splitting of the activities), its reset value is 0.35 and its priority is 0.8 (for example, going to the bathroom has priority 1, so this activity can be interrupted by the toileting activity), and thresholds for different time intervals are defined. Setting a threshold to 1 prevents the agent from filling the need (e.g., between 0 and 7 AM the person cannot eat).

The concepts of "growth" and "satisfaction" are represented as linear functions of time, characterized by angular coefficients that can vary over time (in the next example, every hour the need is increased by 0.267); to introduce a degree of uncertainty and reduce overly deterministic behaviour, a parameter defining uncertainty (in the next example, *epsinc*) is incorporated and applied to the slope of the function.

```
"increments" : [ [ 0, 7, 0.267 ], [ 7, 22, 0.16 ], ...],
"epsinc"     : [ [ 0, 7, 0.013 ], [7, 22, 0.008 ], ...],
"decrements" : 9.6,
"epsdec"     : 0.48
```

To configure the system, an interface was implemented which, while limiting the freedom of configuration to reasonably real scenarios, helps to build all the JSON files by calculating all the coefficients on the basis of the times chosen for each individual person and user characteristics (e.g., she/he takes snacks some times). Average analysis of observed user behaviours is also considered such as meal duration, estimation of bathroom breaks, taking into account that slopes and thresholds vary for day and night, for lunch/dinner, etc.). The limitations do not affect generality, as users can directly modify parameters and time intervals in the configuration files.

The simulator produces *graphical representations* and .csv files describing people's activities, places visited and friendships (discussed later). Figure 1 shows how needs change over time. Each need, color-coded, follows a pattern: growing until it reaches a maximum peak, and then declining when no higher-priority needs are active. For example, the declining line for sleep indicates that the agent is sleeping; this is represented also by the background colour (legend on the right). Coloured vertical bars show the next satisfied needs, corresponding to the decreasing "lines". Throughout the day, various needs emerge and are satisfied.

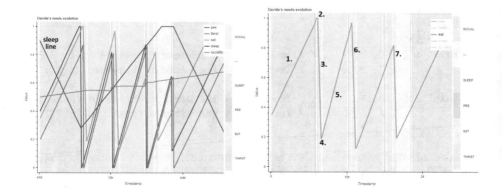

Fig. 1. Plot of needs evolution over a sample day. (Color figure online)

Fig. 2. Eating need evolution over a sample day.

To give a clearer representation of how needs change over time, Fig. 2 reports the evolution of the hunger need.

1. Overnight, the hunger need *grows* according to the incremental parameters.
2. Once it reaches the maximum threshold, its status becomes *ready* to be satisfied.
3. If there are no other needs with higher priority, the need can be *satisfied*, graphically represented by the decrementing.
4. When the minimum threshold is reached, the hunger need of the individual is completely satisfied.
5. Next, the hunger need starts increasing and, as in real-life, in the meanwhile the person executes other actions.
6. The same trend repeats for lunch
7. and for dinner with some randomness to emulate real life.

Sociality Among Individuals. Each character is located in a populated environment where they can interact and socialize, according to an *affinity matrix* that influences the behaviour of the characters, making them move in the environment to meet and stay together or avoid each other.

The *affinity matrix* models the relationship between each pair of individuals. For this model we assumed that the symmetric property holds; this means, for example, that if agent A considers B compatible with himself, the converse is also true. While this is not generally true, in our view it is an acceptable limitation and reduces the complexity of the simulations. The sociality relationship is modelled by two parameters: *compatibility* (in the range $[-1, 1]$ where -1

incompatible, 1 compatible, 0 neutral) between the subjects, which represents the affinity between each pair of subjects, and the initial state of friendship (in [0, 1] - 0 are not friends, 1 already know each other and are friends, 0.5 neutral). An example is the following, where subjects 3 and 1 have an initial positive friendship (equal to 1) which will gradually evolve to negative values due to the compatibility −0.2.

```
{"compatibilities": [ [ 2, 1, 0.4], [ 3, 1, -0.2], ... ]
 "friendships":     [ [ 2, 1, 0.5], [ 3, 1, 1], ... ] }
```

Besides specifying the affinity matrix, each individual is also associated with a sociality need, represented like any other need.

Considering the affinity matrix, sociality between pairs of people evolves and this is graphically represented by heatmaps generated every day. Figure 3 shows an example of a heatmap: green shades indicate friendship, white neutrality, and red shades hostility.

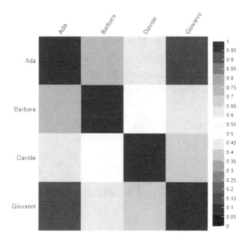

Fig. 3. Heatmap of the evolution of relationships with 4 subjects. (Color figure online)

Weather Conditions and Other Features. The simulator includes also weather conditions that influence behaviour. For example, if it rains, individuals will adapt their behaviour and avoid outdoor locations. We have also expanded the number of needs, making the simulator more versatile and accurate in replicating real-world scenarios: in particular, we considered self-actualization needs, corresponding to the highest levels of Maslow's pyramid [2], such as spirituality and creativity. These needs will move the agents towards places such as the church or the art/music lab (or any other place where these needs may be satisfied).

Representing Isolation. Another aspect taken into consideration by the simulator is the need for isolation, a common behaviour exhibited by Alzheimer's patients. Conscious isolation is what a person chooses voluntarily because he/she wants or because he/she needs it. This need is modelled with a higher priority than the need for sociability. If the time needed to satisfy the isolation need is short (from a few tens of minutes to a few hours), it represents conscious isolation. By setting the maximum threshold to 0, the pathological behaviour of total isolation is induced in the simulated subject.

3.1   Behavioural Drifts

The purpose of the simulator is to allow the production of data that spans long periods of time, capturing the gradual cognitive declines that occur over extended durations. The availability of such datasets would enable researchers to study, develop, and validate algorithms and methodologies for identifying behavioural drifts. To best replicate real-life scenarios, we have gone beyond the current state of the art by emulating the slow and gradual decay of people's behaviour and some disorders caused by physical and/or cognitive pathologies.

Behavioural drift is modelled as the superposition of three different components: i) a step function, which represents an abrupt change, ii) a sinusoidal function, which represents the permanence of cyclical behaviours (e.g., seasonal changes), and iii) a logistic function, which represents a growing and continuous change, represented by the following equation.

$$d(t) = A \cdot H(t - t_0) + H(t - t_1) \cdot \frac{B}{2} \sin(\omega(t - t_1) + \phi) + C \cdot P(t - t_2)$$

with

$$H(t - t_0) = \begin{cases} 0 & \text{for } t - t_0 < 0 \\ 1 & \text{for } t - t_0 \geq 0 \end{cases}$$

and

$$P(t - t_2) = \frac{1}{1 + e^{-k(t - t_2)}}$$

The logistic function has amplitude C, transition centred at time $t2$ and slope of the transition part $k = \frac{C}{\Delta T}$, where ΔT is the duration of the transition of the sigmoidal function.

Their combination allows the representation of the vast majority of behavioural drifts and can be applied to each need separately. The dynamism of the drifts is defined by the parameters; these can allow you to emulate even a very slow drift of behaviour.

4   Experimental Results

The tests have been carried out by simulating hypothetical scenarios that we consider realistic to demonstrate the versatility and flexibility of the proposed approach.

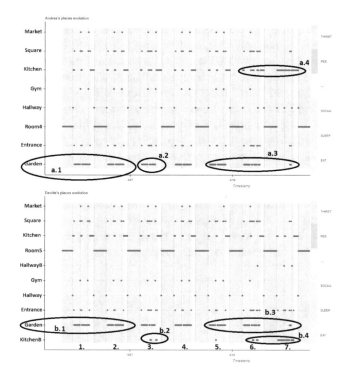

Fig. 4. Places evolutions over time from day 1 to day 7, due to adaptive behaviour.

Adaptive Behaviour of Two Individuals. In this first case study, two friends begin to dislike each other (in the affinity matrix we set the initial friendship to a high value and the compatibility to the negative value -1); as a consequence, they modify their habits to avoid meeting, seeking for new friends in new locations. Figure 4 illustrates the evolution of their relationship and social behaviour over time. In the early days (see a.1 and b.1 in the figure), after fulfilling their physiological needs, the individuals spend most of the time socialising in common areas (the garden in this case). As their relationship sours, from day 3 they begin to adapt their behaviour: one subject starts exploring new places (see b.2), while the other subject keeps spending his time in the same places as usual (see a.2). From days 5 to 7, as the animosity continues to grow there is a decreasing habit for both subjects to go to the common place where they are sure to meet each other (see a.3 and b.3), and consequently, there is an increase in the visits in new different social places (see a.4 and b.4) indicating the establishment of new habits. This simple scenario highlights the simulator's ability to capture real-life behaviours and adapt to different situations, leading to a wide range of possible scenarios.

Apathy and Depression. The second case scenario highlights the simulator's ability to model pathy and depression. Apathy is characterized by the lack of feelings, emotions, interests, concerns or motivations [21,22]. The subject tends not to participate, if not solicited, in social activities and tends to isolate himself. To simulate this, we reduced a subject's social activity and stimulated the subject to isolate himself. After a long simulation, the relationships among the people in the simulated world have become stable with people loving (in green) and people hating (in red), as shown on the left-hand side in Fig. 5. However, there is an individual who does not establish relationships. He is exhibiting apathetic conduct, which is represented graphically through the colour white which indicates the lack of positive or negative feelings towards others.

Moreover, depression in Alzheimer's patients can lead to sleep and appetite disturbances, anxiety, fatigue, and slowed movement [21,22]. Sleep and appetite disturbances can be modelled by configuring the corresponding needs, as shown on the right-hand side in Fig. 5: during the night, the person interrupts the sleep phase (the sleep line decreases) to have small night snacks (the eat lines in a. and b. increase and decrease 3–4 times for each snack).

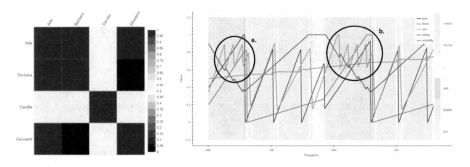

Fig. 5. Left: Relationships at the end of a simulation showing an apathetic subject. Right: Needs evolution in a depressed subject (a. and b. represent sleep and appetite disturbances).

Cognitive Decay. The last scenario simulates cognitive decay. This is a slow and gradual process but, for the sake of presentation, a shorter 45-day simulation was conducted using a step decay model with a noticeable slope. In this simulation, an amplitude value of −100 and an activation value of 0.5 were applied, representing a significant change in behaviour occurring after 15 days.

Figure 6 shows that sociality drops after 15 days affecting both habits of the subject and relationships with others. The behaviour of the subject is clearly changed: if before the person was used to go to specific places to socialise, meet with others and, in general, had a healthier and more active life, after the decay his habits change incrementing loneliness and sedentary life as more then half of the places are no longer frequented.

The new lifestyle of the subject has of course an impact on relationships too. It is possible to see consequences in friendships in the output heatmaps in Fig. 7:

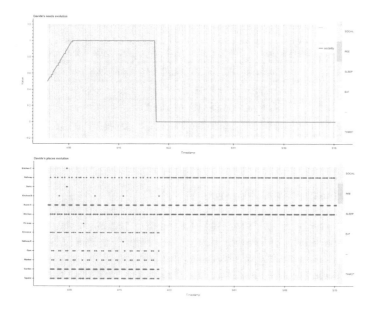

Fig. 6. Evolution of Needs and Places in a subject with cognitive decay.

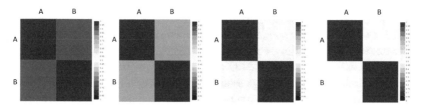

Fig. 7. Evolution of friendship in subject B after beginning of cognitive decay. (Color figure online)

after the beginning of cognitive decay the intense green gradually decreases in shades until it reaches white (without modifying the friendship and compatibilities values).

Simulation Execution Time. Simulation execution time depends on the number of people and on the number of simulated days. Table 1 shows a subset of the measured execution times of the simulator with a processor Intel(R) Core(TM)2 Quad CPU Q8200 2.33 GHz 2.33 GHz by varying the number of people in the virtual environments and the number of simulated days. By computing the interpolation equation derived using least squares, we obtained Time(sec) = 2.83 * People + 94.10 * Days + 12.11. For a large number of people and days of simulation, the error is very low. For example, for simulations with 50 people from 1 day to 200 days, the estimation error (estimated time vs measured time) ranges from −1% to +4%; in general, the maximum estimation error is less than 10%.

Table 1. Some simulation execution times measured in our experiments.

#People	1 day	5 days	20 days	50 days	200 days
0	00:02:09	00:07:55	00:29:31	01:12:43	04:48:39
1	00:02:16	00:08:15			
10	00:02:44	00:08:37	00:30:33		
50	00:04:19	00:10:25	00:34:09	01:22:16	05:17:13
100	00:06:20	00:13:08	00:35:58		

5   Conclusions and Future Work

In this paper, we presented NeeMAS, a simulator based on the concept of *needs* that allows the study of human behaviour, cognitive decay, and behavioural drifts resulting from ageing or the onset of diseases.

The limitations of the simulator are due to the fact that the representation of human behaviour in a simulated environment is a simplification of real life. Consequently, these assumptions and simplifications may not accurately reflect real-world human behaviour in all countless facets and scenarios.

In future work, we plan to broaden the simulator's capacity to model a broader range of scenarios, utilizing real-world data to project user behaviour. Moreover, we plan to improve aspects like relationship dynamics by introducing new parameters for handling complex situations, and the graphical representation of movement patterns, using 2D or 3D visualization.

In conclusion, the simulator presented in this study offers a valuable tool for replicating human behaviour in diverse scenarios and opens up opportunities for future investigations to utilize the simulator in generating data on the effects of various interventions on the behaviour of individuals with Alzheimer's and dementia. The results obtained from this work can be used to develop new algorithms based on movement patterns [23,24] to improve the quality of life for elderly individuals and reduce the economic burden on public and private healthcare systems.

References

1. UNFPA and HelpAge International: Ageing in the Twenty-First Century: A Celebration and A Challenge. United Nations Population Fund (UNFPA) and HelpAge International (2012)
2. Maslow, A.: A theory of human motivation. Psychol. Rev. (1943)
3. Cook, D.J., Crandall, A.S., Thomas, B.L., Krishnan, N.C.: Casas: a smart home in a box. Computer **46**(7), 62–69 (2013)
4. Skubic, M., Alexander, G., Popescu, M., Rantz, M., Keller, J.: A smart home application to eldercare: current status and lessons learned. Technol. Health Care Off. J. Eur. Soc. Eng. Med. **17**(3), 183–201 (2009)
5. Nugent, C.D., Mulvenna, M., Hong, X., Devlin, S.: Experiences in the development of a smart lab. Int. J. Biomed. Eng. Technol. **2**(4), 319–331 (2009)

 6. Yamazaki, T.: The ubiquitous home. Int. J. Smart Home **1**, 17–22 (2007)
 7. Intille, S.S., et al.: Using a live-in laboratory for ubiquitous computing research. In: PERCOM, pp. 349–365. Austria, Innsbruck (2006)
 8. de Ruyter, B., Aarts, E., Markopoulos, P., Ijsselsteijn, W.: Ambient intelligence research in HomeLab: engineering the user experience. In: Weber, W., Rabaey, J.M., Aarts, E. (eds.) Ambient Intelligence, pp. 49–61. Springer, Heidelberg (2005). https://doi.org/10.1007/3-540-27139-2_4
 9. Macal, C., North, M.: Agent-based modeling and simulation, December 2009
10. Helal, A., Cho, K., Lee, W., Sung, Y., Lee, J., Kim, E.: 3D modeling and simulation of human activities in smart spaces. In: 9th International Conference on UIC-ATC 2012, pp. 112–119 (2012)
11. Ho, B., Vogts, D., Wesson, J.: A smart home simulation tool to support the recognition of activities of daily living. In: Proceedings of the South African Institute of Computer Scientists and Information Technologists 2019, SAICSIT '19. Association for Computing Machinery, New York, NY, USA (2019)
12. Bouchard, K., Ajroud, A., Bouchard, B., Bouzouane, A.: SIMACT: a 3D open source smart home simulator for activity recognition. In: Kim, T., Adeli, H. (eds.) ACN/AST/ISA/UCMA -2010. LNCS, vol. 6059, pp. 524–533. Springer, Heidelberg (2010). https://doi.org/10.1007/978-3-642-13577-4_47
13. Bruneau, J., Consel, C.: DiaSIM: a simulator for pervasive computing applications. Softw. Pract. Exp. **43**(8), 885–909 (2013)
14. Park, J., Moon, M., Hwang, S., Yeom, K.: Cass: a context-aware simulation system for smart home. In: 5th ACIS International Conference on Software Engineering Research, Management and Applications (SERA 2007), pp. 461–467 (2007)
15. Nishikawa, H., et al.: UbiREAL: realistic smartspace simulator for systematic testing. In: Dourish, P., Friday, A. (eds.) UbiComp 2006. LNCS, vol. 4206, pp. 459–476. Springer, Heidelberg (2006). https://doi.org/10.1007/11853565_27
16. Lertlakkhanakul, J., Choi, J.W., Kim, M.Y.: Building data model and simulation platform for spatial interaction management in smart home. Autom. Construct. **17**(8), 948–957 (2008)
17. McGlinn, K., O'Neill, E., Gibney, A., O'Sullivan, D., Lewis, D.: Simcon: a tool to support rapid evaluation of smart building application design using context simulation and virtual reality. J. Univ. Comput. Sci. **16**, 1992–2018 (2010)
18. Alshammari, N., Alshammari, T., Sedky, M., Champion, J., Bauer, C.: Openshs: open smart home simulator. Sensors **17**(5) (2017)
19. Synnott, J., Chen, L., Nugent, C., Moore, G.: The creation of simulated activity datasets using a graphical intelligent environment simulation tool. In: 36th Annual International Conference of the IEEE Engineering in Medicine and Biology Society, pp. 4143–4146 (2014)
20. Cabañero, L., Perez-Vereda, A., Nugent, C., Cleland, I., Hervas, R., González, I.: A software tool and a metamodel for digital twins of inhabited smart environments. In: Bravo, J., Ochoa, S., Favela, J. (eds.) UCAm I 2022. LNNS, vol. 594, pp. 747–759. Springer, Cham (2023). https://doi.org/10.1007/978-3-031-21333-5_74
21. American Psychiatric Association et al.: Manuale diagnostico e statistico dei disturbi mentali (dsm-5). Milano, Raffaello Cortina Editore (2014)
22. AlzheimersAssociation, "Depression," 04–07 2021. https://www.alz.org/help-support/caregiving/stages-behaviors/depression
23. Bellini, G., Cipriano, M., et al.: Understanding social behaviour in a health-care facility from localization data: a case study. Sensors **21**(6) (2021)
24. Masciadri, A., Comai, S., Salice, F.: Wellness assessment of Alzheimer's patients in an instrumented health-care facility. Sensors **19**(17) (2019)

Predictive Modeling of Body Shape Changes in Individuals on Dietetic Treatment Using Recurrent Networks

Nahuel Garcia-D'Urso[1], Pablo Ramon-Guevara[1], Jorge Azorin-Lopez[1(✉)], Marc Sebban[2], Amaury Habrard[2,3], and Andres Fuster-Guillo[1]

[1] Department of Computer Technology, University of Alicante,
03690 San Vicente del Raspeig (Alicante), Spain
jazorin@ua.es
[2] University Jean Monnet, UMR CNRS 5516, Laboratoire Hubert-Curien,
42000 St-Etienne, France
[3] Institut Universitaire de France (IUF), Paris, France

Abstract. This paper presents an approach that utilizes deep recurrent neural networks to predict body shape changes in individuals undergoing dietetic treatment. It contributes to computational body modelling by offering a reliable tool that assists healthcare professionals in tailoring recommendations and motivating individuals to achieve their body shape goals. The approach is focused on the regression of body shape parameters over time, which enables the prediction of shape changes using the individual body history. This method has been applied and evaluated using a dataset obtained from 80 individuals undergoing dietetic treatment over an 8-month period. The results demonstrate the effectiveness of the proposed model in accurately predicting body shape transformations resulting from dietetic treatment. The predictive capabilities of the model provide valuable insights for healthcare professionals, enabling them to tailor personalized recommendations for individuals.

Keywords: Body shape · LSTM · Prediction

1 Introduction

Overweight and obesity are defined as abnormal or excessive accumulation of fat that is detrimental to health. The prevalence of overweight and obesity has increased worldwide, tripling in the last three decades [18]. In European Union countries, the prevalence of these diseases in the population over 18 years of age is about 53% according to the latest data from the fourth wave of the 2019

The work was supported by the Spanish State Research Agency (AEI) under grant PID2020-119144RB-I00 funded by MCIN/AEI/10.13039/501100011033 and grant CIAICO/2022/132 funded by Conselleria d'Innovació, Universitat, Ciència i Societat Digital.

J. Bravo and G. Urzáiz (Eds.): UCAmI 2023, LNNS 842, pp. 100–111, 2023.
https://doi.org/10.1007/978-3-031-48642-5_10

European Health Interview Survey [8]. Overweight and obesity contribute to the occurrence of chronic diseases such as arterial hypertension, type II diabetes, cancer, or the development of neurodegenerative diseases such as Alzheimer's or dementia [4,7] leading to an increased risk of premature death, reduced quality of life, and increased frequency of medical visits. In addition, the website COVID has highlighted overweight and obesity as a risk factor for the pandemic [21]. Moreover, the need to bring treatment closer to the patient by promoting personalized medicine has become even more acute. At the European level, the costs associated with obesity are estimated to be between 0.09% and 0.61% of the total annual gross domestic product, corresponding in absolute terms to a cost of 10 billion euros per year [15, 26].

Previous studies have shown that the expectation of weight loss is an important predictor of adherence to nutrition therapy [25]. It has been recognized that expectations of treatment and expectation of positive outcomes are critical factors in motivating the adoption of healthy dietary habits, especially in at-risk populations such as overweight and obese individuals [25] and in the initial phase of adopting such habits [22]. Thus, it has been shown that patients who can set realistic expectations about the efficacy of treatment, through predictive 4D models that allow them to know how their bodies will develop if they actually adhere to a healthy dietary pattern, have greater motivation and adherence to treatment and thus greater weight loss [23]. When patients perceive these expectations as more realistic [9], it enhances the success and satisfaction they derive from dietary and nutritional treatment [23]. Realistic expectations have been shown to significantly boost adherence, making them a key factor influencing treatment effectiveness [17]. In this sense, the development of 4D images would allow giving a realistic character to these expectations by visualizing the patient's own body and its evolution over time through the proposed predictive models.

In this paper, we propose a novel method for predicting changes in body shape using the progression of shape evolution over time in individuals. Our approach is based on recurrent neural network models, specifically Long Short-Term Memory (LSTM) networks, which are capable of learning the temporal patterns and dependencies within shape evolution data. By learning the shape of a parametric model over time, we can regress the unknown shape for the future. In this study, we use the widely used parametric SMPL model, which can synthesize a point cloud representation of a person based on shape and pose parameters. To evaluate the effectiveness, we conducted experiments with 80 individuals who underwent dietary treatment over an 8-month period, showing the predictive capabilities in accurately predicting shape changes resulting from dietary interventions. The results highlight the potential of our approach to assisting healthcare professionals in understanding and predicting the effects of dietary changes on an individual's body shape.

The results of this research could have significant implications for healthcare professionals and individuals seeking effective dietary management. By accurately predicting changes in body shape, our model provides valuable insights for personalized recommendations that allow healthcare professionals to tailor interventions to individual needs. In this way, individuals can make informed decisions about their dietary treatment, leading to more effective outcomes and improved overall well-being.

2   Related Works

2.1   3D Human Body Representation

Parametric models are a class of mathematical models that use parameters, or coefficients, to represent and analyze relationships within a system. These models are characterized by a finite set of parameters whose values are estimated from data to provide a representation of complex structures. These models significantly allow for the creation of realistic representations of the human body by defining adjustable parameters for shape and pose. Nowadays, parametric models are essential tools in a multiple of applications.

In animation and gaming industries, parametric models are indispensable for creating lifelike characters with diverse physical attributes and dynamic movements. They are also pivotal in virtual and augmented reality environments, providing users with immersive experiences through accurately rendered avatars. In fashion and apparel design, these models are employed for virtual try-ons and to simulate the fit of garments on bodies.

Allen's method [1] is an early model using statistical techniques for parametric 3D human body representation, focusing primarily on pose interpolation without accounting for pose-dependent body shape variations. The SCAPE model [2] further develops this approach by using triangular deformations to represent changes in body shape and posture, although it lacks specificity in representing muscle activity and movement-related tissue disturbances.

BlendSCAPE [11] extends SCAPE, optimizing shape and pose registration concurrently through blending techniques. However, it faces compatibility issues with standard graphics packages, limiting its widespread application. SymmetricSCAPE [5], a variant of SCAPE and BlendSCAPE, introduces symmetry constraints, enhancing both the model's accuracy and stability for a more robust 3D human body representation.

The widely utilized SMPL model [16] distinguishes body pose and shape parameters, applying vertex-based skinning and corrective blend shapes for enhanced flexibility (see Sect. 3.2 for more details). Subsequent models, SMPL-H and SMPL-X [20], incorporate additional parameters for hand and face respectively. These models are known to have limitations when deforming the body in areas close to joints. The STAR model [19] achieves better deformations in these areas, providing realistic body deformations with fewer parameters required.

2.2   Neural Networks for Learning Data Sequences

In recent times, predominant methodologies for analyzing data sequences are grounded in neural networks, which have exhibited remarkable proficiency in forecasting data patterns. Their effectiveness is rooted in their skill to learn from complicated and high-dimensional data, which is crucial for tasks like time series forecasting. Specifically, specifically, Recurrent Neural Networks (RNNs) stand out as a specialized category of neural networks adept at managing temporal data [13]. A distinctive feature of RNNs is the recurrent hidden state, functioning as a memory mechanism. This state is refreshed at each step of the data sequence, integrating information from the current input and the previous hidden state. However, RNNs also have their challenges, notably the vanishing gradient problem [3]. During the training process using backpropagation, gradients can shrink exponentially when tracked back through time, particularly over long sequences. This leads to slow updates to the network's weights, difficulting the learning process in the earlier network layers or stopping it entirely. This limitation affects the RNN's ability to represent long-term dependencies in the data.

Addressing the vanishing gradient dilemma, Long Short-Term Memory Networks (LSTM) [12] have emerged as a prominent and efficacious solution. LSTM networks consist of units with a cell state and three gates: input, forget, and output. The cell state transports information throughout the network with little alteration, while the gates control the flow and modification of information within the cell. They offer a robust solution for tasks that require modeling and understanding sequential data with complex temporal dynamics.

Another variant is the Gated Recurrent Unit Networks (GRU) [6]. It was introduced to address the vanishing gradient problem and to simplify the complexity inherent in LSTM. The GRU networks maintain the capacity to capture dependencies in sequential data effectively without the need for a cell state. Instead, a GRU utilize two gates: update and reset. The update gate determines the degree of the past information to be retained, while the reset gate decides the extent of the new information to be incorporated. The simplified architecture of the GRU not only reduces the number of parameters compared to LSTM but also computationally more efficient.

A recent development is Transformers networks [24]. These have achieved considerable success. In contrast to utilizing recursion, Transformer networks use an attention mechanism to weight the importance of different parts of an input sequence so that they can focus on the most relevant parts. Attention mechanisms allow to process entire sequences of data simultaneously, resulting in much faster training times compared to RNN or LSTM. However, Transformers generally demand greater memory capacity, particularly when handling extended sequences.

3   BSP-RNN: Predicting Body Shape Using Recurrent Networks

In this section, the method for predicting body shapes over time is presented. It is based on learning the sequence of previous information about the changes in a set of known body shapes. After the model is learned, it is used to regress the body parameters as predictions.

3.1   BSP-RNN Workflow

Let us assume that we have access to a set of n body models $B = \{B_i\}_{i=1}^n$ supposed to be representatives of body shape parameters over time. Each B_i is composed by a set of body shapes $B_i = \{b_t^i\}_{t=1}^k$ that can change their morphology over k times. A body shape b_t^i is described as a set of parameters that describes a set of captures points of a body $\{p_j^{i,t}(p_j^x, p_j^y, p_j^z) \in \mathbb{R}^3\}_{j=1}^{n_t}$ representing its geometry in the 3D euclidean space.

The proposed method aims to predict the evolution of the body shape over time based on the knowledge of the deformations that the body has undergone in the past. The workflow of our method is described and illustrated in Fig. 1. It is divided into two main stages:

1. An **estimation of body parameters**, which consists in computing the parameters of the body b_t^i of an acquisition of the body i at time t from the 3D point cloud p_t^i.
2. A **prediction of the body shape** that predicts the body shape parameters b_{t+1}^i for time $t+1$ according to the history B_i of the individual i.

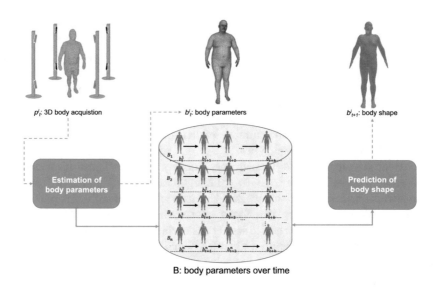

Fig. 1. Workflow of the BSP-RNN according to the two main steps.

3.2   Estimation of Body Parameters

The first stage of our method focuses on estimating parameters from a point cloud p_t^i to a body model. The main goal is to translate the unstructured data related to the point cloud into a model that allows us to capture the underlying structure and variations of the body shape by providing compact representation, generalization, robustness, and most importantly, parameterized control that allows us to generate a wide range of body shapes. Next, the parametric model used in this work is described. Then, the method of fitting the model from the point cloud is presented.

Parametric Model. In this work, the Skinned Multi-Person Linear Model (SMPL) [16] is used as a parametric model. SMPL serves as a framework able to represent both human body shapes and postures. The model encapsulates body shapes and postures within a low-dimensional linear space. This representation is achieved through the application of skinning and blend shapes, with the model parameters learned from an extensive dataset of 3D models of the human body.

Specifically, the SMPL model is comprised of 300 shape parameters, denoted as β, and 72 pose parameters, represented by θ. The set of shape parameters, β, is derived through the application of Principal Component Analysis (PCA) to the bodies in the training dataset. These components provide us with an overall understanding of the patient's body shape. Moreover, the pose parameters, θ, define the position and the orientation of the joints. While the pose parameters provide a detailed representation of the body position, the primary focus of this study is on understanding body shape. Therefore, our analysis predominantly centers on the Shape parameters.

SMPL is ideally suited to our problem because it is compact, continuous, and differentiable. The low dimensionality of the model reduces computational complexity being a good alternative for large-scale analyses. Furthermore, the model's characteristics enable the application of gradient-based optimization techniques, crucial for accurate parameter estimation. The model's continuous representation ensures smooth transitions between various body shapes, enhancing the realism and accuracy of the body representations generated.

Model Fitting. SMPL parameters are extracted from 3D scans p_t^i by using the method provided by Garcia-D'Urso et al. [10]. This method consists in the use of intermediate templates for the prediction of the shape, β, and pose, θ, components of the parameter model. This method contains 3 main phases:

1. **Prior estimation** of a 3D template by the BPS Neural Network: BPS was used to generate an intermediate 3D model. This method encode the point cloud in a vector of distances to a fixed set of points. This array is used as input to a dense network with two blocks of two fully connected layers.
2. **Coarse fitting**: The next step is to minimize the parameters of the SMPL model with to acquire the same pose and the shape than the BPS model.
3. **Fine fitting**: This second minimization was performed to create a parametric model with the same pose and shape than the original model captured with RGB-D sensors. The objective function of this minimization take into account the distance and the normal angle between points.

3.3   Prediction of Body Shape

Following the study of neural networks for learning sequential data in Sect. 2.2 section), transformer networks have some limitations despite their great success as they could not capture the sequential nature of data as effectively as RNN [14]. Furthermore, transformer networks are not only more challenging to implement but also demand larger datasets to leverage their attention mechanisms fully. Taking these factors into account, in this paper, the architecture is based on LSTM networks due to their ability to process long sequences and retain essential information, thereby mitigating the vanishing gradient problem. In addition, the implementation complexity and data requirements for predicting body shapes, make it a balanced approach.

Due to the nature of our data, there exists a variation in the number of dietetic treatment sessions attended by patients. Furthermore, these sessions are not temporally equidistant, with the time intervals fluctuating between a few days to multiple months. This is a problem for neural networks, since they work best with uniformly spaced data. Recurrent neural networks are flexible in terms of the length of the input, but we still need to account for the temporal variability of these sessions.

To solve this problem, the remaining days to the next session are used as another important feature of our proposal, being the network able to predict the expected change for the next session based on this number of days. In other words, we decided to make the neural network predict the daily change instead. This approach allows the network to predict daily body changes, giving us the flexibility to predict any number of days into the future.

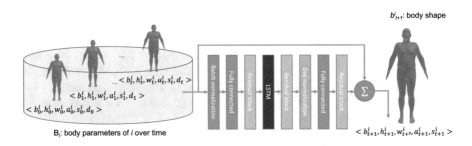

Fig. 2. LSTM-based neural network architecture to predict bodies.

The architecture of the network is shown in Fig. 2. It has as input features a sequence of the β parameters of the SMPL model of b^i at different instant of time (according to the session of the treatment) with the patient's height (h^i), weight (w^i), age (a^i) and a one-hot vector encoding their sex (s^i). Attempts to include body fat percentage and muscle mass percentage did not yield improved results and were excluded to prevent over-fitting. After that, a batch normalization mechanism has been added, which serves to normalize input features and potentially expedite convergence during training. This is succeeded by a linear transformation layer and an associated residual block.

Subsequently, the architecture integrates a Long Short-Term Memory (LSTM) layer, signifying the necessity to capture temporal or sequential intricacies inherent in the dataset. Following the LSTM, another residual block is integrated, and the output is then scaled by a factor corresponding to the intervening days prior to the subsequent session (Day normalization). This scalar multiplication aims to condition the network to predict daily changes. After that, the network employs an additional residual block. To implement a residual nature to the network, the raw input data is added to the output of this block. In this way, we want to learn the differences of changes between sessions being able to predict the daily change, and scaling the output to the number of days until the next session.

Each constitutive residual block in the network is a sequence of an application of the following layers:

- A **layer normalization** mechanism is utilized at the outset to stabilize the activations and facilitate gradient propagation during backpropagation.
- A **linear layer**, which expands the feature dimensions by a designated spreading factor, is employed. In our case, we found a factor of 4 to exhibit good results.
- Subsequently, the **Gated Exponential Linear Unit** (GELU) activation function is applied.
- Another **linear layer** is incorporated, serving to revert the dimensionality post-expansion to its antecedent state.
- Finally, a **skip connection** is integrated, which directly adds the preceding layer's output to the block's input, effectively allowing the block to model and learn incremental residuals.

4   Experiments

In this section, the results obtained from the experiments are discussed. First, we will describe the experimental setup, then evaluate our predictive model's performance, and conclude with an analysis of the generated bodies.

4.1   Experimental Setup

This work uses the dataset that is being developed in the Tech4Diet project. It comprises data from approximately 400 sessions obtained from 80 patients, having each individual from 3 to 11 sets of data over time (being 6 the mean of sessions per patient). The sessions, conducted at irregular intervals, comprised 3D scans of the patient along with various other measurements, including weight, height, body fat percentage, among others. However, substantial data processing was requisite due to the presence of missing measurements and a

plethora of data outliers within the raw data. Following an extensive data cleaning procedure-which involved treating missing data, eliminating invalid data,
and standardizing variable units-approximately 200 sessions were deemed viable
for analysis (Fig. 3).

Fig. 3. 3D body model reconstructions during a patient's weight loss treatment, with
each scan captured roughly a month apart, with a total weight loss of 3.8 kg.

Regarding implementation details, the network was instantiated using
PyTorch. The Adaptive Moment Estimation with Weight Decay (AdamW)
served as the optimizer, utilizing a variable learning rate and weight decay with
Mean Squared Error (MSE) loss functioning as the objective criterion. To mitigate the risk of overfitting, early stopping was employed during the training
phase. Specifically, training was halted if there was no improvement observed in
the validation loss for a span of 50 epochs. Finally, we used a 5-fold and a grid
search algorithm to find optimal hyperparameters.

4.2   Experimental Results

The mean average error, MAE, of the architecture when predicting the body
shape b_{t+1} according to the β parameters of SMPL is 0.064, indicating a very
low error. Figure 4 shows the predictions of the model with respect to the ground
truth for the first 9 β. As the β values are related to the PCA, we can see that
our model is able to obtain favourable predictions in firsts components, with
more errors from β_9, which does not have a big impact on the final result.

Evaluating the quality of generated bodies qualitatively presents a challenge;
however, as depicted in Fig. 5, it is discernible that the model successfully generates shapes resembling the patient. It displays the reconstructed 3D model of a
patient's body, with images from various sessions. Notably, the last image in the
series represents a prediction by the model of the patient's body one month subsequent to the final session, with a time lapse of 74 days from the commencement
to the conclusion of sessions.

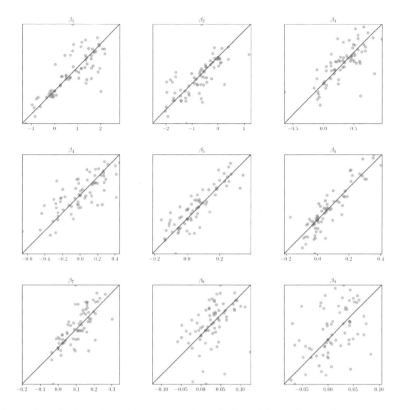

Fig. 4. Predictions (y-axis) vs ground truth (x-axis) for first 9 β parameters.

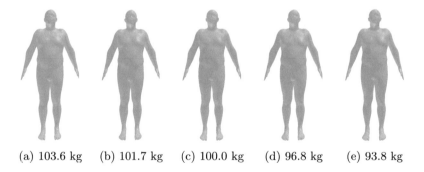

(a) 103.6 kg   (b) 101.7 kg   (c) 100.0 kg   (d) 96.8 kg   (e) 93.8 kg

Fig. 5. 3D body model reconstruction of a patient, with the last image predicting changes one month post-final session. Total time span: 74 days from first to last session.

5   Conclusions

This research presents a novel and effective method for predicting body shape changes in individuals undergoing weight loss treatment. By harnessing the power of machine learning models, specifically the SMPL parametric model combined with LSTM-based neural networks, our predictive approach demonstrated its ability to approximate future body shape changes based on historical data. The predictive capabilities of our proposed model have significant implications for weight loss treatment programmes and healthcare professionals. By providing personalised visualisations of future body shape changes, our model becomes a valuable tool to motivate and engage patients in their weight loss programme.

As future lines, while the model showed robust performance during the validation phase, we acknowledge the challenges posed by the complexity of human body shape change and individual variability in weight loss patterns. To address these limitations, continued refinement and expansion of the dataset promise to further improve the accuracy and practicality of the model.

References

1. Allen, B., Curless, B., Popovic, Z.: The space of human body shapes: reconstruction and parameterization from range scans. ACM Trans. Graph. **22**, 587–594 (2003). https://doi.org/10.1145/1201775.882311
2. Anguelov, D., Srinivasan, P., Koller, D., Thrun, S., Rodgers, J., Davis, J.: Scape: shape completion and animation of people. ACM Trans. Graph. (TOG) **24**, 408–416 (2005). https://doi.org/10.1145/1073204.1073207
3. Bengio, Y., Simard, P., Frasconi, P.: Learning long-term dependencies with gradient descent is difficult. IEEE Trans. Neural Netw. **5**, 157–66 (1994). https://doi.org/10.1109/72.279181
4. Boraxbekk, C.J., et al.: Diet-induced weight loss alters functional brain responses during an episodic memory task. Obes. Facts **8**, 261–272 (2015). https://doi.org/10.1159/000437157
5. Chen, Y., Song, Z., Xu, W., Martin, R.R., Cheng, Z.Q.: Parametric 3D modeling of a symmetric human body. Comput. Graph. **81**, 52–60 (2019)
6. Cho, K., van Merrienboer, B., Gülçehre, Ç., Bougares, F., Schwenk, H., Bengio, Y.: Learning phrase representations using RNN encoder-decoder for statistical machine translation. CoRR abs/1406.1078 (2014). https://arxiv.org/abs/1406.1078
7. Drigny, J., et al.: Effect of interval training on cognitive functioning and cerebral oxygenation in obese patients: a pilot study. J. Rehabil. Med. **46**, 1050–1054 (2014). https://doi.org/10.2340/16501977-1905
8. Eurostat: Overweight and obesity - BMI statistics (2019)
9. Fuster-Guilló, A., Azorín-López, J., Saval-Calvo, M., Castillo-Zaragoza, J.M., Garcia-D'Urso, N., Fisher, R.B.: RGB-D-based framework to acquire, visualize and measure the human body for dietetic treatments. Sensors **20**(13) (2020). https://doi.org/10.3390/s20133690. https://www.mdpi.com/1424-8220/20/13/3690
10. Garcia-D'Urso, N.E., Azorin-Lopez, J., Fuster-Guillo, A.: Accurate estimation of parametric models of the human body from 3D point clouds. In: García, P., et al. (eds.) SOCO 2023, pp. 236–245. Springer, Cham (2023). https://doi.org/10.1007/978-3-031-42536-3_23
11. Hirshberg, D.A., Loper, M., Rachlin, E., Black, M.J.: Coregistration: simultaneous alignment and modeling of articulated 3D shape. In: Fitzgibbon, A., Lazebnik, S., Perona, P., Sato, Y., Schmid, C. (eds.) ECCV 2012. LNCS, vol. 7577, pp. 242–255. Springer, Heidelberg (2012). https://doi.org/10.1007/978-3-642-33783-3_18

12. Hochreiter, S., Schmidhuber, J.: Long short-term memory. Neural Comput. **9**, 1735–80 (1997). https://doi.org/10.1162/neco.1997.9.8.1735
13. Hopfield, J.J.: Neural networks and physical systems with emergent collective computational abilities. Proc. Natl. Acad. Sci. **79**(8), 2554–2558 (1982)
14. Katrompas, A., Ntakouris, T., Metsis, V.: Recurrence and self-attention vs the transformer for time-series classification: a comparative study. In: Michalowski, M., Abidi, S.S.R., Abidi, S. (eds.) AIME 2022. LNCS, vol. 13263, pp. 99–109. Springer, Cham (2022). https://doi.org/10.1007/978-3-031-09342-5_10
15. Lehnert, T., Sonntag, D., Konnopka, A., Riedel-Heller, S., König, H.H.: Economic costs of overweight and obesity. Best Pract. Res. Clin. Endocrinol. Metab. **27**, 105–115 (2013). https://doi.org/10.1016/j.beem.2013.01.002
16. Loper, M., Mahmood, N., Romero, J., Pons-Moll, G., Black, M.J.: SMPL: a skinned multi-person linear model. ACM Trans. Graphics (Proc. SIGGRAPH Asia) **34**(6), 248:1–248:16 (2015)
17. Olson, E.A., Visek, A.J., McDonnell, K.A., DiPietro, L.: Thinness expectations and weight cycling in a sample of middle-aged adults. Eat. Behav. **13**(2), 142–145 (2012)
18. World Health Organization: World health organization. global strategy on diet, physical activity and health (2004)
19. Osman, A.A.A., Bolkart, T., Black, M.J.: STAR: a sparse trained articulated human body regressor. In: European Conference on Computer Vision (ECCV), pp. 598–613 (2020). https://star.is.tue.mpg.de
20. Pavlakos, G., et al.: Expressive body capture: 3D hands, face, and body from a single image. In: Proceedings IEEE Conference on Computer Vision and Pattern Recognition (CVPR), pp. 10975–10985 (2019)
21. Renzo, L.D., et al.: Eating habits and lifestyle changes during COVID-19 lockdown: an Italian survey. J. Transl. Med. (2020). https://doi.org/10.1186/s12967-020-02399-5
22. van Stralen, M.M., De Vries, H., Mudde, A.N., Bolman, C., Lechner, L.: Determinants of initiation and maintenance of physical activity among older adults: a literature review. Health Psychol. Rev. **3**(2), 147–207 (2009)
23. Teixeira, P., Going, S., Sardinha, L., Lohman, T.: A review of psychosocial pretreatment predictors of weight control. Obes. Rev. **6**(1), 43–65 (2005)
24. Vaswani, A., et al.: Attention is all you need (2017)
25. Willmott, T., Parkinson, J.: Motivation, opportunity, and ability: understanding new habits and changes adopted for weight management. Int. J. Consum. Stud. **41**(3), 291–298 (2017)
26. Withrow, D., Alter, D.A.: The economic burden of obesity worldwide: a systematic review of the direct costs of obesity. Obes. Rev. **12**, 131–141 (2011). https://doi.org/10.1111/j.1467 789X.2009.00712.x

Understanding Students' Perspectives About Human-Building Interactions in the Context of Smart Buildings

Allan Berrocal[ID], Luis Quesada[ID], Kryscia Ramírez-Benavides[ID],
and Adrian Lara[(✉)][ID]

Centro de Investigaciones en Tecnologías de la Información y Comunicación,
Universidad de Costa Rica, San José, Costa Rica
{allan.berrocal,luis.quesada,kryscia.ramirez,adrian.lara}@ucr.ac.cr
http://citic.ucr.ac.cr

Abstract. Smart buildings provide a variety of sensor-based services to support and enhance the quality of human activities. Advanced technologies such as robotics are increasingly added to smart buildings' ecosystems, creating a need to incorporate affective computing techniques to augment the quality of human-building, and human-robot interactions. To better understand user's needs and expectations about human-building interactions, we conducted a pilot study using a mixed methods approach combining short surveys and controlled laboratory activities. We recruited 66 participants and collected several data elements characterizing their perceptions and expectations about smart building services. This paper presents preliminary evidence showing acceptance of specific human-building interaction methods based on ambient-sensors information such as in-context voice, behavior, and emotion, recognition. We also identified a need for educational activities to promote the understanding of smart building concepts and their impact in modern society. These results can be leveraged to assist the design of future services that include human-building and human-robot interactions.

Keywords: Smart buildings · Human-Building Interaction (HBI) · Society impact

1 Introduction

Modern buildings integrate and process data from multiple sensors to provide valuable services to its users. Due to the wide availability of complex sensors, and advanced computing power embedded in modern buildings, they are referred to as smart buildings [1]. Smart buildings' services vary along many dimensions [5] such as purpose and level of sophistication (e.g. automatic door operations, air quality control, energy savings, ambient customization, among others). Finally, smart buildings have started considering affective computing techniques [6] to enhance the quality of human-building, and human-robot interactions.

J. Bravo and G. Urzáiz (Eds.): UCAmI 2023, LNNS 842, pp. 112–118, 2023.
https://doi.org/10.1007/978-3-031-48642-5_11

This paper reports our findings while designing a smart laboratory for our students. We asked ourselves: How familiar are our students with smart buildings? What are their expectations of a smart building? What are the opportunities for affective computing [6] scenarios with our students?

This paper summarizes the results of a first study with $N = 66$ participants where we aim to assess and investigate the following aspects: Prior knowledge or awareness about smart buildings, perceptions about smart building services advertised as *intelligent*, perceived importance of a number of human-building interaction methods, perceptions about smart buildings including self interest, importance for society, and implementation difficulty.

2   Related Work

Table 1 shows several studies addressing perception of smart buildings, with privacy being the most common topic addressed. The majority of these studies work with users that are familiar with or live inside smart buildings. Our study is slightly different in that we deal with students with little or no experience on the topic. Thus, our focus is on what they know about or expect from smart buildings.

Table 1. Related work summary

Authors	Topic	Participants	Year
Zheng et al. [10]	Privacy	11	2018
To et al. [9]	Importance of sustainable features	494	2018
Marky et al. [4]	Privacy	15	2020
Rokooei et al. [7]	Perception of smart buildings	43	2022
Taher et al. [8]	Privacy	20	2022
Krawczyk et al. [2]	Thermal comfort perception	73	2022
Li et al. [3]	Privacy	24	2023

3   Methods

Our protocol consisted of gathering students in a classroom where they completed three activities. In the first activity, students completed a survey with two sections. The first one collected demographic characteristics (gender and age), and the second one inquired about three categories of information: (a) prior knowledge about smart buildings, (b) perceptions about building services and interaction methods, and (c) perceptions about relevance and complexity of smart buildings. Table 2 shows the survey questions utilized in this study.

In the second activity, students were shown a pre-recorded video[1] explaining concepts such as ubiquitous computing and robotics. The video format was meant to reduce bias by ensuring that all students received exactly the same information.

Finally, in the third activity, participants completed a short survey with three questions from the first activity. Doing so allowed us to measure differences in perceptions after being part of the second study activity.

Table 2. Survey Questions

Question	Category
1. How much do you know about smart buildings?	Knowledge
2. Should the following building services be called intelligent? a. Access control mediated by card readers b. Automatic light control (on and off) c. Automatic room temperature control d. Automatic room ventilation control e. Interaction with services by voice commands f. Analysis of people circulation in the building	Services and interaction methods
3. How important are the following human-building interaction methods? a. Interacting with the building through voice recognition b. Interaction with the building through gesture recognition c. Recognition of behaviors and emotional states through building sensors d. Recognition of people to provide context-aware services	
4. Are smart buildings important for society?	Relevance
5. Are you interested in smart buildings?	and
6. What is the technical difficulty of implementing smart buildings?	complexity

Participants. Students enrolled in the Introduction to Computer Science course. Traditionally, these students have little knowledge of smart buildings and IoT. Some of them have prior programming knowledge, but the majority of them have very little experience in computer science in general. We were able to work with four out of eight groups offered by the department. All groups had between 10 and 25 students.

---

[1] https://www.youtube.com/watch?v=FY2aV4OmHYA.

We conducted the activities in a customized laboratory to study perceptions about smart buildings. This laboratory house sensor technologies and Internet of Things (IoT) devices (see Table 3) to explore their integration into daily life. Orchestrated by a Home Assistant server and further supported by servers like InfluxDB and Grafana, the laboratory infrastructure facilitates comprehensive data collection, manipulation, and visual analytics. It measures various environmental parameters and allows interaction with devices like lights, projectors, and cast devices based on sensor data or via voice commands through smart speakers like Amazon's Alexa.

Table 3. Sensors and IoT devices inside the laboratory.

Device	Feature(s)
Aqara Camera Hub G3	Real-time video streaming, Human tracking, Face recognition, Gesture recognition, Sound detection
Aqara Temperature sensor	Temperature, Humidity, Atmospheric pressure
Aqara Motion sensor	Motion detection, Illuminance
Aqara Door sensor	Open/close detection
Blink camera	Real-time video streaming, Motion detection
Amazon Air Quality Monitor	Indoor air quality, Temperature, Humidity, Particulate matter, Carbon monoxide, Volatile organic compounds
Customized application (Raspberry pi + sensors)	Carbon dioxide, Distance, Vibration detection, Sound detection
Linksys router	Network usage
Amazon Echo Show/Dot	Voice recognition

4   Results

A total of $N = 66$ participants completed both surveys of the study; one before, and one after watching the video (see Sect. 3). The gender distribution was 15% female, 83% male, and 2% other. In terms of age, 65% were between 16–18 years old, 29% between 19–23, 3% between 24–30, and 3% between 31–40.

Student Perceptions about Smart Buildings. First of all, in question No. 1 of the survey, most students (65%) reported limited knowledge or awareness about the concept of smart buildings. More concretely, 32% of respondents said they knew nothing at all, while 33% said they knew very little. Moreover, about a fourth of the students (26%) declared neutral knowledge, and only 9% said they knew enough about the concept. Summarizing, in this particular study, participants were predominantly less knowledgeable about the concept of smart buildings.

In question No. 2, we presented a list of rather simple, sensor-based, and commonly implemented smart building services. We wanted to assess the degree to which students agree to call these services *intelligent*. As shown in Fig. 1, aside from the first service (access control card readers), we found that a majority of students are in favor (agree or totally agree) to call these types of services *intelligent*.

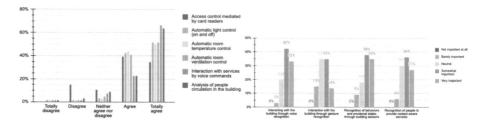

Fig. 1. Intelligent building services.   **Fig. 2.** Human-building interaction methods.

In question No. 3, we presented a list modern human-building interaction methods, and we wanted to assess how important were these interaction methods for students. Figure 2 shows that most students consider these interaction methods somewhat important (38% on average), or very important (27% on average). On the other hand, human-building interaction via gesture recognition was rated as very important 14% of the time, which is about 50% less often than other interaction methods. Additionally, 35% of the participants said they were neutral regarding the importance of this particular interaction method.

Education for Awareness Shift. As explained in Sect. 3, study participants contributed responses to questions No. 4, 5, and 6 twice; one before and after taking part in the educational activity (No 2.) of the study protocol. Based on these two data points, we are able to identify response shifts from participants (i.e. difference of the response from the first and second surveys).

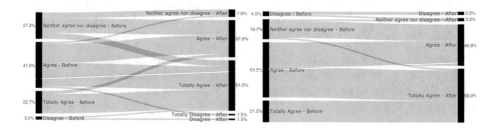

Fig. 3. Importance for Society   **Fig. 4.** Personal Interest

In Fig. 3 one can see that a large percentage of participants in the first survey said they disagree (3%), or neither agree nor disagree (27.3%) that smart buildings are important for society (question No. 4). However, after activity No. 2, most of those participants expressed higher agreement totaling 89.4% between the agree, and totally agree categories.

Similarly, regarding participant's level of personal interest in smart buildings (question No. 5), Fig. 4 shows that the combined percentage of disagreement and neutral shrank from 24.2% to 3% after activity No. 2. Moreover, the combined percentage of agree and totally agree increased from 75.7% to 94%.

5   Discussion

Based on the results, we can see that this particular sample shows limited knowledge about the concept of smart buildings. This information is useful in at least two ways: (a) to introduce these concepts more carefully during the introductory courses of the major, and (b) to feedback downstream educational levels, particularly technical high schools in our country.

Another aspect we observed in this study is the tendency to accept the term *intelligent* as a qualifier for simple, yet sophisticated, sensor-based automation services widely present in smart buildings (Fig. 1). Our task is to explore whether or not such a tendency will prevail when we survey other populations.

From this study, we also collected valuable data that may influence user's acceptability of smart building services based on the type of human-building interaction method offered by such services (Fig. 2). We learned that recognizing voice, behaviors and emotional states through sensors are all important aspects to account for when designing interaction methods. Similarly, provisioning buildings with ubiquitous context-aware services may be seen as acceptable human-building interaction methods.

Finally, when we look at the student responses before and after watching the video in this study (see Sect. 3), we can infer that a short class about smart buildings had a noticeable impact as an educational tool. The observed tendency is to increase the perceived level of personal interest as well as the perceived value for society after engaging in the activity.

6   Conclusions and Future Directions

The goal for this study was to pilot part of the methodology of a larger aim to investigate people's perceptions regarding smart buildings. The long term goal is to understand, document and implement relevant affective computing techniques [6] in the context of the design of human-building and human-robot service-interactions.

From our results, we can conclude that there is a need for educational activities to promote the understanding of smart building concepts and their impact in modern society. We also confirm a positive acceptance of human-building interaction methods leveraging ambient-sensors information such as recognition of voice, behaviors, emotions, and context.

In future work, we will continue with the larger study surveying individuals from more educational levels, age brackets and occupations. Additionally, we plan to pilot specific implementations of affective human-building interactions in a controlled laboratory environment.

References

1. Buckman, A., Mayfield, M., Beck, S.B.M.: What is a smart building? Smart Sustain. Built Environ. **3**(2), 92–109 (2014)
2. Krawczyk, N., Wojciechowska, P.: Analysis of indoor environment perceptions in the smart building. J. Phys. Conf. Ser. **2339**, 012019 (2022). IOP Publishing
3. Li, B., Tavakoli, A., Heydarian, A.: Occupant privacy perception, awareness, and preferences in smart office environments. Sci. Rep. **13**(1), 4073 (2023)
4. Marky, K., Zimmermann, V., Stöver, A., Hoffmann, P., Kunze, K., Mühlhäuser, M.: All in one! user perceptions on centralized IoT privacy settings. In: Extended Abstracts of the 2020 CHI Conference on Human Factors in Computing Systems, pp. 1–8 (2020)
5. Mofidi, F., Akbari, H.: Intelligent buildings: an overview. Energy Build. **223** (2020)
6. Picard, R.W.: Affective computing. Pattern Anal. Appl. **1**(1), 71–73 (1998)
7. Rokooei, S., Karji, A.: Exploring construction students' perceptions of smart buildings. In: ASC 2021. 57th Annual Associated Schools of Construction International Conference, vol. 2, pp. 284–292 (2021). https://doi.org/10.29007/rwtr
8. Tawer, R., Mehrnezhad, M., Morisset, C.: I feel spied on and I don't have any control over my data: user privacy perception, preferences and trade-offs in university smart buildings. IN: Socio-Technical Aspects in Security, STAST 2022 (2022)
9. To, W.M., Lai, L.S., Lam, K.H., Chung, A.W.: Perceived importance of smart and sustainable building features from the users' perspective. Smart Cities **1**(1), 163–175 (2018)
10. Zheng, S., Apthorpe, N., Chetty, M., Feamster, N.: User perceptions of smart home IoT privacy. Proc. ACM Hum.-Comput. interact. **2**(CSCW), 1–20 (2018)

IoT-Driven Real-Time Glucose Monitoring: Empowering Diabetes Care and Prevention

José L. López Ruiz[1]([✉])[ID], Juan F. Gaitán Guerrero[1][ID],
Carmen Martínez Cruz[1][ID], David Díaz Jimenez[1][ID], Jesús González Lama[2][ID],
and Macarena Espinilla[1][ID]

[1] Department of Computer Science, University of Jaén, Campus Las Lagunillas,
A3-140, 23071 Jaén, Spain
{llopez,cmcruz,ddjimene,mestevez}@ujaen.es, jfgg0010@red.ujaen.es
[2] Maimonides Biomedical Research Institute of Cordoba, Reina Sofia University
Hospital, 14004 Córdoba, Spain
jegonla@telefonica.net

Abstract. The widespread use of Internet of Things (IoT) devices has revolutionized monitoring systems, driving advancements in various societal domains. This study presents a cutting-edge IoT-based architecture for real-time glucose monitoring in the context of Internet of Medical Things, emphasizing its ability to enable timely and accurate sample collection. The primary objective of this system is to continuously monitor glucose levels using the Freestyle Libre 3 sensor, making a significant impact on the health management of patients affected by diabetes and proactively preventing the disease's onset. To ensure the system's effectiveness, a comprehensive quantitative evaluation is conducted, focusing on both battery life and the seamless collection of real-time samples. This meticulous assessment guarantees the system's reliability, efficiency, and ability to deliver vital health data promptly. Additionally, the contribution includes an evaluation of the system's alignment with the Sustainable Development Goals (SDGs), demonstrating its potential contributions to broader social, economic, and environmental objectives. This research showcases the transformative potential of IoT technology in healthcare, offering unprecedented opportunities for continuous health monitoring and proactive intervention.

Keywords: Internet of Medical Things · Real-time glucose monitoring · Monitoring System · Diabetes · Healthcare technology · Freestyle Libre 3 sensor

1 Introduction

Nowadays, the Internet of Things (IoT) is widely employed across various domains. IoT, as a technology enabling communication among diverse

This result has been partially supported by PID2021-127275OB-I00 funded by MCIN/AEI/10.13039/501100011033 and by "ERDF A way of making Europe".

Internet-connected devices, has transformed the provision of numerous services, leading to enhanced quality.

The prevalence of IoT is evident in numerous fields, including home [14], city [11], agriculture [3], and notably, healthcare [2,6]. For instance, in home settings, such systems can be utilized to ensure real-time location tracking [10] of patients, preventing undesirable incidents like falls.

IoT technology serves as the foundation for monitoring systems, aiming to control specific characteristics through smart devices and/or sensors. These systems collect data, perform data cleansing, store it persistently, and even employ intelligent processing to extract valuable insights. In the realm of healthcare, this knowledge extraction supports diagnostics and disease monitoring by healthcare professionals.

While these systems facilitate various societal aspects, there is a growing need to analyze their alignment with Sustainable Development Goals (SDGs), which some studies have overlooked, as indicated by Espinosa et al. [19].

Presently, diabetes is becoming increasingly prevalent worldwide. According to the World Health Organization (WHO), diabetes is a chronic disease [22] caused by either insufficient insulin production by the pancreas or ineffective utilization of insulin by the body. Consequently, patients with diabetes experience hyperglycemia, an uncontrolled increase in blood sugar levels that can lead to severe complications if left untreated.

In 2021, the International Diabetes Federation [8] reported that 537 million adults (aged 20–79) were living with diabetes (1 in 10 patients). This number is projected to increase to 643 million by 2030 and 783 million by 2045. Europe alone accounts for 61 million adults with diabetes (1 in 11), of which 36% remain undiagnosed. Healthcare costs related to diabetes amounted to USD 189 billion in 2021, with 1.1 million diabetes-related deaths reported.

Diabetes is commonly classified into three types. The first type is gestational diabetes, where patients experience hyperglycemia with blood sugar levels above normal but below those diagnosed for diabetes. The second type is type 2 diabetes mellitus, which can cause high blood sugar levels due to the body's ineffective use of insulin. In these cases, treatment through pills or even insulin injections is often required. Lastly, type 1 diabetes mellitus involves deficient insulin production in the body, necessitating daily insulin administration.

Monitoring systems, combined with IoT, can be a valuable tool to enhance the quality of life for patients with diabetes, aid in diagnosing the condition, and even prevent its onset. This leads to reduced mortality and healthcare costs associated with diabetes. Thus, this paper introduces a real-time monitoring system based on IoT devices, designed to control glucose levels for patients using the Freestyle Libre 3 sensor [1]. In addition, it includes a web application to access the latest glucose data received and any historical data of the patient. Compared to other applications such as Nightscout, it allows the data of any patient who wants to use the system to be sent and consulted. Also, the robustness of the system is evaluated in relation to battery life and samples collected, a crucial feature for patients with diabetes.

Finally, evaluating this proposed system against the Sustainable Development Goals (SDGs) is of utmost importance as it allows us to assess its potential positive impact on broader social, economic, and environmental aspects. Additionally, this evaluation presents the alignment of the proposed system with specific SDGs, enabling us to understand how it addresses key sustainability issues and contributes to achieving the global agenda for a more sustainable and inclusive future.

The following are the sections comprising the document. Firstly, Sect. 2 provides a brief literature review of related works. Next, in Sect. 3, the proposed glucose monitoring system is presented. Subsequently, in Sect. 4, a quantitative evaluation of the system is conducted concerning battery life and collected samples. Finally, in Sect. 5, an assessment of the system's alignment with the SDGs is performed, and in Sect. 6, the conclusions and the future works of the contribution are presented.

2   Related Works

In the healthcare field, numerous works [2] aim to monitor and address various health issues with the ultimate goal of improving people's quality of life. Within this domain, IoT is often referred to as the Internet of Medical Things (IoMT). One such research by Bhardwaj et al. [6] presents a system designed to identify COVID-19 patients through the monitoring of blood pressure, heart rate, oxygen level, and temperature.

Regarding the topic addressed in this research, some works focus on diabetes [16]. On one hand, some opt for non-invasive technologies using sensors based on split ring microwave resonators [5] or smartphone photoplethysmogram (PPG) [23]. However, these experimental technologies may not provide reliable samples and lack a real-world applicable architecture and sampling method.

On the other hand, literature includes works using invasive sensors. In these cases, the research focuses on the methodology for intelligent processing but does not address the continuous real-time data acquisition. For example, the study by Whelan et al. [21] evaluates the experience of diabetes patients with the Freestyle Libre 2 sensor, but this version of the sensor does not provide continuous data and requires patient interaction. Moreover, Nasser et al. [12] propose a prototype for continuous glucose monitoring, which does not include a real system.

These models are based on machine learning [12,20] or deep learning [12] and are primarily used for two different types of predictions. The first type predicts glucose levels in the future to anticipate harmful hyperglycemia and hypoglycemia. Glucose prediction is usually performed in a relatively short future period (between 30 and 60 min). For example, Nasser et al. [12] provide a prediction for the next 30 min using a deep learning model.

The second type detects whether a person has diabetes or not [15,20]. Other characteristics are often used in addition to glucose levels to determine whether a person is diabetic. In the work of Padhy et al. [15] a total of 14 characteristics are used, such as age, gender, family history, physical activity, blood pressure, among others.

Finally, all these research works lack evaluation from a sustainable perspective. While some of the mentioned works contribute to sustainability in energy, industry, economic growth, and even healthcare, none of them explicitly address the SDGs.

3   Proposed Real-Time Glucose Monitoring System

In this section, we propose an IoT-based monitoring system for real-time glucose control.

First and foremost, it is essential to define what glucose is and how it is measured. Blood glucose refers to the sugar present in the blood at a given moment and is measured in milligrams of sugar per deciliter (mg/dL).

Currently, patients with diabetes, especially those with type 1 and type 2 diabetes mellitus, use a glucometer to measure their blood sugar levels. A glucometer is a device that measures capillary blood glucose using a drop of blood obtained from a finger. This process involves two elements: the lancet and the test strip. The lancet is used to prick the finger and extract a small drop of blood, while the test strip reacts with the blood to determine the glucose level. Finally, the glucometer analyzes the test strip to provide the blood glucose reading in mg/dL. All these elements are illustrated in Fig. 1.

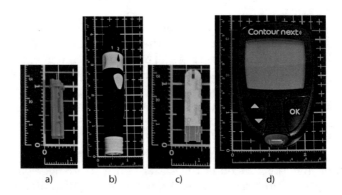

Fig. 1. Devices needed to measure blood glucose: a) needle, b) lancet, c) test strip and d) glucometer (own image).

Blood glucose measurement is a painful process, especially for patients with type 1 diabetes mellitus, which may need to be repeated up to 10 times a day. Additionally, it incurs a significant expense in non-reusable materials, such as lancets and test strips. Therefore, this process could benefit from a more sustainable perspective.

3.1 Proposed Glucose Sensor

Currently, there are commercial sensors available that allow us to measure glucose at regular intervals. Unlike traditional glucometers that measure capillary glucose, these devices use glucose from interstitial fluid, which originates from exchanges between cells in the tissue and blood.

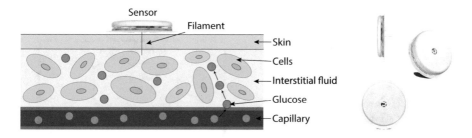

Fig. 2. Example of sensor placement on a person's skin (own image).

The main difference is that it has a delay of 5 to 10 min compared to capillary glucose readings, and both measurements are only coincident during stable moments. For instance, when a user consumes a meal, the values will not coincide and will exhibit a slight delay.

These types of sensors are composed of an electrode or filament. This element is placed subcutaneously using an applicator and measures interstitial glucose values through an enzymatic reaction, providing readings at regular intervals. An example of the sensor applied to a person's skin can be observed in Fig. 2.

In this research work, the commercial sensor Freestyle Libre 3 [1] is utilized, and its characteristics are outlined in Table 1.

Table 1. Characteristics of the glucose sensor.

Size (diameter × depth)	$21 \times 2,9$ mm
Battery life	14 days
Reusable	No
Frequency of data transmission	5 min
Type of connectivity	Bluetooth Low Energy (BLE)
Approximate cost	59.91€

3.2 Architecture of the Proposed System

To monitor the user wearing the sensor, it is necessary to establish an architecture for sending and persistently storing the data. The system's architecture consists of two layers: the fog layer and the cloud layer, as illustrated in Fig. 3.

In the fog layer, the glucose sensor is attached to the user's body, along with a fog node. It is recommended to place the glucose sensor on the upper arm, in an area with minimal muscle activity to avoid any discomfort during movements. This sensor sends a glucose sample every 5 min.

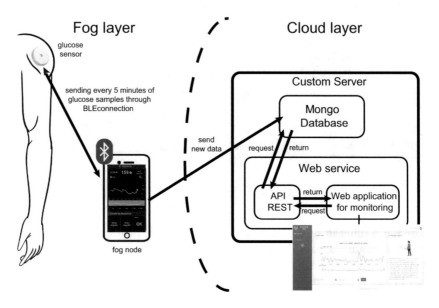

Fig. 3. Proposed Architecture for the Monitoring System Communication (own image).

On the other hand, there is the fog node, which is a smartphone equipped with Near-Field Communication (NFC) and Bluetooth Low Energy (BLE) capabilities. The first step is to initiate communication with the sensor using NFC technology by placing the smartphone close to the sensor. Once this connection is established, the sensor will continuously send glucose data without requiring any further action. All the data is received, processed, and sent to a RESTful service through the Diabox application [7], which is exclusively available for Android devices.

To establish communication with the RESTful service deployed in the cloud, it is necessary to follow the same request schema as Nightscout [13]. This service collects the data sent by the fog node and persistently stores it in a MongoDB database.

Each glucose sample is stored in the MongoDB database as a JSON format document. An example of such a document is shown in Fig. 4. For each sample, the most relevant fields include:

- **sgv**: indicates the glucose value in mg/dL.
- **dateString**: indicates the time instant at which the sample was collected.
- **utcOffset**: offset in minutes to be applied to the date. In this case the actual date would be: "2022-12-24T16:41:17".

```
▼ {
    date : 1671896477434 🕐
    dateString : 2022-12-24T15:41:17.434Z
    sgv : 115
    delta : 8.007
    direction : FortyFiveUp
    type : sgv
    filtered : 127000
    unfiltered : 127000
    rssi : 100
    noise : 1
    sysTime : 2022-12-24T15:41:17.434Z
    utcOffset : 60
}
```

Fig. 4. Example of a document in MongoDB for each sample collected (own image).

Additionally, a web application has been developed to display glucose data for patients. Unlike the Nightscout service, this application allows centralized remote access to glucose data for anyone who wishes to send their glucose data through our platform. To identify the patient, the security code requested in the Nightscout tool is used. In our case this 12-character alphanumeric code (e.g. wyQ9e9jFY5rt) is to identify the patient. This code is provided when the user registers and is unique. Although the system can support a large number of patients, the prototype has been tested with a small group of people, as the system has limited resources.

This web application offers two types of monitoring: real-time glucose and historical data for the user. The real-time monitoring provides an immediate view of the data from the last hour, along with the date of the last sent sample and the smartphone battery status used for data transmission. The real-time graph also includes a slider to view data from the past 24 h, and it automatically updates with new glucose data. This type of monitoring is illustrated in Fig. 5.

The second type of monitoring allows users to query historical glucose data. These queries enable filtering time ranges between two dates. If the start and end days are the same, it will display the historical data for that particular day only. Additionally, a linguistic summary of the most relevant events within that time span is included. This second type of monitoring is illustrated in Fig. 6.

In both cases, glucose data is visualized through line graphs, which users can interact with using various actions: scrolling on the X-axis, zooming, a slider to narrow down the time range with different step options (5 min, 1 h, or no limit), reset to the initial state, and exporting the graph as a PNG image. Moreover, three horizontal lines are included to represent the glucose levels defined by the American Diabetes Association (ADA) [4]: normal (grey), medium (orange), and high (red).

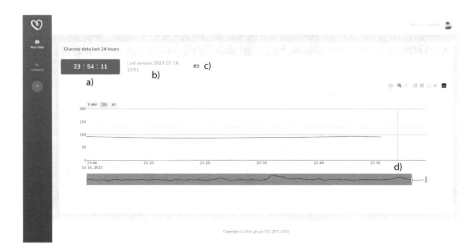

Fig. 5. Web application interface for the first type of monitoring: a) Current time, b) Time instant when the last sample was received, c) Remaining smartphone battery and d) Vertical bar showing the current minute (own image). (Color figure online)

Fig. 6. Web application interface for the second type of monitoring: a) Available actions, b) Slider to modify the time range, c) Query form between two dates and d) Linguistic summary of the filtered time span (own image). (Color figure online)

4   Quantitative Evaluation of the Effectiveness of the Proposed Real-Time Glucose Monitoring System

In this section, a quantitative evaluation of the effectiveness of the proposed real-time glucose monitoring system will be conducted. This evaluation will take into account several factors such as the number of missed samples, the actual sensor battery duration, and the overall system performance. To test the architecture, a sensor was used on a user, and samples were collected until the sensor's battery was depleted (continuously for 14 days).

The data collection period spans from December 24, 2022 (15:36 h) to January 7, 2023 (13:21 h), defined based on the first and last received samples. Overall, the sensor has operated for 99.33% of the time set by the manufacturer (14 days).

Regarding the samples, considering a sampling frequency of 5 min, 4005 samples were expected within that time frame. The system has successfully collected a total of 91.3% of the expected samples. Out of the 8.7% of missed samples, 7.1% were intentionally omitted to test whether the sensor could store data temporarily until reconnecting with the fog node. Thus, the system effectively captured 98.4% of the total samples. Additionally, it was observed that at times, the sensor readings experienced slight delays, leading to a partial loss of expected samples.

Based on the results, it can be concluded that the sensor performed as expected according to the manufacturer's specifications (99.33%). Furthermore, the connection and sample transmission have been robust and reliable (98.4%), which is crucial for type 1 diabetic patients, as they require a dependable system at all times. One limitation is that the sensor cannot store samples when the fog node is unavailable.

These findings are summarized in Fig. 7 and the data used for the evaluation is presented in Table 2.

Table 2. Data used for evaluation.

Time period of data collection			
Start	2022-12-24 15:36	*End*	2023-01-07 13:21
Number of estimated samples			4005
Number of samples collected			3656

Testing time periods	
Start	*End*
2023-1-3 00:50	2023-1-3 07:48
2023-1-4 22:35	2023-1-5 07:29
2023-1-5 23:32	2023-1-6 07:23
Number of samples lost	284

Example of samples with altered frequency		
Timestamp 1	*Timestamp 2*	*Time between samples (min)*
2022-12-25 17:17	2022-12-25 17:24	7
2022-12-25 18:05	2022-12-25 18:13	8
2022-12-25 22:58	2022-12-25 23:04	6

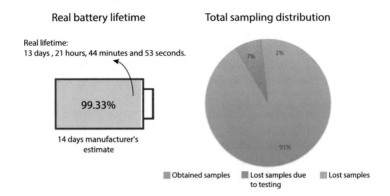

Fig. 7. Summary of the results obtained in the quantitative assessment (own image).

5   Alignment with Sustainable Development Goals

In 2015, the United Nations (UN) established the 2030 Agenda as an urgent call for action by all countries. This agenda includes 17 SDGs, 169 targets, and 232 indicators. Therefore, it is essential to evaluate this work from a sustainable perspective [17].

This type of evaluation is crucial for the implementation of the system in the real world, because it allows us to know how it affects our society and the environment. It also provides another type of metric to compare related systems. In the literature, there are studies [9,18] that propose methodologies to evaluate this type of system in a comprehensive way. In contrast, this paper makes a generalist evaluation of this system by listing the main SDGs with which we consider the monitoring system to be aligned.

Firstly, we consider the system proposed in this document aims to replace the traditional method of glucose measurement using a glucometer. On one hand, we believe that the presented proposal establishes a new service in the health-care domain to assist patients and provide a means for disease management by healthcare professionals. Hence, we consider that it aligns with *SDG 3: Good Health and Well-being*.

On the other hand, an patientes with type 1 diabetes is estimated to use the glucometer 10 times a day. By substituting this method with the proposed system, they would avoid using 140 needles and test strips over a 14-day period. Therefore, we also consider that the system aligns with *SDG 12: Responsible Consumption and Production*.

6   Conclusions and Futures Works

This document presents a real-time glucose monitoring system for patients with diabetes. The proposed system utilizes an invasive glucose sensor that reads glucose levels from the interstitial fluid. By enabling real-time sample collection, this innovative system empowers healthcare professionals and patients alike to make informed decisions, leading to better health outcomes and improved quality of life. To do so, the sensor is connected to a smartphone via NFC and BLE, and it continuously sends glucose readings without requiring any action from the user. Additionally, the glucose values are persistently stored in a MongoDB database. These data are retrieved and displayed in a web application, enabling real-time visualization of glucose levels, historical data with temporal filtering capabilities, and linguistic summaries of the selected time frames. The system underwent a 14-day evaluation and achieved a capture rate of 98.4% of the expected samples. Furthermore, the sensor's battery lasted 99.33% of the manufacturer-specified duration. Therefore, we believe that the system meets the expected requirements and needs. Finally, by considering the SDGs, we can better gauge the significance of this system in promoting health and well-being (SDG 3) and supporting responsible production and consumption practices (SDG 12), and thus guiding future developments towards a more sustainable world.

Our future work in this area will focus on providing real-time descriptive summaries of blood glucose levels to raise awareness among patients about the impact of their dietary choices and exercise habits on their health. By integrating data from the monitoring system with nutrition and activity logs, the proposed system aims to offer personalized feedback and insights to patients, empowering them to make informed decisions and adopt healthier lifestyle choices. The descriptive summaries will not only facilitate self-monitoring but also foster a deeper understanding of the relationship between glucose levels and daily habits, encouraging patients to take a proactive approach to managing their condition and ultimately improving their overall well-being. This aspect of the system holds immense potential for promoting patient engagement and adherence to treatment plans, leading to more effective diabetes management and enhanced health outcomes.

References

1. Abbott: Freestyle Libre 3 sensor. https://www.freestyle.abbott/es-es/productos/freestylelibre-3.html
2. Abdulmalek, S., et al.: IoT-based healthcare-monitoring system towards improving quality of life: a review. Healthcare 10(10), 1993 (2022)
3. Almalki, F.A., Soufiene, B.O., Alsamhi, S.H., Sakli, H.: A low-cost platform for environmental smart farming monitoring system based on IoT and UAVs. Sustainability 13(11), 5908 (2021)
4. American Diabetes Association: Glucose levels for the diagnosis of diabetes. https://diabetes.org/diabetes/a1c/diagnosis
5. Baghelani, M., Abbasi, Z., Daneshmand, M., Light, P.E.: Non-invasive continuous-time glucose monitoring system using a chipless printable sensor based on split ring microwave resonators. Sci. Rep. 10(1), 1–15 (2020)
6. Bhardwaj, V., Joshi, R., Anshu, M.: IoT-based smart health monitoring system for COVID-19. SN Comput. Sci. 3(2), 1–11 (2022)
7. Bubblandevteam: Diabox Application. https://github.com/Bubblandevteam/diabox.git

8. International Diabetes Federation: Diabetes around the world in 2021. https://diabetesatlas.org/
9. López, J.L., Espinilla, M., Verdejo, A.: Evaluation of the impact of the sustainable development goals on an activity recognition platform for healthcare systems. Sensors **23**(7), 3563 (2023). https://doi.org/10.3390/S23073563
10. López Ruiz, J.L., Verdejo Espinosa, A., Montoro Lendínez, A., Espinilla Estévez, M.: OBLEA: a new methodology to optimise bluetooth low energy anchors in multi-occupancy location systems. JUCS J. Univ. Comput. Sci. **29**(6), 627–646 (2023)
11. Mali, P., Patil, A.S., Gavade Pratibha, S., Mane Mrunal A., Patil Aniket A.: IoT based battery monitoring system for electric vehicle. Int. J. Adv. Res. Sci. Commun. Technol. 37–43 (2022)
12. Nasser, A.R., et al.: IoT and cloud computing in health-care: a new wearable device and cloud-based deep learning algorithm for monitoring of diabetes. Electronics **10**(21), 2719 (2021)
13. Nightscout contributors: Nightscout RESTful service framework. https://github.com/cosm0naut/nightscout/blob/deploy/swagger.yaml
14. Ahsan, M., Based, A., Haider, J., Rodrigues, E.: Smart monitoring and controlling of appliances using LoRa based IoT system. Designs **5**(1), 17 (2021)
15. Padhy, S., Dash, S., Routray, S., Ahmad, S., Nazeer, J., Alam, A.: IoT-based hybrid ensemble machine learning model for efficient diabetes mellitus prediction. Comput. Intell. Neurosci. **2022** (2022)
16. Rodriguez León, C., Villalonga, C., Muñoz Torres, M., Ruiz, J., Baños, O.: Mobile and wearable technology for the monitoring of diabetes-related parameters: systematic review. JMIR Mhealth Uhealth **9**(6), e25138 (2021). https://mhealth.jmir.org/2021/6/e25138
17. United Nations: Sustainable Development Goals (2015). https://sdgs.un.org/goals
18. Verdejo, A., Espinilla, M., López, J.L., Jurado, F.: Assessment of sustainable development objectives in smart labs: technology and sustainability at the service of society. Sustain. Cities Soc. **77**, 103559 (2022)
19. Verdejo, A., López, J.L., Mata, F.M., Estevez, M.E.: Application of IoT in healthcare: keys to implementation of the sustainable development goals. Sensors **21**(7), 2330 (2021)
20. Verma, N., Singh, S., Prasad, D.: Machine learning and IoT-based model for patient monitoring and early prediction of diabetes. Concurr. Comput. Pract. Exp. **34**(24), e7219 (2022)
21. Whelan, M., Orme, M., Kingsnorth, A., Sherar, L., Denton, F., Esliger, D.: Examining the use of glucose and physical activity self-monitoring technologies in individuals at moderate to high risk of developing type 2 diabetes: randomized trial. JMIR Mhealth Uhealth **7**(10), e14195 (2019)
22. World Health Organization: Diabetes disease. https://www.who.int/news-room/fact-sheets/detail/diabetes
23. Zhang, G., et al.: A noninvasive blood glucose monitoring system based on smartphone PPG signal processing and machine learning. IEEE Trans. Industr. Inform. **16**(11), 7209–7218 (2020)

Edge IoT System for Wearable Devices: Real-Time Data Processing, Inference, and Training for Activity Monitoring and Health Evaluation

M. Lupión[1]([✉])[iD], F. Romero[1,2][iD], L. F. Romero[2][iD], J. F. Sanjuan[1][iD], and P. M. Ortigosa[1][iD]

[1] Department of Informatics, ceiA3, University of Almería, 04120 Almería, Spain
{mlupion,jsanjuan,ortigosa}@ual.es
[2] Department of Computer Architecture, University of Málaga, Málaga, Spain
{fr,felipe}@uma.es

Abstract. Developing a medical device or solution involves gathering biomedical data from different devices, which can have different communication protocols, characteristics and limitations. Thus, deploying a test lab to record experiments can be challenging, requiring the synchronisation of the source signals, processing of the information, storing it and extracting conclusions. In this work, we face this problem by developing an edge Internet of Things (IoT) system composed of a Raspberry Pi and an NVIDIA Jetson TX2 device (integrating an NVIDIA Pascal GPU). The information from two biomedical devices (Biosignals Plux and Polar Verity Sense) is synchronised and fused, interpolating the information and extracting features such as mean and standard deviation in real-time. In parallel, the Jetson TX2 device is able to execute a Deep Learning (DL) model in real-time as new data is received using the Message Queuing Telemetry Transport (MQTT) protocol. Also, an online learning approach involving a loss function that takes into account past predictions is proposed, as well as a density-based clustering algorithm that selects the most representative samples of the most repeated class. The system has been deployed in the Smart Home of the University of Almería. Results show that the proposed fusion scheme accuracy represents the intrinsic information of the received data and enables the DL model to run in real time. The next steps involve the deployment of the system in a hospital, in order to monitor epilepsy patients, create a robust dataset and detect epileptic seizures in real-time.

Keywords: Edge processing · Data Fusion · Real-time inference

1 Introduction

The design of a wearable device that can predict or detect physical activity, human behaviour, situation, or health condition requires knowing in advance

© The Author(s), under exclusive license to Springer Nature Switzerland AG 2023
J. Bravo and G. Urzáiz (Eds.): UCAmI 2023, LNNS 842, pp. 131–142, 2023.
https://doi.org/10.1007/978-3-031-48642-5_13

the physical variables that may change when before or during the event[6]. For example, some epileptic patients suffer from repetitive muscle spasms, which can be detected using Inertial Motion Units (IMUs) and surface electromyography (sEMG) sensors [15]. In the early stages of research experiments, to measure the physical signals, different biomedical kits or devices such as Shimmer [4] are used. These provide data with very high precision and frequency (achieving 500 Hz), which needs to be processed, cleaned, and stored.

Due to the complexity of the human body, estimating or detecting patterns from different biomedical signals of a particular individual may require numerous devices with various technologies or data formats. There are several studies that have used Neural networks (NNs) to estimate human biomedical signals. For example, a study used Long Short-Term Memory (LSTM) NNs to automatically detect sleep apnea events in respiratory signals [14]. Another study proposed a lightweight CNN-based anomaly detection framework for IoT devices to detect abnormal power consumption behaviour [7]. Overall, NNs have shown promise in estimating human biomedical signals. However, to train NNs in such a complex ecosystem, it is necessary to carry out proper integration to correlate variables and estimate different situations of interest, such as gestures or movements that may pose a risk to the health of elderly individuals.

When developing health solutions, the information from the patients is confidential and must be processed and stored locally to avoid data leaks. Furthermore, the local processing and communication of the data allows the system to have a lower latency [12], required as most of the applications should be run in real time. Low-cost System-on-chip (SoC) devices such as Raspberry Pi have been proposed to receive, fuse, process, clean, and store sensor data [5]. However, when having different sources of sensors, high frequencies and images in some use cases, the computation capabilities of these SoCs are not enough [3].

For this, an edge IoT system that can obtain, process, and store data from different sources of sensors at high frequencies, as well as train and perform the execution of a Machine Learning (ML) model in real-time is proposed. First, physical variables such as skin temperature, Electrocardiogram (ECG), Peripheral oxygen saturation (SpO_2), Electrodermal activity (EDA), and sEMG are transmitted by Bluetooth; Photoplethysmography (PPG), accelerometer, and gyroscope are sent using MQTT. Second, the Raspberry Pi device stores the data, synchronises, aggregates the signals, and calculates features. Third, the main gateway device consists of an NVIDIA Jetson X2 that performs inference and online training of an ML model.

To evaluate the efficiency and usefulness of the proposed architecture, a system that monitors the activity of users to evaluate their cardiovascular health has been deployed at the University of Almería. In this case, a simple DL model has been deployed to demonstrate a successful integration.

The remainder of this paper is organised as follows: Sect. 2 analyses the existing literature, containing state-of-the-art and solutions similar to ours; Sect. 3 describes the proposed solution. After that, the experimental setup and results are exposed in Sect. 4. Finally, conclusions and ongoing works are explained in Sect. 5.

2   Related Works

2.1   Wearable Devices and Biomedical Kits

Nowadays, the most common wearable devices such as smartwatches incorporate a wide range of sensors that enable the acquisition of physiological data of the user. Sensors such as accelerometer, gyroscope, PPG, SpO_2 and EDA can be found in low-cost devices such as Xiaomi Smart Band and Apple Watch [10], allowing the detection of health problems in users that wear these devices.

Researchers and developers aim to develop health solutions making use of these sensors. Nevertheless, most companies do not offer raw data in Software Development Kits (SDKs) or Application programming interfaces (APIs). For this reason, in the field of biomedical engineering, researchers and associate companies have created research kits providing a wide variety of sensors, together with APIs and wearable devices that enable the capture of signals without any restriction.

In literature, the most used biosignal kit are the shimmer device [4] and biosignal plux [2]. Both of them incorporate sensors such as sEMG, ECG and EDA, wich are not easily found in smartwatches such as those mentioned before. The main drawback of the kits is that the sensors are connected by wires and users cannot wear them in real-life scenarios. To handle this, smartwatches have been created. The main device is the Empatica E4 wristband [8], having one day of autonomy, and contains most of the sensors mentioned before. However, the price is an economic obstacle (more than 1,000 euros).

Therefore, although the number of biomedical devices and devices incorporating health sensors that are being created is increasing, there is no complete solution that enables researchers to access the raw data and deploy algorithms on them to monitor users in real life. The real-time streaming of the sensors is mostly offered by commercial solutions, but the integration and synchronisation between different sources of sensors is not well addressed (or addressed in research-oriented solutions that do not work in real-life environments [1]).

2.2   Data Fusion from Heterogeneous Devices

The field of edge NN systems for biomedical signal analysis and data fusion has seen significant advancements in recent years. Researchers have explored various approaches to integrating data from heterogeneous devices and enabling real-time edge processing. In [13], researchers investigated the use of low-cost SoC devices like Raspberry Pi for running ML models on the edge, highlighting their capabilities and limitations.

Some works deal with the problem of running ML models on the edge in order to detect several anomalies in real-time such as epileptic seizures [11]. In the previous work, although the use of different sensors is presented, no fusion of different sources is proposed, only the ECG is used. This problem is faced in [9]. The authors propose a fusion scheme of different sensors (sEMG, strain, accelerometers and inclinometers) in order to classify the patients' motion intent. As the frequency is highly elevated (2 kHz in the case of sEMG), neural processors gather the information from the sensors, calculate features (mean, variance, number of slope sign changes and histogram), execute a partial NN from the information, and the results are aggregated with the other NN processors, reducing the latency of the data transmission. However, the system is constrained by the number of data sources it includes (it requires at least as many processors as data sources the system has). Another drawback is the storage of the data, which is not handled in works involving pre-defined datasets.

Therefore, in our work, we propose two main modules to process the data provided by the sensor devices: the first one synchronizes devices, stores the data and calculates the features. The second one is in charge of running the ML model in real-time.

3 Matherials and Methods

This section defines the overall architecture of the proposed system, and then, the different phases are more deeply described.

3.1 System Architecture

The chosen architecture (Fig. 1) for this work aims to address data collection from three different technology sources (smartband, biosignal kit and environmental sensing) in order to store and preprocess them on a central server for integration. A Raspberry Pi has been selected to store confidential information locally, preventing data leaks and allowing lower latency in communications. On this server, the variables undergo resampling to synchronise the data at the same frequency and phase. The goal is to use them as input to a NN deployed on a unit with a Graphical Processing Unit (GPU) (NVIDIA Jetson TX2). NN protocols have been employed for direct communication between the various sources and the server. Both the integrated data and the network's predictions are stored in an Influx database located in the Raspberry Pi device. As for communication with the Biosignal and Polar devices, two auxiliary devices are required: a laptop and a tablet. In the case of the Biosignal device, the software that allows a real-time streaming is only available in Windows, macOS and Linux. As for the Polar device, the API to access the raw data is offered in Android and iOS.

3.2 Biosignal Devices

Biosignal Plux. This biosignal device [2] has five different sensors: an ECG; an sEMG, which measures muscle activity; an EDA sensor; a body temperature sensor; and a SpO_2 sensor.

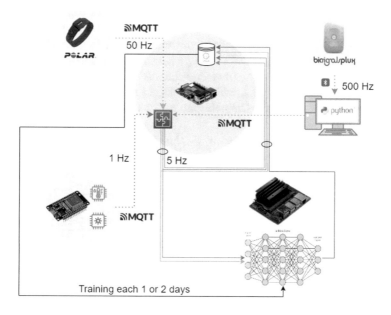

Fig. 1. System architecture

The kit allows the sampling rate of the data to be changed between 10 Hz and 4000 Hz. This device establishes a Bluetooth connection with a laptop and the information from the sensors is streamed directly through it using a Lab Streaming Layer (LSL) implemented in Python. In this case, the device requires a dedicated connection with the PC in order to transmit the information in real-time. Nevertheless, NN devices such as Raspberry Pi cannot handle this connection with the Biosignal device because the installation of the software is not allowed on these architectures.

Polar Verity Sense. Polar Verity Sense device is an optic cardiac frequency sensor that provides one of the best heart pulse accuracies among the wearable devices available in the market. It also incorporates an IMU sensor that provides accelerometry and gyroscope data. The most important feature is the incorporation of an open SDK that enables researchers and developers to extract the raw data from the sensors. Its development platforms are Android and iOS. Using them, the frequency of the sensors can be modified, as well as the programming of offline recordings when the device is powered on. To use the developer tools, a mobile application has to be developed. In our case, Android was selected due to the availability of Android devices in our working lab. The developed application displays a user interface to allow the user to change the data frequency and the data that has to be streamed. When the user starts the streaming, the data is sent to an MQTT server deployed in the Raspberry Pi device.

ESP32. An ESP32 device connected to the WiFi network is used to measure brightness, ambient temperature and humidity. These ambient conditions can affect the physiological variables of the patients. An LDR photoresistor is used to measure luminosity and a DHT11-type sensor is used to measure humidity and temperature. Each measurement is published in real-time using the MQTT protocol, and the Raspberry Pi, which acts as a broker and client, subscribes to the data and stores it in InfluxDB using an agent of Telegraf. The frequency can be adjusted using MQTT messages.

In this experiment, the frequency of the streamed data from both devices mentioned before is shown in Table 1.

Table 1. Frequency of the different signals

Device	Variable	Frequency (Hz)
Biosignal Kit	SpO$_2$	500
	ECG	500
	EDA	500
	sEMG	500
	Temperature	500
Polar Verity Sense	PPG	55
	Accelerometer	52
	Gyroscope	52
ESP32	Temperature	2
	Humidity	2
	Luminosity	2

3.3 Data Resampling and Synchronisation

The resampling and synchronisation of data are necessary for three reasons:

- The maximum publication frequency of MQTT messages is much lower than the sampling rate of devices like Biosignals or Polar. Therefore, in order to send them remotely, the information must be simplified or downscaled. Thus, Biosignals and Polar stream the data to a controller device (laptop and tablet respectively). In these devices, the data are temporarily stored and sent to the Raspberry Pi in a fixed timestamp set by the user. In our case, this is set to 2 s.
- Since the server is a Raspberry Pi, its storage capacity is more limited, necessitating a reduction in the amount of data processed per second.
- By synchronising the data at the same frequency, the complexity of processing in the NN is reduced, as it avoids the need to handle data with different frequencies and phases. This enables more efficient execution of the model and more effective utilisation of hardware resources, such as the GPU on the Jetson TX2 device.

As the sampling frequencies of each data source are very different, the Raspberry Pi device incorporates a methodology that recalculate values for the desired resampling frequency in two different ways (Fig. 2). These are:

- If the sampling frequency is lower than the desired one, data interpolation is performed for the chosen timestamps.
- If the sampling frequency is higher, to avoid losing too much information, the mean and standard deviation are calculated for windows of data, ensuring that the results are centred around the target timestamps.

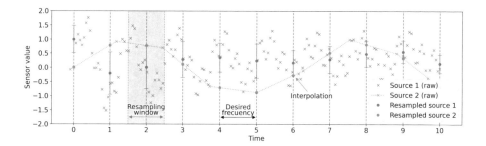

Fig. 2. Data resampling

3.4   AI Models to Fuse the Information of the Sensors

In this section, two lightweight NN architectures are proposed to fuse the information from the signals provided by the sensors.

Recurrent (Long Short-Term Memory). The first configuration contains recurrent blocks that explore the temporal properties of the input 1D signals in order to classify the window. Long short-term blocks find patterns in the signals incorporating forget operations that discard the useless information. In this case, a stateless approach is followed, as the signal history of the patient does not have to be taken into account, only the selected window of time.

After the main temporal patterns have been extracted in the recurrent block, these are fused in a layer and then a fully connected part is incorporated. The fully connected part maps the temporal features to the output layer. The architecture is illustrated in Fig. 3.

Convolutional. The second configuration makes use of convolutional layers in the initial part of the NN. These layers obtain spatial features in the input signal. More precisely, 1D convolutional layers are used in this case. In the first layer, the signals are processed in an independent way. After this, the NN contains as many new characteristics as the number of filters in the first layer.

Finally, same as before, after the recurrent layers, the final regression fully connected layers are added.

It is important to remark that the convolutional part of the NN does not have to be designed from scratch. In the literature, there are convolutional architectures[1] that stand-out because of their state-of-the-art performance in several tasks such as object recognition and image segmentation.

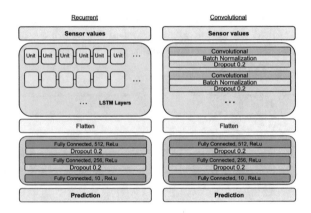

Fig. 3. NN Architectures

3.5 Online Training in NVIDIA Jetson TX2

The inference process involves the execution of the NN with the input data. Large NNs require better computation capabilities, which can be found in clusters or laptops. Problems having images as input data usually require deeper NN architectures to have a good performance. However, in our case, the data have been resampled and pre-processed, and the size of the window that is classified does not usually last more than 5–10 seconds. Thus, the NNs will not contain millions of parameters.

In these cases, NNs can be run in inference mode in devices such as Raspberry Pi, using Tensorflow Lite. The difference between running in the CPU of these devices and powerful GPUs such as Tesla V100 or RTX 3900 is not noticeable. Nevertheless, the training process cannot be carried out in NN devices such as Raspberry Pi. The solution to this could be sending the data to the cloud. However, devices connected to Internet connections inside hospitals have a lot of limitations, making impossible in some cases to send information to private servers on the cloud. For this reason, there exist development boards such as NVIDIA Tesla TX2, integrating a GPU Volta TX100, with an 8 GB memory that enables the training of models in the edge.

In our system, the processed data is streamed in real-time to the NVIDIA Jetson TX2, and the model is executed on it. After the inference process, the data is stored in a .csv file, together with the prediction. If the annotation of the data is also available in real-time, it is also annotated.

---

[1] https://keras.io/api/applications/.

In problems such as epilepsy seizure detection, sleep apnea or heart problems, the events that present the anomaly are fewer than the normal state of the patient. Therefore, retraining the model as soon as new labelled alert samples are available is key in the online learning of the system.

This work will be further evaluated in epilepsy seizure detection, which can lead to 3-4 seizures maximum every day. Thus, as soon as the doctor annotates an epileptic seizure, the .csv file in the Jetson TX2 is updated, and the model is retrained. The annotation is carried out using EDF files. In the online training of the model, two steps are followed: first, the backbone is frozen and the last layers are fine-tuned. Second, the backbone is unfrozen and the last layers are also fine-tuned, but considering a very small learning rate.

This process allows the model to learn from new samples while maintaining the knowledge. However, the complexity of the training would increase as new data are stored. Thus, a new schema is proposed. First, a Density-Based Clustering is performed on the previous data, identifying different clusters on the most repeated class. The most representative samples are selected and the dataset is balanced. Second, as the model performs inference on new data, the data which is incorrectly classified in real-time has more importance in the loss function of the NN in a factor established by the user.

These techniques allow the selection of the most representative samples to online train the model, avoiding the increase in training time as time passes.

4 Experimental Results

In this Section, the experimental setup and results are discussed. This system will be deployed in the hospital to detect epileptic seizures in real time, but the experiments have not been started yet. Thus, in order to validate the system, it was deployed in the Smart Home of the University of Almería.

4.1 Dataset Details

A total of three users participated in the experiments, two men and one woman. In this work, changes in the values of the biosignal variables need to be detected, so the users were required to perform different physical and daily living activities in the Smart Home. These activities and their duration are the following:

- Physical exercises (20 min): Jumping, burpees, squats.
- Stretching (5 min).
- Resting (5 min).
- Walking (10 min).
- Sweeping, washing windows and organising furniture (5 min).

As the number of users is only three, the dataset cannot be used to train stable and powerful DL models that can be deployed in real environments. As mentioned before, the main goal of this work is to validate the proposed IoT architecture and pre-processing approach in order to deploy it in real experiments in hospitals.

4.2   Hardware Infrastructure

The design and testing of the different NNs have been carried out in the cluster of the University of Almería. In this cluster, a node with 2 NVIDIA TESLA V100 GPUs was used. The operative system of the node is CentOS 8.2 (OpenHPC 2), with DDR4 3200 MHz RAM.

However, the execution of the models in the Smart Home is performed in the NVIDIA Jetson TX2 device. It incorporates the GPU NVIDIA Pascal (GTX 1000, 8 GB memory), 128 bits CPU ARM, and 8 GB LPDDR4.

4.3   Results

The data from the different devices have been aggregated in the Raspberry Pi. Figure 4 shows the difference between the original source of data represented by an x and the synchronised and aggregated data, represented by a dot. Also, three different signals are displayed: the brightness, in blue; the sEMG of the Biosignals plux kit in orange, and the acceleration of the Polar device in green. These signals have been chosen for representation since they belongs to different data sources what proves the effectiveness of the method for integration.

The unsynchronised data (marked with "x") are the ones received by the database through the Telegraf agent. The brightness data is collected directly from the ESP32, so it corresponds exactly to the raw data extracted from the sensors, i.e., data with the "raw" frequency of this device, which is lower than the chosen frequency, in this case, 1 Hz. On the other hand, the agent also receives resampled data as indicated in Sect. 3.3, belonging to the Biosignals and Polar devices, which have higher sampling frequencies than the chosen one.

Subsequently, the Raspberry Pi retrieves all this stored data from the database, reads it, and interpolates it to align with the chosen timestamps based on the new resampling frequency. This way, it is ensured that the data represented by a point coincides with the same timestamps regardless of the data source, and furthermore, it does not seem to affect the ground truth. The standard deviation of the data has also been interpolated to avoid losing too much information.

Regarding the different devices, the most challenging step is attaching the different sensors from the biosignal kit to the user. When the user performs physical activity, the electrodes and attachers may come loose from the user's skin, providing incorrect values. However, in medical centres, the experiments do not involve such physical activities (in the case of epilepsy seizure detection, sleep apnea or heart problems, as mentioned before). In these cases, users are only required to stay in bed, eat, walk inside the room or go to the WC.

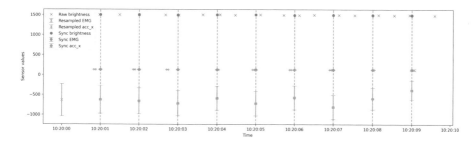

Fig. 4. Comparison between unsynchronised and synchronised sources.

5 Conclusions and Future Works

In this work, several biomedical devices have been integrated into an NN edge system that is able to receive and process the data in real-time. The system is composed by an aggregator NN device, a Raspberry Pi, and a NVIDIA Jetson TX2.

Due to the different frequency of data, these are synchronised using the same frequency, set by the user and problem requirements. If the data frequency is fewer, the interpolation is proposed. Furthermore, instead of using all the data in the Artificial Intelligence (AI) models, two features have been calculated: mean and standard deviation.

In addition, two NNs have been proposed to fuse the information from the sensors and extract some conclusions from the biosignal plux kit variables (EDA, sEMG and SpO$_2$), the wearable (PPG, accelerometer and gyroscope) and ambiental device's data. An online training of the model is implemented in the Jetson TX2 device, taking into account the inference error to leverage the importance of the past samples, and a density-based clustering to reduce the number of samples of the majority class.

Results show that the proposed synchronisation and window scheme allows the processing, synchronisation, storage, feature extraction of the signals in real-time, as well as the execution and train of the AI models in the NVIDIA Jetson TX2 device.

Future works involve the deployment of the system in a hospital, involving patients and medical staff. More NN architectures will be compared, and the online training approach assessed.

Acknowledgements. This work has been funded by the projects R+D+i PID2021-123278OB-I00 and PDC2022-133370-I00 from MCI-N/AEI/10.13039/501100011033/ and ERDF funds; and the Department of Informatics of the University of Almería. M. Lupión is a fellowship of the FPU program from the Spanish Ministry of Education (FPU19/02756).

References

1. imotions. https://imotions.com/. Accessed 19 Sept 2023
2. PLUX Biosignals. https://www.pluxbiosignals.com/. Accessed 13 July 2023
3. Blanco-Filgueira, B., García-Lesta, D., Fernández-Sanjurjo, M., Brea, V.M., López, P.: Deep learning-based multiple object visual tracking on embedded system for IoT and mobile edge computing applications. IEEE Internet Things J. **6**(3), 5423–5431 (2019). https://doi.org/10.1109/JIOT.2019.2902141
4. Burns, A., et al.: Shimmer^{TM}-a wireless sensor platform for noninvasive biomedical research. IEEE Sens. J. **10**(9), 1527–1534 (2010)
5. Greco, L., Percannella, G., Ritrovato, P., Tortorella, F., Vento, M.: Trends in IoT based solutions for health care: moving AI to the edge. Pattern Recogn. Lett. **135**, 346–353 (2020)
6. Guk, K., et al.: Evolution of wearable devices with real-time disease monitoring for personalized healthcare. Nanomaterials **9**(6), 813 (2019)
7. Lightbody, D., Ngo, D.M., Temko, A., Murphy, C.C., Popovici, E.M.: Host-based intrusion detection system for IoT using convolutional neural networks. In: 2022 33rd Irish Signals and Systems Conference (ISSC), pp. 1–7 (2022)
8. McCarthy, C., Pradhan, N., Redpath, C., Adler, A.: Validation of the empatica e4 wristband. In: 2016 IEEE EMBS International Student Conference (ISC), pp. 1–4. IEEE (2016)
9. Otseidu, K., Jia, T., Bryne, J., Hargrove, L., Gu, J.: Design and optimization of edge computing distributed neural processor for biomedical rehabilitation with sensor fusion. In: 2018 IEEE/ACM International Conference on Computer-Aided Design (ICCAD), pp. 1–8 (2018). https://doi.org/10.1145/3240765.3240794
10. Raja, J.M., et al.: Apple watch, wearables, and heart rhythm: where do we stand? Ann. Transl. Med. **7**(17) (2019)
11. Sakib, S., Fouda, M.M., Fadlullah, Z.M.: A rigorous analysis of biomedical edge computing: an arrhythmia classification use-case leveraging deep learning. In: 2020 IEEE International Conference on Internet of Things and Intelligence System (IoTaIS), pp. 136–141 (2021). https://doi.org/10.1109/IoTaIS50849.2021.9359721
12. Shi, W., Cao, J., Zhang, Q., Li, Y., Xu, L.: Edge computing: vision and challenges. IEEE Internet Things J. **3**(5), 637–646 (2016)
13. Utomo, D., Hsiung, P.A.: Anomaly detection at the IoT edge using deep learning. In: 2019 IEEE International Conference on Consumer Electronics - Taiwan (ICCE-TW), pp. 1–2 (2019). https://doi.org/10.1109/ICCE-TW46550.2019.8991929
14. Van Steenkiste, T., Groenendaal, W., Deschrijver, D., Dhaene, T.: Automated sleep apnea detection in raw respiratory signals using long short-term memory neural networks. IEEE J. Biomed. Health Inform. **23**(6), 2354–2364 (2019). https://doi.org/10.1109/JBHI.2018.2886064
15. Verdru, J., Van Paesschen, W.: Wearable seizure detection devices in refractory epilepsy. Acta Neurol. Belg. **120**, 1271–1281 (2020)

Analysing Requirements Specification Languages for Self-adaptive AAL Systems

Inmaculada Ayala[1,2]([envelope]) [ID], Mercedes Amor[1,2] [ID], and Lidia Fuentes[1,2] [ID]

[1] Universidad de Málaga, Málaga, Spain
{ayala,pinilla,lff}@lcc.uma.es
[2] ITIS Software, Málaga, Spain

Abstract. Ambient Assisted Living (AAL) systems are usually deployed in complex environments and should behave autonomously. So, self-adaptation is an essential concern of these systems that should be part of their requirements specification. The modelling of self-adaptation requirements is a challenging task because it is strongly related to other system requirements. The achievement of self-adaptation requirements typically implies prioritisation or denial of different requirements, so it is crucial to have mechanisms for their modelling and analysis. We study the suitability of requirement specification languages for self-adaptation in the context of a project to develop AAL systems. Specifically, we compare specifications made in RELAX, Tropos4AS and GODA. Our study with 14 people concludes that understanding these specifications is similar for the three languages but poses some difficulties for users. Specifically, users perceive the understanding of RELAX specification requires less effort and makes them feel less frustrated.

Keywords: AAL · self-adaptation · requirements · robotics

1 Introduction

Ambient Assisted Living (AAL) systems are critical applications for the sustainability of the healthcare system in many developed countries. The complexity of the development and deployment of these technologies comes from many different sources like the diversity of the stakeholders involved, the technical challenges of the technology used and their requirements in terms of dependability [14]. One of the solutions to realise dependability is self-adaptation, a quality of software systems that can evaluate their functioning and adjust their behaviour accordingly to optimise its performance, recover from failures, configure its functionality, or secure some part of an entire system [10].

Self-adaptation of AAL systems is challenging because many human aspects are involved. Usually, self-adaptive systems change their behaviour by considering technical factors like the system's workload, the battery's level, and the failure of some components, to mention a few. In addition, a self-adaptive AAL system should consider the patient's current condition, the number of people in

J. Bravo and G. Urzáiz (Eds.): UCAmI 2023, LNNS 842, pp. 143–154, 2023.
https://doi.org/10.1007/978-3-031-48642-5_14

the room, the scheduling of the nursing home, etc. [3]. While engineers entirely decide on technical aspects, the human elements of the adaptation should come from the joint work of the engineers and health personnel. As in any software project, the requirements specification phase is fundamental to agree on what the final system should provide regarding system and self-adaptation functionalities.

Self-adaptive systems should be aware of their requirements [5]. While initial works on self-adaptation manage this concern as something ad-hoc. There is an agreement that self-adaptation should be included since the beginning of the system's conception [4,12]. The motivation for this is the strong relationship between self-adaptation and system requirements. Self-adaptation typically affects the other requirements of the system, relaxing or dropping them, giving more priority to some requirements than others. There are several approaches for self-adaptation requirement specification [2,4,12,16]. These works are generally based on some early-requirement specification language (i.e., i*, KAOS, GRL) extended to support self-adaptation specification.

In this work, we study the suitability of methodologies for self-adaptation requirement specification in the context of a project to develop AAL systems. Specifically, we compare requirements specifications made in RELAX [16], Tropos4AS [12] and GODA [11]. We have selected these languages because they are well-known approaches that follow different modelling approaches. The goal of this comparison is to identify the effort of users to understand these specifications. AAL systems demand a combination of proactive and reactive self-adaptive behaviour. In addition, stakeholders belong to starkly different backgrounds, and the validation of requirements is of the utmost importance as human lives are involved in these systems. To illustrate the use of these languages, we specify the requirements of a case study of an assistive robot in a nursing home. The project stakeholders have to answer questions about these requirement specifications. We study aspects like comprehension and perceived workload.

This paper is structured as follows: Sect. 2 describes the case study used to illustrate our proposal, Sect. 3 introduces the requirement specification languages and explains their use for the mentioned case study, Sect. 4 provides the experimental design, we report and analyse the results of the experiment in Sect. 5 and Sect. 6 presents some related work. The paper finishes with conclusions and some ideas for future work.

2   Case Study

To compare the work of FLAGS, RELAX and Tropos4AS, we use a case study of an assistive robot in a nursing home. The goal of this robot is to pick up the menu selection for the next day of the residents (see Fig. 1). To do so, it has information about each resident's scheduling, the daily plan of the nursing home, software for face recognition and LIDAR sensors to know the distance to possible obstacles. The robot uses voice and screen to obtain information from residents. It has to adapt the screen and the discourse to the specific condition of the resident. The robot should adjust its behaviour to the activity that is taking place in the room that it is going to enter. These activities affect the number of people in the room, but some require that the robot disturb as little as possible. In addition, technical aspects affect the robot's behaviour, like the battery level and the position of the obstacles. A conjunction of proactive and reactive strategies determines the behaviour of the robot. It should be proactive because it has to predict how to behave in a room before entering, and it has to be reactive to act according to the resident's condition, the position of the obstacles and the current battery level.

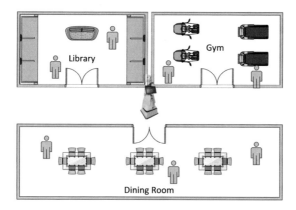

Fig. 1. Case study of the menu selection.

The sources of uncertainty of this case study come from the ability of the robot to recognise contexts that cause adaptation. They are the face of a specific resident, the number of people in a room, the obstacles using sensors or bumpers and the battery level. To simplify the case study, we do not consider uncertainty in the outcome of self-adaptation actions.

3   Self-adaptive Requirements Specification

In this section, we provide an overview of the three requirements specification languages, and we give the specification of the requirements in them. All these languages have formal semantics that permit validating the requirements specification. However, here, we focus on the modelling perspective and how these languages capture stakeholders' expectations of the system.

3.1   RELAX

RELAX [16] is a structured natural language for requirements specification, including operators explicitly designed to capture uncertainty. Textual requirements languages prescribe system behaviour using model verbs such as SHALL or WILL, which define the functionality that the system must always provide. However, in self-adaptive systems, these functions may not always be provided. RELAX has operators to distinguish between requirements that should never change and those that can be relaxed in some contexts. In addition, this language can specify constraints on how these requirements can be relaxed using operators depicted in Table 1. So, we rewrote the requirement specification for our case study as follows:

Table 1. RELAX operators

RELAX operator	Description
SHALL	A requirement must hold
MAY...OR	A requirement specifies one or more alternatives
EVENTUALLY	A requirement must hold eventually
UNTIL	A requirement must hold until a future position
BEFORE, AFTER	A requirement must hold before or after a particular event
IN	A requirement must hold during a particular time interval
AS EARLY, LATE AS POSSIBLE	A requirement specifies something that should hold as soon as possible or should be delayed as long as possible
AS MANY, FEW AS POSSIBLE	A requirement specifies a countable quantity but the exact count may be relaxed

- **R1:** The robot SHALL search for nursing home residents to gather about the menu preferences for tomorrow UNTIL rest time AS EARLY AS POSSIBLE.
- **R2:** The robot SHALL update its interface taking into account the condition of the resident in front of it.
- **R3:** The robot SHALL update its sound level to the condition of the resident in front of it and the activity of the room.
- **R4:** BEFORE entering a room, the robot SHALL adapt its speed to the current activity that is taking place.
- **R5:** The robot SHALL adapt the speed to the number of people in the room and their position.
- **R6:** The robot SHALL adapt its trajectory to avoid obstacles.
- **R7:** The robot MAY adapt its speed and trajectory, considering its battery level OR, IN case its battery becomes empty, to stay in a safe location and notify the responsible person.
- **R8:** IN case the resident is not in the expected location, the robot searches for another resident that has not been contacted that day.

RELAX supports the specification of the uncertainty sources using the keywords **ENV**, which defines a set of properties that define the system's environment, **MON**, which establishes a set of properties that the system can monitor, **REL**, which defines the relationship between the **ENV** and **MON** properties, and **DEP** that identifies the dependencies between the relaxed and invariant requirements. For our case study, the **ENV** comprises the identity of the resident in front of the robot, the activity that is taking place in a room, the expected location of a resident, the number of people in a room, the distance to the obstacles and its battery level. The **MON** comprises the camera, the resident's schedule, the nursing home's daily plan, the LIDAR sensor and the battery sensor. Accordingly, the **REL** component considers that the camera provides the identity of the resident in front of the robot, the scheduler offers information about the expected location of a resident, the daily plan of the nursing home provides information about the current activity in a room, the camera provides the number of people in a room, and the battery sensor provides information about the battery level. The **DEP** attribute states that **R4**, **R5**, **R6** and **R7** negatively impact **R1**.

3.2 Tropos4S

Tropos4AS [12] is a goal-oriented requirement specification language that uses the methodology of Tropos [13]. This language considers three dimensions for the requirements specification: an extended goal model, the environment model that contains essential elements that affect goal satisfaction, and the failure model that assists requirements engineers in eliciting undesirable states of affairs and possible recovery procedures.

In goal-oriented requirements engineering, like Tropos4AS, models capture the stakeholder's objectives and define a system's high-level functional and non-functional goals and alternatives. Goals are delegated from stakeholders to the system actor and can be decomposed into additional actors or sub-systems called agents. Goals are decomposed and delegated to other actors or operationalised by tasks.

A goal model in Tropos4S contains different elements that support how agents achieve goals through their interaction with the environment. In our case study, the *Robot* agent (see Fig. 2) has the goal *Menu preferences of each resident gathered* and *Battery safe navigation*. The condition to fulfil the first goal is *All menu preferences gathered*, and it is operationalized through the plans *Determine resident location, Go to resident location, Locate resident, Go to resident, Present options, Capture preferences* and *Search next resident*. Some plans require artefacts to be executed; for example, *Present options* requires *Speaker* and *Screen*. Some plans should be performed taking into account soft goals; for example, *Go to resident location* is affected by *Obstacle avoidance* and *Caution*. Uncertainty is managed in Tropos4AS using the failure scenarios, which model the behaviour of the system when something unexpected happens. For example, for *Locate Resident*, we consider the failure *Resident not found* with the error *The Resident is not in the expected location* and the selected recovery plan *Search next resident*.

3.3 GODA

GODA [8,11] is a goal-oriented requirements engineering framework for runtime dependability analysis targeting self-adaptive systems. Here, we focus on [8], which proposes an assurance process that supports taming different sources of

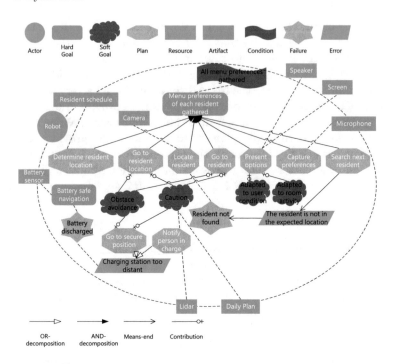

Fig. 2. Case study of the menu selection in Tropos4AS.

uncertainty. GODA uses Contextual Goal Models, which extend classical goal models with contexts, which are partial states of the world relevant to a goal. When the context changes, this may influence the (quality of the) goals and the means of achieving them.

The goal model of our case study (see Fig. 3) is modelled using the extension of piStar tool proposed by authors in [8]. GODA extends this tool with contextual information and modelling guidelines for uncertainty. The case study requirements have been modelled using the goals *Menu preferences of residents gathered* and *Battery safe navigation*. We add goal *Cautious navigation* because the iStar 2.0 standard does not permit associating an intentional element of the type quality with a resource. So, we have modelled this requirement as a goal operationalized by the goals *Daily plan aware trajectory* and *Obstacle free trajectory*. These goals are decomposed into tasks that need resources *Daily plan* and *Lidar*.

GODA has different mechanisms to model uncertainty. Taking into account our case study, we focus on uncertainty in the environment. GODA considers uncertainty in the execution context and noise in sensing. The execution context is modelled by adding elements representing the execution context in which the robot operates. So, we have elements like *Battery level*, *Daily plan* or *Resident schedule*. The noise in the sensing is modelled using the guideline that the task of collecting data is decomposed into three sub-tasks, *Read data*, *Filter data*

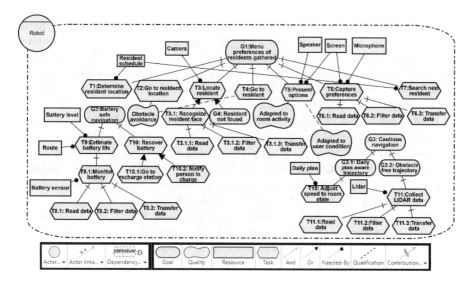

Fig. 3. Case study of the menu selection in GODA using piStar.

and *Transfer data*. We use this guideline for *Monitor battery*, *Recognize Resident face*, *Capture preferences* and *Collect LIDAR data*. Finally, to capture failure scenarios like in Tropos4AS, we use the support of GODA for non-deterministic behaviour. For those tasks that can fail, we have added a root task and decomposed several into tasks using Or-decomposition. For example, *Recover battery* is Or-decomposed into *Go to recharge station* and *Notify person in charge*. The first task represents successful behaviour, while the second one represents failure. We follow a similar approach to model the failed behaviour for *Locate resident*.

4   Experimental Design

We have designed the experiments in this section following the guidelines proposed in [1]. Our experiments aim to analyse the differences in RELAX, Tropos4AS and GODA for evaluation concerning their effects on the accuracy of understanding the models from the viewpoint of researchers in the context of experiments conducted in the Campero project. The materials used in the experiment are the case study in Sect. 2, the requirements specification in the three modelling languages from Sect. 3, a set of questions to check the understanding of the models, a NASA-TLX questionnaire, and a demographic questionnaire.

The procedure that we have followed has been as follows. All the project participants know the case study to be modelled, so we focus on the modelling languages. We had a short presentation on the three requirements specification languages, and after that, we sent all the materials for the experiments. Participants had three days to answer the three questionnaires, and we suggested having different sessions and focusing on one modelling language in each one.

We assess the understanding of the information of the models using the following questions: (a) What is the robot's primary goal?; (b) Which tasks require contextual information to be performed?; (c) What tasks are context dependent?; (d) What does the robot do in case of failure? Mention at least one issue; (e) List the sensors used by the robot; (f) List the information artefacts used by the robot; (g) What elements of the specification are related to proactiveness?; and (h) Could you mention something in the specification related to uncertainty?

There are slight differences in the questions depending on the modelling language. Each question has a maximum score of one point, and the answer can be correct (1 point), partially correct (0.5 points) or incorrect. To assess the perceived effort required to understand the models, we use the NASA-TLX questionnaire[1] in the form of an editable text document. This questionnaire is a tool for measuring and conducting a subjective mental workload assessment to determine a participant's mental workload while performing a task. Participants should grade between 1 to 10 the mental demand, the physical demand, the temporal demand, the performance, the effort and the frustration linked to the performance of a specific task. In our case, the understanding of the goal models is checked with the questionnaire. The answers to these questions are available in the link https://drive.google.com/file/d/1OXNEvSce-M-xR9xX1AzjbvDK3MDsXBht/view?usp=sharing.

Taking into account the goal of our study our research question is: *Does a difference in the requirement specification language influence the understanding of the specification?*

5   Analysis of Results

The number of participants in the experiments has been 14; their age is between 25 and 51 years, with an average of 36. Every participant speaks Spanish. Most of them work in academia, so a 63% have a doctorate, and a 27% have a master's degree. The field of study was related to technical areas, but three participants worked on health-related topics, and one was not technical. A percentage of 54.5% of the participants have some experience with requirement specification, and 72.7% have some experience with modelling languages.

Table 2 presents the descriptive statistics for the grades obtained in the understanding (maximum score of 8) and NASA-TLX (top score of 10) questionnaires for the different languages. In the *Language* column, we specify which specification language we are considering (RELAX, Tropos4AS or GODA), and we set the studied variable (i.e., Understanding, Mental Demand, ...) in the *Variable* column. We present the mean, standard deviation, skewness (i.e., Pearson's second coefficient) and Kurtosis. Regarding understanding, the mean of the tests for different languages is around 5, with the highest standard deviation for RELAX. The values of Skewness and Kurtosis suggest that the results for Tropos4AS and GODA are nearly symmetrical, while the results for RELAX are highly skewed.

---

[1] https://humansystems.arc.nasa.gov/groups/tlx/downloads/TLXScale.pdf.

We have Kurtosis values similar to zero, so we can conclude that the data distribution is close to the normal distribution. Regarding the NASA-TLX workload metrics, we see that the *Performance* variable, which captures the feeling of successfully performing a task, is the one that concerns the users the most. Not surprisingly, the physical demand scores the lowest values. In general, results seem similar, but values for *Effort* and *Frustration* (lower for RELAX). Skewness and Kurtosis for NASA-TLX suggest symmetry in the results.

Table 2. Descriptive statistics for understanding questionnaires

Language	Variable	Mean	S.D.	Skewness	Kurtosis
RELAX	Understanding	4.75	1.72	−0,86	−1.3
	Mental demand	5.5	1.74	0	−0.09
	Physical demand	3.57	2.68	0.08	−1.22
	Temporal demand	5.79	1.76	−0,36	0.27
	Performance	6.14	2.66	−0,97	−0.63
	Effort	3.21	0.97	−0,88	0.37
	Frustration	2.86	1.23	−0,35	−0.52
Tropos4AS	Understanding	4.96	0.91	−0.15	−0.12
	Mental demand	5.29	2.61	−0.06	0.33
	Physical demand	2.62	1.80	1.39	1.02
	Temporal demand	4.86	2.21	−0.29	−0.87
	Performance	6.5	1.91	−0,77	0
	Effort	5.86	1.61	0.14	−0.27
	Frustration	4.64	2.13	−0,34	−0.50
GODA	Understanding	4.85	1.21	−0.36	−0.38
	Mental demand	6.31	1.97	−1.05	−1.29
	Physical demand	3.62	2.10	−0,55	−1.35
	Temporal demand	5.85	1.52	−0,30	−0.81
	Performance	6.15	2.08	0.22	−1.12
	Effort	5.69	2.14	−0,43	0.31
	Frustration	4.77	1.88	−0.37	−1.71

To answer our research question, we analyse if there is a statistically significant difference between the results of the experiments for the three languages. We have performed a Welch t-test analyzing the variables of Table 2. Table 3 summarises the results for *Understanding, Effort*, and *Frustration* comparing pairs of languages, values of p-value lower than 0.05 indicate a statistically significant difference. We see statistical differences in *Effort* and *Frustration* comparing RELAX with Tropos4AS and RELAX with GODA, but not for the other NASA-TLX metrics. There is no statistical difference in the level of *Understanding*. The answer to our research question is: A difference in the requirement specification language does not influence the understanding of the specification, but it affects the perceived effort and frustration of the users. Specifically, users perceive the understanding of RELAX specification requires less effort and make them feel less frustrated. This is especially interesting because RELAX is the only textual language analysed.

Table 3. Results for the Welch t-test (Test statistic, Criticial two tail, p-value)

		RELAX	Tropos4AS	GODA
Understanding	RELAX		(−0.41,2.09,0.69)	(−0.17,2.07,087)
	Tropos4AS	(−0.41,2.09,0.69)		(0.28,2.07,0.78)
	GODA	(−0.17,2.07,087)	(0.28,2.07,0.78)	
Effort	RELAX		(−5.25,2.08,**0.00**)	(−3.83,2.11,**0.00**)
	Tropos4AS	(−5.25,2.08,**0.00**)		(0.23,2.07,0.82)
	GODA	(−3.83,2.11,**0.00**)	(0.23,2.07,0.82)	
Frustration	RELAX		(−2.71,2.08,**0.01**)	(−3.10,2.09,**0.01**)
	Tropos4AS	(−2.71,2.08,**0.01**)		(−0.16,2.06,0.87)
	GODA	(−3.10,2.09,**0.01**)	(−0.16,2.06,0.87)	

The presented results raise concerns about external validity, which assesses the generalizability of experiment outcomes. The study made here targets teams working on AAL projects and their level of comprehension of requirements specifications. In this regard, we consider that the number of subjects and the profile of the participants is an excellent example of this kind of team.

6   Related Work

The representation of self-adaptation at the requirement phase has brought the attention of different works. In the recent literature, different studies propose languages to support self-adaptation explicitly. Still, none provides a comparative analysis between different languages when used in the same scenario.

Some works provide comparative analysis related to adaptation and dynamic behaviour linked with uncertainty. The work in [9] focus on message sequence charts (MSC) for automotive domain requirement, which are used for specifying interaction sequences and the interaction-based behaviour of such reactive systems. This work reports on a comparative analysis to investigate three approaches for MSC-specification generation. Results show that the different approaches applied lead to correct yet MSC specifications that exhibit different characteristics and are thus suited for other requirements engineering purposes. With a similar goal, the work in [15] studies the efficiency of TROPOS graphical representations vs. textual ones in modelling and presenting software requirements. The work explores the impact of using structured textual vs. graphical representations while performing requirement comprehension tasks. Results observe no statistically significant difference in terms of accuracy. However, it is spent more time and effort working with the graphical representation, although this extra time and effort does not affect accuracy. Results emphasise that training can significantly improve the efficiency of working with graphical representations. However, TROPOS structured textual representation is more efficient when performing the requirements comprehension tasks.

The approach in [6] addresses how understanding an adaptive system's requirements is crucial to correctly developing them. The study evaluates specifications of three adaptation semantics formalised using A-LTL and KAOS. Although this work offers a means to incorporate the adaptation semantics into the goal-oriented requirements specifications of an adaptive system, it is not compared with other adaptation semantics description languages or methodologies. The work in [2] reviewed 52 publications and assessed methods that integrate goal models with SysML models (or use them separately) to support runtime self-adaption, considering the context of the social cyber-physical system. The review answers many questions of broad interest to researchers and practitioners considering using goal models, SysML models, or both in SCPSs or self-adaptive systems contexts. Although there has been much improvement in the last decade, the principal results show that improving the usability and scalability of goal/SysML integrations for adaptive systems with proper tool support is necessary. As for [7], it presents a comparative study of Web methodologies for handling adaptation requirements as functional requirements.

7   Conclusions

In this work, we have presented the results of a study on the suitability of methodologies for self-adaptation requirement specification in the context of a project to develop AAL systems. Specifically, we have focused on understanding requirement specifications and perceived workload. To do so, we have specified the requirements of an assistive robot in a nursing home in three specification languages that explicitly support this issue. Then, fourteen people related to the project answered questions about their understanding of the models and their perceived workload using the NASA-TLX questionnaire. The conclusion is the level of understanding is the same, but there are statistical differences regarding perceived effort and frustration. In any case, the values for understanding are around five, being eight the top score; this suggests that these languages have aspects that could be improved to make them more understandable to users.

In future work, we plan to extend these experiments, assessing the effort to create requirement specifications or the speed of doing tasks related to design and understanding. In addition, our experiments' results suggest difficulties in understanding these specifications, so another interesting future line of research would be to improve them to facilitate their use.

Acknowledgements. This work is partially funded by projects TED2021-131739B-C21, PDC2022-133597-C42 and IRIS PID2021-12281 2OB-I00 funded by the Gobierno de España and FEDER funds, by the Junta de Andalucía (Spain) under contract QUAL21 010UMA and project DISCO B1-201212 funded by Universidad de Málaga.

References

1. Gralha de Almeida, A.C.: Quality evaluation of requirements models: the case of goal models and scenarios. Ph.D. thesis, Universidade Nova de Lisboa (2019)
2. Anda, A.A., Amyot, D.: Self-adaptation driven by SysML and goal models - a literature review. e-Informatica Softw. Eng. J. **16**(1) (2022)
3. Ballesteros, J., Ayala, I., Caro-Romero, J.R., Amor, M., Fuentes, L.: Evolving dynamic self-adaptation policies of mhealth systems for long-term monitoring. J. Biomed. Inform. **108**, 103494 (2020)
4. Baresi, L., Pasquale, L., Spoletini, P.: Fuzzy goals for requirements-driven adaptation. In: 18th IEEE International Requirements Engineering Conference, pp. 125–134 (2010)
5. Bencomo, N.: Requirements for self-adaptation. In: Lämmel, R., Saraiva, J., Visser, J. (eds.) GTTSE 2011. LNCS, vol. 7680, pp. 271–296. Springer, Heidelberg (2013). https://doi.org/10.1007/978-3-642-35992-7_7
6. Cheng, B.C., Goldsby, H.: Goal-oriented modeling of requirements engineering for dynamically adaptive system. In: 14th IEEE International Requirements Engineering Conference (2006)
7. Escalona, M.J., Koch, N.: Requirements engineering for web applications - a comparative study. J. Web Eng. **2**(3), 193–212 (2004)
8. Félix Solano, G., Diniz Caldas, R., Nunes Rodrigues, G., Vogel, T., Pelliccione, P.: Taming uncertainty in the assurance process of self-adaptive systems: a goal-oriented approach. In: IEEE/ACM 14th International Symposium on Software Engineering for Adaptive and Self-Managing Systems (2019)
9. Keller, K., Brings, J., Daun, M., Weyer, T.: A comparative analysis of ITU-MSC-based requirements specification approaches used in the automotive industry. In: System Analysis and Modeling. Languages, Methods, and Tools for Systems Engineering (2018)
10. Kephart, J., Chess, D.: The vision of autonomic computing. Computer **36**(1), 41–50 (2003)
11. Mendonça, D.F., Nunes Rodrigues, G., Ali, R., Alves, V., Baresi, L.: GODA: a goal-oriented requirements engineering framework for runtime dependability analysis. Inf. Softw. Technol. **80**, 245–264 (2016)
12. Morandini, M., Penserini, L., Perini, A., Marchetto, A.: Engineering requirements for adaptive systems. Requirements Eng. **22**(1), 77–103 (2017)
13. Penserini, L., Perini, A., Susi, A., Mylopoulos, J.: High variability design for software agents: extending tropos. ACM Trans. Auton. Adapt. Syst. (2007)
14. Purohit, P., Khanpara, P., Patel, U., Kathiria, P.: IoT based ambient assisted living technologies for healthcare: concepts and design challenges. In: 6th International Conference on I-SMAC, pp. 111–116 (2022)
15. Sharafi, Z., Marchetto, A., Susi, A., Antoniol, G., Guéhéneuc, Y.G.: An empirical study on the efficiency of graphical vs. textual representations in requirements comprehension. In: 21st International Conference on Program Comprehension (ICPC), pp. 33–42 (2013)
16. Whittle, J., Sawyer, P., Bencomo, N., Cheng, B.H.C., Bruel, J.M.: Relax: a language to address uncertainty in self-adaptive systems requirement. Requirements Eng. **15**(2), 177–196 (2010)

A Deep Learning and Probabilistic Approach to Recognising Activities of Daily Living with Privacy Preserving Thermal Sensors

Matthew Burns[1]([✉]) [ID], Chris Nugent[1] [ID], Sally McClean[1] [ID], Javier Medina Quero[2] [ID], and Aurora Polo-Rodríguez[3] [ID]

[1] School of Computing, Ulster University, Belfast, Co. Antrim, Northern Ireland
m.burns2@ulster.ac.uk
[2] Department of Computer Engineering, Automation and Robotics Higher Technical School of Computer Engineering and Telecommunications, University of Granada, Granada, Spain
[3] Department of Computer Science, University of Jaén, Campus Las Lagunillas, 23071 Jaén, Spain

Abstract. A privacy preserving approach for recognising Activities of Daily Living (ADLs) is proposed within this work. Low-resolution thermal sensors and Convolutional Neural Network (CNN) models were utilised for detecting positions within a smart environment and classifying human poses. A Hidden Markov Model (HMM) was implemented for which the position and pose data acted as the model's observable information. An average F-score of 0.8171 was achieved for the poses on a test dataset. From a separate test dataset, the times in which each ADL began and ended were estimated with a maximum of 30 s between estimations and ground truth. Each ADL was correctly classified from the test dataset. Further discussion on the results are presented in this article.

Keywords: Activity Recognition · Deep Learning · Computer Vision · Thermal Sensors

1 Introduction

Recognising activities is a complex problem with a common solution that involves the use of sensors to collect data [1, 2]. Such a solution can be particularly useful for developing an understanding of older adults' long-term health care [3]. Accurate recognition of activities has facilitated earlier detection of age-related diseases such as dementia and Alzheimer's [4, 5]. ADL analysis has also shown to aid in the progress of developing platforms that aim to provide independent living for elderly people [6]. To expect a person's family or carer's to be able to provide necessary and continuous monitoring of ADLs would not be practical. An automated solution through the use of sensors, whether they are wearable or installed within an environment, can address this problem. It should, however, remain important to consider the privacy of a home's inhabitant when aiming to analyse ADLs. Subsequently, the type of sensor data collected when performing ADL recognition becomes significant as, for example, privacy preservation is negatively impacted through the use of video cameras as they can be considered to be too intrusive for a home's inhabitant [7].

Using a range of sensors and machine learning approaches, ADLs have been able to be successfully recognised [2, 8, 9]. Some approaches include the use of image or video data due to the high recognition accuracies that can be achieved [10]. There can, however, be privacy-related concerns when collecting image or video data within one's home, therefore, it is important to consider the privacy offered by a smart environment [11]. More specifically, it is the extent to which image data captures a person's identity or the environment in which they inhabit that requires attention [12]. Thermal cameras have been used to offer privacy preservation while they monitor an environment [13–16]. For the study presented in this article, low resolution Thermopile Infrared Sensors (TISs) were used for a privacy-preserving solution to detecting and recognising ADLs as, with such sensors, image data can be collected while no discernible characteristics of either inhabitant nor environment are visible.

Previous work has been completed by the authors using the aforementioned low-resolution thermal sensors for the determination of falls and an inhabitant's pose and location within a smart environment [17–20]. The approach presented in this study aims to utilise low resolution thermal data alongside a probabilistic HMM to accurately recognise ADLs whilst preserving privacy. Seminal research has presented the concept that three fundamental problems should be used to describe HMMs: the Likelihood problem, the Decoding problem and the Learning problem [21]. Given the descriptions of the three problems, the work presented in this study corresponds with problem two: The Decoding problem. The Viterbi algorithm was presented as the means for solving this problem. The Decoding problem has been described as *"Given the observation sequence $O = O_1 O_2 \ldots O_T$, and the model λ, how do we choose a corresponding state sequence $Q = q_1 q_2 \ldots q_T$ which is optimal in some meaningful sense (i.e., best explains the observations)?"* [21].

The remainder of this article is structured as follows: the methodology for the use of thermal sensors and an HMM for the classification of ADLs is presented in Sect. 2; The design of the experiment is presented in Sect. 3; The results of the experiments are discussed in Sect. 4; and finally, conclusions are drawn in Sect. 5.

2 Methodology

An HMM was the probabilistic model chosen for the approach due to its ability to use observable events to detect events that are not clearly observable from the data input. In the context of data used in this research, the poses and positions of an inhabitant were directly observable from the TIS frames, however, the ADL being performed by the inhabitant was not. The HMM's hidden states were represented by one of three ADLs: Making a Tea/Coffee, Watching TV or Preparing a Microwavable Meal. The observations from each state represented sub-activities that were necessary for completing the respective ADLs. For example, Making a Tea/Coffee required the completion, in any order, of the sub-activities: Using the Kettle, Using the Cupboard and Using Fridge. By inferring sub-activities from sequences of TIS frames, sub-activity sequences were used for the Viterbi algorithm's input. The Viterbi algorithm's output was the ADLs which were most likely to have been completed in order to produce the sequence of sub-activities that were used as the input.

2.1 Deployment of Thermopile Infrared Sensors Within a Smart Environment

Thermal data was captured with the TIS for both training and testing. Privacy was maintained as the data did not include any discernible characteristics of the inhabitant. Throughout the kitchen, five TISs were deployed and each offered a unique lateral perspective. One TIS was setup opposite the kitchen counter within a shelving unit. The other TISs were deployed in each of the room's corners. Only one inhabitant was present in the kitchen during the experiment. The environment is illustrated in Fig. 1.

The TISs labelled *Corner 1 TIS, Corner 2 TIS, Corner 3 TIS* and *Corner 4 TIS* will be referred to as C1, C2, C3 and C4, respectively. The fifth TIS will be referred to as NO-TIS (Nearest Object – Thermopile Infrared Sensor). Each TIS was installed in a location that ensured it was possible to detect the inhabitant whenever they were in the same vicinity as the TIS. A lateral perspective of the kitchen counter and table was given by the NO-TIS, subsequently facilitating the estimation of which object the inhabitant was closest to at any given time. The nearest object estimation process was also supported by C1 as it helped determine whether the inhabitant was near the kitchen counter, prior to consulting the NO-TIS for which specific object was closest.

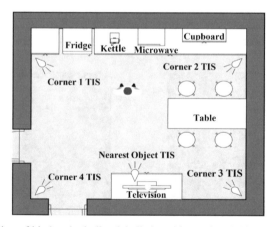

Fig. 1. Illustration of kitchen including labelled positions of each kitchen object and TIS.

2.2 Classifying Poses

From C1, C2, C3 and C4 500 pose instances were recorded for four pose classes: *Arms Down, Arm Forward, Bend* and *Sitting*. The CNN architecture and approach used for pose classification in [19] was used again in this study with no modifications made to the model's infrastructure. A TIS image of each pose class is shown in Table 1 with photographs of the same poses included for reference.

Table 1. Example TIS image and photograph of each pose class.

Arm Forward	Arms Down	Bend	Sitting

2.3 Determining the Nearest Object

Once a pose prediction was made, the nearest kitchen object was determined so that the sub-activity could be inferred. As there was no depth information captured by the TISs, the NO-TIS could not be used to determine the nearest object alone. The example presented in Fig. 2 demonstrates how the inhabitant could have been using the cupboard, yet from the perspective of NO-TIS their centroid was closer to the table, therefore, causing confusion on which object was likely to be closer. To overcome this issue, the Euclidean distance between the inhabitant's centroid and the pre-determined position of the kitchen counter was calculated using the perspective of C1. If the inhabitant was deemed near the kitchen counter, NO-TIS was used to determine the Euclidean distances between the inhabitant and objects on the counter. Otherwise, NO-TIS was used to only calculate distances between the inhabitant and the table or television.

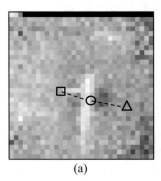 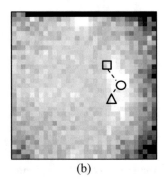

(a)   (b)

Fig. 2. (a) The perspective provided by C1 of the inhabitant using the cupboard. (b) The perspective offered by NO-TIS of the same interaction with the cupboard. The square represents the approximate location of the cupboard and the triangle represents the approximate location of the edge of the table that was closest to the inhabitant. The circle represents the centroid of the inhabitant.

2.4 Inferring Sub-activities

The approach to sub-activity inference in [18] was used in this study with slight modifications. The revised process is presented in Fig. 3. Once a pose was classified and the nearest object was calculated, the pair of labels were compared against the known pose and nearest object pairings to infer the sub-activity for the TIS frame. The diagram highlights that the inference of the majority of sub-activities was only possible if the *Arm Forward* pose was classified. Due to the importance of classifying this pose whenever the inhabitant interacted with the objects on the counter, C1 and C2 were positioned to monitor only the kitchen counter.

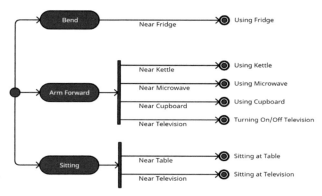

Fig. 3. Visualisation of the sub-activity inference process where the superscript indicates the ADL(s) each sub-activity was linked with where a = *Making Tea/Coffee*, b = *Preparing a Microwavable Meal* and c = *Watching Television*.

2.5 Using a Hidden Markov Model to Represent Activities of Daily Living

The ADLs selected for detection and recognition in this study were chosen as they were deemed necessary to be classified in other ADL recognition studies [6, 22–25]. The selected ADLs were defined, in this research, as sequences of known sub-activity performances. The approach to ADL detection and recognition entailed the development of a stochastic model known as an HMM. One of the components of an HMM is the discrete-time process first order Markov Chain (MC), which consists of a state sequence. For an HMM the states are hidden, however, each state produces observable information.

The sequence of states was represented by S where there were M potential states. The probability of transitioning from state i to state j was quantified as a_{ij}. As any state could transition to or from any other state, there were a potential $M \times M$ state transitions, which were represented with transition matrix A. The equation used to determine the probability of transitioning from state i to state j is presented in Eq. 1:

$$a_{ij} = P(S_k = j | S_{k-1} = i) \tag{1}$$

where a_{ij} is the probability of state i transitioning to state j and S_k represents the HMM's state at time k.

An HMM assumes that the observation sequence O occurs concurrently with the sequence of states and that O is dependent on S. The states of the HMM were hidden, however, the observations emitted from each state were not and so the probability of an observation j being emitted from state i could be calculated. The observable data from each state were assumed to be one of K values and so the emission probability matrix B had a size of $M \times K$. The equation for calculating the probability of a state producing a given observation is shown in Eq. 2:

$$b_{ij} = P(O_k = j|S_k = i) \tag{2}$$

where b_{ij} is the probability of observation j being generated from state i and O_k represents the observation detected at time k.

The HMM used in this methodology is presented in Fig. 4. The model considered the ADLs as the three hidden states due to the ADLs not being directly observable from a recorded TIS frame. The inhabitant's pose and the nearest object were observable from the TIS data and so the sub-activities inferred using this information represented each hidden state's observations.

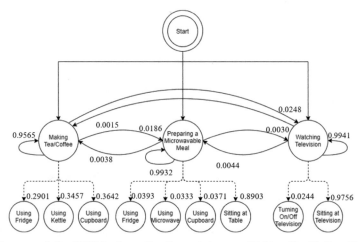

Fig. 4. Diagram of the HMM where the three states are *Making Tea/Coffee, Preparing a Microwavable Meal* and *Watching Television*. The values next to the arrows connecting the states indicate the likelihood of transitioning from each state to another. Connected downwards from each state are the observations where the values indicate the likelihood of the observations being generated from their respective states.

To define the transition and observation probabilities, ten instances of each of the three ADLs were first executed and recorded with the TISs. Each thermal frame was manually labelled with the pose, sub-activity and ADL being conducted. For the frames that had both a sub-activity and ADL label, the particular sub-activity and ADL indicated the observation and the state of the frame, respectively. With this information, both an observation and state sequence were created for the thirty ADL executions. The state sequence was iterated over and using Eq. 1, the state transition probabilities presented in

Fig. 4 were calculated. Similarly, the observation probabilities in Fig. 4 were calculated by iterating over the observation sequence and using Eq. 2.

3 Experimental Design

The proposed approach involved detecting a predetermined number of observations from TIS frames captured during ADLs, which could then be used as input for the Viterbi algorithm. Prior to testing the methodology, a validation process was conducted to select an observation sequence threshold which would facilitate the most accurate ADL predictions. The threshold indicated the number of observations that were to be detected before the Viterbi algorithm estimated the most likely ADL. This involved repeatedly detecting ADLs from a validation dataset, using different observation sequence thresholds. The validation data consisted of two performances of each ADL captured from each TIS. The threshold value used to obtain the highest ADL recognition performance was selected for the test dataset.

The test dataset consisted of three performances of each ADL. Consecutive performances of the same ADL were included in the dataset to investigate how such a scenario would be handled. Once a pose was classified and the nearest object to the inhabitant was calculated, inference of a sub-activity was attempted. Inference of a sub-activity meant that an observation for the HMM was detected and so the observation was appended to an observation sequence. The same process was repeated for each frame until the length of the observation sequence was equal to the observation threshold.

The observation sequence and the transition and emission probability matrices were used with the Viterbi algorithm. The output was the sequence of hidden states that the model most likely transitioned through to generate the observation sequence. The sequence of predicted states was the same length as the input observation sequence and so the mode of the state sequence was calculated. The ADL that corresponded to this state was selected as the ADL that was most likely performed over the duration of the sequence. The predicted ADL's beginning was estimated as the frame containing the sequence's first observation. If an ADL prediction was different from the prediction made for the previous observation sequence, it was assumed that the end of the previous ADL came with the end of its respective observation sequence.

4 Experimental Results and Discussion

This Section presents the results achieved with the test datasets.

4.1 Pose Classification Results

Each corner TIS was associated with a separate CNN model, therefore, any given pose was classified using four CNNs. The pose classification approach was used in [19]. For any given frame, the position of the inhabitant dictated which CNN outputs were used. If multiple CNNs were required to predict a pose, the *Most Confident* prediction selection technique used in [19] indicated the CNN whose prediction would be chosen. Sensitivity,

Table 2. Classification results achieved from using the trained CNNs to classify each pose from the test dataset.

Pose	Sensitivity	Precision	Specificity	F-score
Arm Forward	0.4949	0.8750	0.9910	0.6323
Arms Down	0.9415	0.9121	0.8858	0.9266
Bend	0.6800	0.7727	0.9938	0.7234
Sitting	0.9861	0.8927	0.9400	0.9371
All Poses (Average)	0.7756	0.8632	0.9526	0.8171

precision, specificity and F-score were used to measure pose classification performance and the results are presented in Table 2.

The classification of *Arm Forward* was significantly better whenever the arm was fully extended forward during the use of the cupboard and the television. When interacting with the other kitchen objects, however, the extension of the arm was not as apparent and so the CNNs did not often predict the pose correctly. Whenever the pose was misclassified, the sub-activity being performed was unable to be inferred. The majority of *Arm Forward* misclassifications were predicted as *Arms Down* due to the misdetection of the extended arm, as highlighted by the confusion matrix in Table 3.

Table 3. Confusion matrix generated from the test dataset by the trained CNNs.

True Class	Predicted Class			
	Arm Forward	*Arms Down*	*Bend*	*Sitting*
Arm Forward	49	39	2	9
Arms Down	6	467	2	21
Bend	–	4	17	4
Sitting	1	2	1	283

As a result of the low classification scores for *Arm Forward* and *Bend,* multiple sub-activity instances were not accurately inferred. The misclassifications for both *Arm Forward* and *Bend* were not, however, detrimental to the ADL classification accuracy due to the nature of the approach. Only two instances of a sub-activity involving *Arm Forward* or *Bend* needed to be inferred in order to consider the sub-activity as evidence for the most likely ADL. As each sub-activity involving *Arm Forward* was performed twice during an ADL and multiple frames were captured of each sub-activity, a high classification rate of the poses was not vital for ultimately recognising an ADL. While the CNNs did not classify every instance of *Arm Forward* and *Bend*, the models were able to classify a sufficient number of instances for both poses. While successful in achieving accurate ADL recognition, increasing the training dataset size would improve pose recognition capabilities.

4.2   Activities of Daily Living Recognition Results

The ADL test dataset sequence consisted of three performances of each of the three ADLs and The ground truth is visualised in the chart presented in Fig. 5 (a). This chart indicates the order in which the ADLs were performed, their labels and the frames in which each ADL began and ended. Each ADL class is represented with a different colour to highlight the ADL that was performed within the range of frames on the X axis of the chart. The results chart is presented in Fig. 5 (b) where the predictions for the ADL performances with respect to the ADL classes, beginnings and endings are shown.

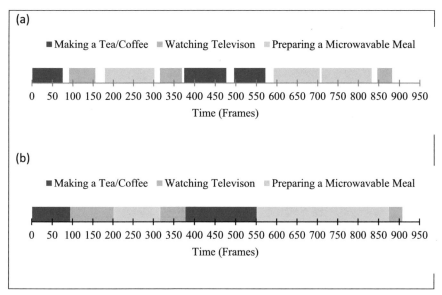

Fig. 5. (a) Chart presenting the ground truth for the ADL detection and classification test in which the ADL labels are highlighted along with the beginning and ending of each ADL instance (b) Chart presenting the results for the ADL detection and classification test in which the ADL labels are highlighted along with the beginning and ending of each ADL instance.

Each coloured partition represents an ADL prediction and the X axis represents the sequence of frames captured during the ADL performances. Comparing both charts with one another indicates that while the correct number of ADLs was not predicted, each ADL that was predicted, was done so with 100% accuracy. On both occasions whenever the same ADL was performed consecutively, the second instance of the ADL was not differentiated from the first. The first instance of the ADL was instead predicted to have lasted for a duration similar to both instances combined.

The estimates for the ADL beginnings and endings were considerably similar to the ground truth as shown in Table 4. In the two cases where the same ADL was performed consecutively, the absolute error for the estimation of the ADL's beginning was calculated by comparing the estimated beginning frame number with the ground truth beginning frame number of the first ADL instance. The absolute error for the estimated ADL ending was then calculated by comparing the estimated frame number with the ground truth ending frame number of the second instance of the consecutively performed ADLs.

Table 4. Comparison between the ground truth and the results from testing the ADL recognition approach with the test dataset where Drink = Making a Tea/Coffee, Meal = Preparing a Microwavable Meal and TV = Watching Television.

Ground Truth ADL	Ground Truth Frame Range	Observation Sequence ADL Prediction	Observation Sequence Frame Range	ADL Beginning Estimate Absolute Error	ADL Ending Estimate Absolute Error
Drink	1–75	Drink	1–74	0	1
TV	92–156	TV	75–181	17	25
Meal	181–304	Meal	182–296	1	8
TV	319–366	TV	287–357	32	9
Drink	373–475	Drink	358–531	15	40
Drink	495–571				
Meal	592–705	Meal	532–854	60	20
Meal	711–834				
TV	849–884	TV	855–887	6	3

The largest absolute error was produced by the estimation for the ending of the last performance of *Preparing a Microwavable Meal* in which the absolute error was only 60 frames. As the TISs operated at approximately eight frames per second, in an online scenario, this estimate would have been incorrect by approximately eight seconds. A value of 237 was calculated by accumulating the absolute error for each ADL beginning and ending estimation. This value equated to circa 30 s.

The approach's act of combining consecutive ADL performances into single instances of the ADLs has not been considered a limitation. The aim to accurately determine the label and duration of an ADL at any given time was accomplished. It is proposed that in a more realistic context, if the inhabitant was making two cups of tea/coffee, both drinks would likely be made in a concurrent manner, rather than a consecutive one and so it would not be necessary to determine if the ADL was conducted twice. As the approach relied on detection of a minimum number of sub-activities, or observations, multiple performances of a sub-activity would only further ensure high predictive performance. If such a scenario did occur in which the same ADL was performed multiple times one after another, the approach has proven to be capable of accurately predicting an ADL for as long as it is being performed.

5 Conclusions

This research has presented an accurate and privacy preserving approach to recognising ADLs. Low-resolution thermal data was used, in which discernible human characteristics were not visible. The size of the pose training dataset was limiting as it had an impact on the pose classification accuracies. Another limitation was the lack of real-time analysis. Arguably, it is not as imperative to produce ADL analysis as immediately as it would be for other systems and so predictive performance was prioritised. While the limitations in this research do not have a significant detrimental impact on the results, there is still scope for future work. The next steps for this research will involve the expansion of the pose dataset. Real-time recognition of ADLs will also be investigated through the use of a Raspberry Pi and integrated TIS so that data collection and classification may be achieved with a self-contained unit.

References

1. Köping, L., Shirahama, K., Grzegorzek, M.: A general framework for sensor-based human activity recognition. Comput. Biol. Med. **95**, 248–260 (2018)
2. Nguyen, H., Lebel, K., Bogard, S., Goubault, E., Boissy, P., Duval, C.: Using inertial sensors to automatically detect and segment activities of daily living in people with Parkinson's disease. IEEE Trans. Neural Syst. Rehabil. Eng. **26**, 197–204 (2018)
3. Alam, M.A.U., Roy, N.: GeSmart: a gestural activity recognition model for predicting behavioral health. In: 2014 International Conference on Smart Computing, Hong Kong, China, pp. 193–200. IEEE (2014)
4. Galasko, D., et al.: An inventory to assess activities of daily living for clinical trials in Alzheimer's disease. Alzheimer Dis. Assoc. Disord. **11**, 33–39 (1997)
5. Bucks, R.S., Ashworth, D.L., Wilcock, G.K., Siegfried, K.: Assessment of activities of daily living in dementia: development of the Bristol activities of daily living scale. Age Ageing **25**, 113–120 (1996)
6. Debes, C., Merentitis, A., Sukhanov, S., Niessen, M., Frangiadakis, N., Bauer, A.: Monitoring activities of daily living in smart homes: understanding human behavior. IEEE Signal Process. Mag. **33**, 81–94 (2016)
7. Rantz, M.J., et al.: Sensor technology to support aging in place. J. Am. Med. Dir. Assoc. **14**, 386–391 (2013)
8. Ding, D., Cooper, R.A., Pasquina, P.F., Fici-Pasquina, L.: Sensor technology for smart homes. Maturitas **69**, 131–136 (2011)
9. Pontes, B., Cunha, M., Pinho, R., Fuks, H.: Human-sensing: low resolution thermal array sensor data classification of location-based postures. In: Streitz, N., Markopoulos, P. (eds.) DAPI 2017. LNCS, vol. 10291, pp. 444–457. Springer, Cham (2017). https://doi.org/10.1007/978-3-319-58697-7_33
10. Mo, L., Li, F., Zhu, Y., Huang, A.: Human physical activity recognition based on computer vision with deep learning model. In: 2016 IEEE International Instrumentation and Measurement Technology Conference Proceedings, Taipei, Taiwan. IEEE (2016)
11. Cook, D.J.: How smart is your home? Science **335**, 1579–1581 (2012)
12. Caine, K.E., Rogers, W.A., Fisk, A.D.: Privacy perceptions of an aware home with visual sensing devices. Proc. Hum. Factors Ergon. Soc. Annu. Meet. **49**, 1856–1858 (2005)
13. Pittaluga, F., Zivkovic, A., Koppal, S.J.: Sensor-level privacy for thermal cameras. In: 2016 IEEE International Conference on Computational Photography (ICCP), Evanston, IL, USA, pp. 1–12. IEEE (2016)

14. Griffiths, E., Assana, S., Whitehouse, K.: Privacy-preserving image processing with binocular thermal cameras. Proc. ACM Interact. Mob. Wearable Ubiquitous Technol. **1**, 1–25 (2018)
15. Guettari, T., et al.: Thermal signal analysis in smart home environment for detecting a human presence. In: 2014 1st International Conference on Advanced Technologies for Signal and Image Processing (ATSIP), Sousse, Tunisia, pp. 334–339. IEEE (2014)
16. Hevesi, P., Wille, S., Pirkl, G., Wehn, N., Lukowicz, P.: Monitoring household activities and user location with a cheap, unobtrusive thermal sensor array. In: Proceedings of the 2014 ACM International Joint Conference on Pervasive and Ubiquitous Computing, pp. 141–145. ACM, New York (2014)
17. Burns, M., Morrow, P., Nugent, C., McClean, S.: Gesture recognition with thermopile sensors. In: Irish Machine Vision and Image Processing Conference Proceedings, pp. 89–96. Irish Pattern Recognition and Classification Society, Belfast, Northern Ireland (2018)
18. Burns, M., Morrow, P., Nugent, C., McClean, S.: Fusing thermopile infrared sensor data for single component activity recognition within a smart environment. J. Sens. Actuator Netw. **8**, 1–16 (2019)
19. Burns, M., Cruciani, F., Morrow, P., Nugent, C., McClean, S.: Using convolutional neural networks with multiple thermal sensors for unobtrusive pose recognition. Sensors **20**, 1–26 (2020)
20. Quero, J., Burns, M., Razzaq, M., Nugent, C., Espinilla, M.: Detection of falls from non-invasive thermal vision sensors using convolutional neural networks. Proc. West Mark. Ed. Assoc. Conf. **2**, 1–10 (2018)
21. Rabiner, L.R.: A tutorial on hidden markov models and selected applications in speech recognition. Proc. IEEE **77**, 257–286 (1989)
22. Liciotti, D., Frontoni, E., Zingaretti, P., Bellotto, N., Duckett, T.: HMM-based activity recognition with a ceiling RGB-D camera. In: Proceedings of the 6th International Conference on Pattern Recognition Applications and Methods, vol. 1, pp. 567–574. Science and Technology Publications, Porto, Portugal (2017)
23. Chen, B., Fan, Z., Cao, F.: Activity recognition based on streaming sensor data for assisted living in smart homes. In: 2015 International Conference on Intelligent Environments, Prague, Czech Republic, pp. 124–127. IEEE (2015)
24. H. Medjahed, D. Istrate, J. Boudy, and B. Dorizzi, "Human Activities of Daily Living Recognition Using Fuzzy Logic For Elderly Home Monitoring," in *2009 IEEE International Conference on Fuzzy Systems*, IEEE, Jeju Island, South Korea (2009)
25. Cheng, B.-C., Tsai, Y.-A., Liao, G.-T., Byeon, E.-S.: HMM machine learning and inference for activities of daily living recognition. J. Supercomput. **54**, 29–42 (2010)

Using Synthetic Data to Improve the Accuracy of Human Activity Recognition

Majid Liaquat⬤, Chris Nugent(✉) ⬤, and Ian Cleland(✉) ⬤

School of Computing, Ulster University, Belfast, UK
{majid-m2,cd.nugent,i.cleland}@ulster.ac.uk

Abstract. Approaches to Human Activity Recognition are typically data driven, classifying activities performed by humans based on low level sensor data from ambient and/or wearable sensors. Collecting this data is a challenging task, being time consuming to collect and label the data, whilst also being expensive to deploy the acquisition systems. The use of synthetic data offers a potential solution to these challenges. This research is based on generating synthetic data, with the aim of improving the performance of HAR. The data used in this study is generated based on a well-known open dataset, Mobile Health (mHealth) dataset, which was collected with wearable sensors for 12 activities. Firstly, the real data was pre- processed and classification methods such as Decision Tree, Gaussian Naïve Bayes, and Support Vector Machine were applied. Three different synthetic generation techniques Synthetic Data Vault Probabilistic Autoregressive (SDV-PAR), Time-series Generative Adversarial Network (TGAN), and Conditional Tabular Generative Adversarial Network (CTGAN) were subsequently used. In comparison to real data, which achieves an accuracy score of 0.9725, the synthetic data generated by CTGAN achieved an accuracy score of 0.8373.

Keywords: Synthetic data · Sensor Data · Tabular Data · Human Activity recognition

1 Introduction

Human Activity Recognition (HAR) data are collected by monitoring various activities performed by humans. These data are collected by sensors such as wearable sensors, smartphones, cameras, Radio Frequency Identification tags and motion capture systems [1]. The collection of HAR data is challenging, time-consuming and costly. Additionally, data leakage and privacy concerns are prevalent, which can prevent the use of the data for training machine learning models.[1]

To overcome these challenges the use of synthetic data has been proposed. Synthetic data is artificial data that is generated with the same underlying characteristics as the real data [2]. HAR data are collected via sensors during physical activities such as walking, standing, and running [3]. The data collected by these sensors is typically imbalanced.

---

[1] This research has been partially funded by the ARC (Advanced Research and Engineering Centre) project, funded by PwC and Invest Northern Ireland.

© The Author(s), under exclusive license to Springer Nature Switzerland AG 2023
J. Bravo and G. Urzáiz (Eds.): UCAmI 2023, LNNS 842, pp. 167–172, 2023.
https://doi.org/10.1007/978-3-031-48642-5_16

Furthermore, classification can be challenging given differences in how individuals perform the activities, in addition to interclass similarity with distinct activities being very similar (e.g. walking vs walking upstairs) [4].

To overcome these challenges, methods of synthetically generating HAR data can be investigated. Using synthetic data, removes the requirement to collect large amounts of labelled data. Given that the synthetic data is artificially generated it can help reduce privacy concerns, making it easier to share and use the data for research purposes [5].

This research is based on generating synthetic data to support HAR. We use real data to recognise activities and then generate synthetic data using various methods, comparing the generated datasets to the real data to determine the most accurate generation method for the activities.

2   Background and Related Work

Traditionally, data has been collected for HAR research using wearable or smart home sensors in various domains. In [6], Qu et al. proposed a new method to overcome the challenges of HAR radar-based data collection. They highlighted that the collection and labelling of this data were expensive and time-consuming by using Wasserstein refined a generative adversarial network with gradient penalty (WRGAN-GP) [6]. In [7], the synthetic data generated for sensor-based activity recognition data with Conditional Generative Adversarial Networks (CGAN) was enhanced by combining a convolutional layer with multiple fully connected networks in the generator's input and discriminator's output of the CGAN. They proposed visual evaluation, similarity measure, and usability evaluation for the quality of the generated data. Anjum et al. collected data on depression-symptomatic activities through indoor and outdoor human activities [8]. They used the GAN to augment the original sensor dataset. The synthetic data was generated at a frequency of 1 Hz. For evaluation, Long Short-Term Memory (LSTM) and Gated Recurrent Unit (GRU) models were employed. The study achieved higher accuracy with the LSTM model.

DeOliveira et al. explored the traditional HAR data generation methods by using Conditional Tabular Generative Adversarial Networks (CTGANs) to synthesise mobile sensor data, which showed more realism to the real-world data [3]. They compared three GAN classes: vanilla, conditional and controllable GANs, and used F1-score as evaluation metrics. They proposed HAR-CTGANs, which generated more realistic data than the above-mentioned GANs.

The aforementioned research projects used different techniques of GAN to generate synthetic data in the context of HAR. In [3], the CTGAN's method was used to generate synthetic data. Although all the works performed different methods within the state of art to generate realistic synthetic data, there were still some gaps like comparison of synthetic data to real-world data and using different approaches of machine learning models to generate more realistic data.

In the current work, we use different methods to generate synthetic data to improve the performance of HAR. We focused on generating synthetic data with different techniques (SDV-PAR, Time GAN, and CTGAN) and combined their performance by applying the training of real and synthetic data in different approaches. For reducing the high dimension data during evaluation, t-Distributed Stochastic Neighbour Embedding (t-SNE) was utilised [9].

Our aim is to address the gap in augmenting real data with synthetic data for HAR, which represents the state-of-the-art in sensor data processing. Furthermore, we tested the effects of using combined real and synthetic data using machine learning models to enhance the training and performance.

3 Methodology

For conducting this research, the mHealth (Mobile Health) dataset was chosen [10, 11]. The reason for choosing this public data was due to it being collected from different sensors placed on different parts of the body and it is based on various activities in the context of HAR. The mHealth dataset was based on activities which were collected by 10 volunteers using the Shimmer2 wearable sensor. These sensors were placed on the volunteer's chest, left ankle, and right wrist with elastic straps, [11]. The data was collected at a 50 Hz sample rate, which is considered appropriate for capturing human activity. There were four activities: *'waist bends forward, frontal elevation of arms, knee bending, and jumping front and back'* each performed 20 times, and other eight with a duration of 60 s. These data was collected from the accelerometer, magnetometer and 2- lead of electrocardiogram (ECG).

In the original dataset there were a large number of null classes labelled with '0', that were not part of the activities, hence they were removed. Outliers in the dataset were detected by using a normal distributed standard deviation method which helped to clean the dataset. Furthermore, the dataset was split as dependent and independent (X, y), respectively, and then the feature selection method 'mutual_info_classif' [13] from scikit-learn was applied. Initially, the top 5 features, then the top 10 features, were tested with machine learning models, and the performance of the tested data was recorded (10 features included with the CTGAN approach).. The features selected were based on the mutual information scores of the applied method. The data was further split into train and test, 75% for train and 25% for test, with a random sample of 42 to ensure that the results were reproducible.

Three methods were implemented with different hyperparameters, and comparison evaluations were undertaken with real data to obtain more accurate synthetic data. The pre-processed data (clean data) was subsequently used for generating synthetic data with the following techniques.

The first method implemented was SDV-PAR [14]. The generated data was then loaded into the programme, split in the same manner as the real data, and applied to machine learning classifier models. The data was also normalised before testing with the models.

Time GAN [15], was implemented as the second method for generating synthetic data. The module was set to "lstm" and "gru" with different hyperparameters. The final method used in this research was CTGAN [16], which was applied to generate synthetic data.

4   Implementation and Results

The findings we derived by employing the aforementioned mentioned classifiers on both real and synthetic datasets are discussed in this Section. The results are presented as accuracy scores, allowing a comparative analysis of classifier performance across different techniques and configurations.

Table 1. Accuracy scores of machine learning models with real and synthetic datasets.

Data Source	DT AS	GNB AS	SVM AS
Real Data	0.9723	0.8423	0.9725
SDV-PAR Synthetic Data	0.1243	0.2401	0.2200
Time GAN LSTM	0.1774	0.3145	0.2799
Time GAN GRU	0.0946	0.1588	0.2799
CTGAN 5 Features (1000 epochs)	0.6285	0.5957	0.7136
CTGAN 5 Features (1500 epochs)	0.6491	0.6178	0.7561
CTGAN 5 Features (3000 epochs)	0.1789	0.1133	0.1430
CTGAN 10 Features (1000 epochs)	0.5364	0.7189	0.8181
CTGAN 10 Features (1500 epochs)	0.5552	0.7319	0.8373
CTGAN 10 Features (3000 epochs)	0.1246	0.1689	0.2181

Overall, by analysing the accuracy scores of different models in Table 1, it shows the data generated with 1500 epochs with the CTGAN gives a higher accuracy score as the same number of data that was split for train and test and that is similar data as real data as shown in Fig. 1.

Table 2. Combined accuracy scores (real and synthetic) with classification models.

Classification Models	Combined Accuracy	Real Data Accuracy
Decision Tree Accuracy	0.9554	0.9723
Gaussian NB Accuracy	0.7517	0.8423
SVM Accuracy	0.9268	0.9725

Table 2 presents the accuracy scores of machine learning models using a combination of real and synthetic data, which were split into 75% for training and 25% for testing. The Decision Tree (0.9554) achieved the closest accuracy compared to the real data (0.9723), then the SVM (0.9268) and the GNB (0.7517).

In Fig. 2, a t-SNE visualisation was performed by comparing the clean data and generated data with CTGAN at 1500 epochs. The t-SNE is a non-linear visualization technique that maps high-dimensional data into a lower-dimensional space for easier visual interpretation [10].

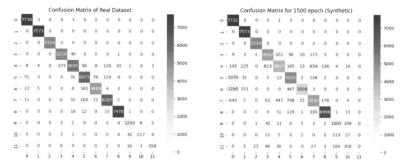

Fig. 1. Confusion matrix of real data and synthetic data with 1500 epochs.

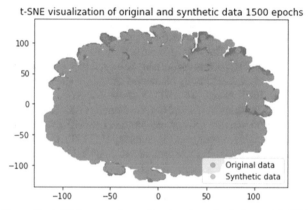

Fig. 2. Shows the t-SNE comparison of real and synthetic data with 1500 epochs.

5   Discussion and Conclusions

We used synthetic data to help improve HAR due to lack of data by comparing the effectiveness of various data generation methods and their impact on machine learning models' performance. The dataset we used which was collected from 10 volunteers performing 12 different activities [12]. The results (Table 1) indicate that the synthetic data generated using CTGAN with 1,500 epochs yielded a better accuracy score of 0.8373 compared to other methods. Additionally, when combined, we achieved an accuracy score of 0.9554 using a decision tree model, which is close to the real-world data accuracy score of 0.9723. Nevertheless, there are still some gaps that should be covered as using synthetic data generators are limited, and a single dataset has been used. Also, a limited number of hyperparameters were applied to each single method, which can be improved by using more resources and research, in addition to comparison with different evaluation matrices for a more accurate result.

References

1. Pires, I.M., Hussain, F., Marques, G., Garcia, N.M.: Comparison of machine learning techniques for the identification of human activities from inertial sensors available in a mobile device after the application of data imputation techniques. Comput. Biol. Med. **135**, 104638 (2021)
2. Murtaza, H., Ahmed, M., Khan, N.F., Murtaza, G., Zafar, S., Bano, A.: Synthetic data generation: state of the art in health care domain. Comput. Sci. Rev. **48**, 100546 (2023)
3. DeOliveira, J., Gerych, W., Koshkarova, A., Rundensteiner, E., Agu, E.: HAR-CTGAN: a mobile sensor data generation tool for human activity recognition. In: IEEE International Conference on Big Data (Big Data), pp. 5233–5242 (2022)
4. Dahmen, J., Cook, D.: SynSys: a synthetic data generation system for healthcare applications. Sensors (Basel), **19**(5). 1181 (2019)
5. Rajendran, M., Tan, C.T., Atmosukarto, I., Ng, A.B., See, S.: SynDa: a novel synthetic data generation pipeline for activity recognition. In: IEEE International Symposium on Mixed and Augmented Reality Adjunct (ISMAR-Adjunct), pp. 373–377 (2022)
6. Qu, L., Wang, Y., Yang, T., Sun, Y.: Human activity recognition based on WRGAN-GP-synthesized micro-doppler spectrograms. IEEE Sens. J. **22**(9), 8960–8973 (2022)
7. Jimale, A.O., Mohd Noor, M.H.: Fully connected generative adversarial network for human activity recognition. IEEE Access. **10**, 100257–100266 (2022)
8. Anjum, F., Alam, S., Bahadur, E.H., Muhammad Masum, A.K., Rahman, M.Z.: Deep learning for depression symptomatic activity recognition. In: 2022 International Conference on Innovations in Science, Engineering and Technology (ICISET), pp. 510–515 (2022)
9. Anowar, F., Sadaoui, S., Selim, B.: Conceptual and empirical comparison of dimensionality reduction algorithms (PCA, KPCA, LDA, MDS, SVD, LLE, ISOMAP, LE, ICA, t-SNE). Comput. Sci. Rev. **40**(100378), 100378 (2021)
10. Banos, O., et al.: mHealthDroid: a novel framework for agile development of mobile health applications. In: Pecchia, L., Chen, L.L., Nugent, C., Bravo, J. (eds.) Ambient Assisted Living and Daily Activities. IWAAL 2014. LNCS, vol. 8868, 91–98. Springer, Cham (2014). https://doi.org/10.1007/978-3-319-13105-4_14
11. Banos, O., et al.: Design, implementation and validation of a novel open framework for agile development of mobile health applications. Biomed. Eng. Online, **14**(Suppl 2), p. S6 (2015)
12. UCI Machine Learning Repository: MHEALTH Dataset Data Set. http://archive.ics.uci.edu/ml/datasets/mhealth+dataset
13. sklearn.feature_selection.mutual_info_classif. https://scikit-learn.org/stable/modules/generated/sklearn.feature_selection.mutual_info_classif.html
14. Zhang, K., Patki, N., Veeramachaneni, K.: Sequential models in the synthetic data vault. arXiv preprint arXiv:2207.14406 (2022)
15. Yoon, J., Jarrett, D., van der Schaar, M.: Time-series Generative Adversarial Networks. In: Advances in neural information processing systems, vol. 32 (2019)
16. Kalyan, V., Xu, L., Skoularidou, M., Cuesta-Infante, A.: Modeling tabular data using conditional GAN. In: Advances in neural information processing systems, vol. 32 (2019)

Smart-Lab IoT Research Platform with Modern Open Source Components

Jordan Vincent[(⊠)], Joseph Rafferty, Matthew Burns, and Chris Nugent

Ulster University, York St, Belfast BT15 1ED, UK
vincent-j1@ulster.ac.uk

Abstract. Research efforts in smart environments involves building and maintaining smart labs with a diverse range of sensors facilitate research. These are often set up with bespoke applications to meet the specific needs of the lab being designed. However, with these custom applications come bespoke problems. One of these is the engineering effort required to maintain the system and cope with change. Time spent maintaining and adapting these bespoke systems takes away from time the research they are intended to facilitate. We aimed to produce a greenfield development of the Ulster University's smart lab infrastructure looking at the best-of-breed open source software to reduce the engineering and maintenance overhead. This paper documents the resulting setup using an open-source application stack and presents an example of its in-practice use for data collection. This work demonstrates that the open-source ecosystem has evolved such that bespoke application stacks need not be required for smart lab provisioning.

Keywords: Smart lab · Home Assistant · Kubernetes · Research Data Platform · Internet of Things

1 Introduction

Institutions engaging in smart environment and connected healthcare research need to build and maintain "smart labs" outfitted with a diverse range of sensors to monitor those within the environment and facilitate data collection and analysis with the aim of developing and testing new systems, algorithms, and models. These labs are often bespoke, such as the School of Computing's previous system Sensor Central [12] However, with these bespoke setups come bespoke problems. In parallel to the development of these systems there have been open-source projects with overlapping goals. For example the open-source home-automation ecosystem needs wide sensor integration overlapping with smart-labs which have the same requirement. These open projects are under much more active development and maintenance today than they were previously, where open source solutions existed but their expected longevity was unknown. In this work investigate a greenfield deployment of the School of Computing smart lab infrastructure reflecting on the experience with Sensor Central and looking at adopting open-source software to build the application stack. Using popular, documented, public, open-source systems where possible also has the advantage of

J. Bravo and G. Urzáiz (Eds.): UCAmI 2023, LNNS 842, pp. 173–184, 2023.
https://doi.org/10.1007/978-3-031-48642-5_17

fitting with the research life-cycle - it's less sensitive to researchers moving on to other projects or to other institutions entirely. Additionally, the collaborative nature of open source projects opens up opportunities to facilitate impact of the research. Updated subsystems, algorithms, or features that come out of research can be passed back to the community who provided the platform the research was built on.

2   Related Work

Sensor Central, the preceding system, was built to overcome some functional deficiencies in the alternatives. These included commercial license restrictions, limited sensor type support, and challenges achieving the needed scalability [13]. Sensor Central dealt with these deficiencies and provided the base for many studies over the years. The primary identified drawback through its use has been its bespoke nature. At times this required researchers to implement the necessary changes and software implementation themselves to support varying experiments and sensors. This could take significant time, particularly if the researcher involved was not already proficient with the language, frameworks, and infrastructure the system was developed with.

Comparing IoT systems universally is difficult. Requirements can vary greatly by use-case and there is a lack of widely applicable comparison frameworks. Noting this challenge, *Mijuskovic et al.* [8] identified a set of requirements after reviewing related research studies before presenting a framework and using it to compare 5 cloud IoT systems. Their analysis and list of requirements is not directly applicable as it focuses on cloud services. Their areas of comparison and requirements has however been used as a baseline with the most relevant requirements and aspects for this use case selected. Broadly the areas chosen for comparison are; Authentication, Data-at-Rest Encryption to protect participant data, Database support for storage, Analytics support, APIs to access data, Event Management/Visualisation, Protocol Support (support for many sensors and platforms), Freedom of Use, Performance/scalability, Costs, and Development Support (apis/sdks/documentation). All compared platforms support sufficient authentication methods, database support, development support, and freedom of use. This leaves Data-at-rest encryption, Protocol Support (out-of-the-box sensor support), Analytics support, APIs, Event Management and Visualisation, Performance/scalability, and Costs for discussion.

Home-automation solutions have been compared by *Setz et al.* [15]. They ranked Home Assistant (HA) [6] 1st and OpenHAB [11] 3rd in their initial selection. These two have been chosen for discussion. The second ranked choice, Domoticz, was excluded as they noted there was little documentation on the underlying details of the project. In their feature comparison they found Open-HAB and HA to be broadly comparable; with HA slightly ahead in features and user experience, and ahead by popularity. Considering the first requirement for discussion, data-at-rest encryption, neither OpenHAB or HA provides this, it would need to be provided at a lower level. They both support a variety of sensor types and can connect to them through many different online services providing exceptional protocol support. They both have built in event-driven triggers, but neither is designed to support intensive analytics (OLAP) type workloads out-of-the-box. OpenHAB was included in the original Sensor Central related work, it was noted that it did not have the ability to forward data-sets to independent processes, one of the original Sensor Central requirements [13]. Both platforms now support streaming access to the event bus, covering API support. The home-automation focus has given them easily adaptable UI systems to cope with the very different and individualised setups they cater to, providing good event management/visualisation. Their primary disadvantage is scalability, focusing on single smart-home environments with low throughput. From the perspective of cost, if the system runs locally they are free open-source applications.

IoT offerings covering different parts of the IoT chain such as messaging, device management, and analytics from the two major cloud service providers such as Amazon Web Services [1] and Microsoft Azure [7] have also been considered. They are not open-source but the infrastructure could be used to host open-source systems. These services focus on creating individual components for others to architect solutions with. There is not a single product that will meet all of any set of requirements. For all areas of data-at-rest encryption, analytics, event management, and scalability there are individual services suitable to meet the criteria. The manual architecture of systems components however requires specific cloud service knowledge and represents a return to bespoke system design. Consequently, cloud services are not likely to represent a decrease in design or maintenance effort. Properly architected however, these cloud services can be expected to out scale the previously discussed home automation systems by orders of magnitude. There are also alternate all-in-one service providers that require less development than building a system from the building blocks of the major cloud providers such as Thingspeak and LineMetrics. These can be expected to almost eliminate upkeep of infrastructure, but offer limited adaptability beyond currently supported services and sensors without returning to manual upkeep as the lab changes. Thus they're considered to have limited protocol support out of the box as custom integrations would need to be created and maintained. They do however support UI customisation and Thingspeak additionally supports custom analytics. The costs of these services are recurring and not necessarily consistent or fixed.

A brief summary is presented in Table 1 which shows a contrast between the home-automation solutions and cloud solutions. Home-automation solutions focus on protocol support and user experience, cloud offerings on infrastructure and scalability. Scalability however is relative. A smart lab will be above the level of data throughput expected of a "standard" smart-home as they will have an above average number of sensors including some high throughput sensors. However, it does not reach the industrial scale the cloud IoT offerings are targeted at. As such, with properly chosen architectural design it is possible to mitigate the scaling disadvantages sufficiently making scalability a relevant but

Table 1. Table comparing potential platforms. Data-at-rest encryption support for Thingspeak and LineMetrics is not known.

	OpenHAB	HA	AWS	Azure	Thingspeak	LineMetrics
Data encryption			Y	Y	?	?
Protocol Support	Y	Y				
Analytics			Y	Y	Y	
Suitable APIs	Y	Y	Y	Y	Y	Y
Event mgmt/vis	Y	Y				Y
Scalability			Y	Y	Y	Y
Costs	Free	Free	Recurring	Recurring	Recurring	Recurring
Engineering effort	Low	Low	High	High	Low	Low

lower priority criteria. Cloud options require an ongoing cost commitment from the school independent of funded research projects, which is not cleanly compatible with the research life-cycle. This model could leave the lab infrastructure vulnerable between funding gaps. A hybrid solution is also possible, running HA or OpenHAB on cloud services. However, a notable concern with cloud services or a hybrid approach is data protection - cloud protection policies can change causing significant overheads when storing data from study volunteers. The data protection and privacy implications must be assessed and evaluated for each study and ethics application to see if they are suitable for the type of data being collected. That fails to insulate researchers from the system design instead forcing them to confront it with every study application. Ultimately it is adding to their workload instead of decreasing it. The authors consider the cloud data protection concerns, engineering effort, and protocol support to be the primary criteria for making a decision. That leaves the choice between HA and OpenHAB.

2.1 Solution Requirements in the Context of Research

The overall goal of the system is to integrate with the research life-cycle. The system will be continuously evolving, used by multiple projects concurrently, and used by researchers new to the projects or institution. Therefore, components require good documentation to reduce overhead when on-boarding new researchers. Using open-source components helps fulfil this as large successful open source projects require good documentation to facilitate the large collaboration. HA and OpenHAB are well documented with tutorials and other resources that new researchers can use to gain initial knowledge and experience with the systems, and the projects have active communities.

In addition to the requirements discussed in related work, research oriented systems introduce a number of specific challenges arising from the research life-cycle. Previously these were not all catered for when systems were evaluated prior to creating Sensor Central [13]. Individual components need to update, evolve, and be replaced to suit heterogeneous software needs of concurrent research projects without causing incompatibilities. This is a common software engineering challenge which is often addressed with containerisation [18] in modern systems architecture. Additionally, experiment specific changes from UI updates through to new subsystems will be transient necessitating high levels of deployment automation to avoid creating a burdensome maintenance cost. Failing to cater to this would limit the scope for research, as more time, effort, and funding would need to be devoted to supporting technical work instead of facilitating the research activity itself. The system needs to meet the following additional criteria:

a) Data must be encrypted-at-rest to reduce ethics overhead
b) The system must support concurrent use of the sensors
c) The individual subsystems must be contained from each other to facilitate deployment of subsystems while minimising scope for incompatibilities.
d) The system must abstract the details and intricacies of the lab network.
e) The system must allow for automated installation, update, and removal of transient subsystems supporting different experiments.
f) The system must support multiple concurrent UIs to cater to the needs of different experiments.

Requirement (a) can be satisfied with standard full-disk encryption. Requirements (b) and (f) are satisfied by the choice of HA or OpenHAB and (c) is satisfied by containerisation. Having excluded cloud options which are known for their automation and scalability, requirements (d) and (e) are more complicated. Abstracting network architecture from interconnected applications requires several co-operating components. This abstraction is important to ensure the applications do not become tightly coupled to details of the network, which would increase the technical cost of supporting different experiments with different requirements. A new subsystem may require a high amount of RAM or require GPU acceleration, and the addition of new hardware and network modifications to support it. Deployment and abstraction of containerised systems is a modern infrastructure challenge that can be handled by orchestration systems [18]. A primary purpose for them is to facilitate and automate the setup and decommissioning of application components abstracting as many details as possible from the application. This allows it to see the same environment everywhere. There are good and well-maintained orchestration systems with communities backed by cloud enterprises around these challenges such as Kubernetes [2]. Orchestration systems are also built to support clusters of servers, with differing resources.

3   Framework for an Open-Source IoT Research Platform Ecosystem

An orchestration system as the base of the application infrastructure meets the network and server abstraction requirements. It will handle the networking setup, provide internal DNS for locating services, and provide proxy services to direct network traffic to the correct location within the cluster. Additionally, they automate allocation of resources. The ability to dynamically add new nodes to the system also maximises it's ability to adapt to the changing needs of AI and IoT research.

There are a variety of orchestration systems. Options considered for this are Kubernetes, Docker compose, Docker Swarm, and OpenShift. Docker compose on it's own is limited to a single server and thus excluded. Among the three cluster scale options, Kubernetes, Docker Swarm, and OpenShift, Kubernetes is the large-scale leader with managed offerings by the major cloud providers. Together these are now much more mature, stable, and capable platforms than were available at the time of the previous lab setup. As a result it's now much more reasonable to build on top of these projects than it was previously. Kubernetes is the most popular [16]. To give an indication of it's enterprise backing; the three largest cloud providers (total market share 65% [17]) provide managed Kubernetes offerings, AWS, Azure, and GCP. Kubernetes was itself originally developed by Google before being open-sourced and transferred to the Cloud Native Computing Foundation which is itself a Linux Foundation project. Kubernetes also provides ease of getting deployments up for testing through cloud services or tools such as Docker Desktop for testing. This allows researchers interacting with the lab to easily create their own replicas of the system to test any changes. Easy deployment of test environments with minimal technical knowledge is key to minimising researcher time spent on non-research tasks. OpenShift can be seen as a RedHat customised and supported version of Kubernetes. As it is RedHat supported, officially supported platforms for OpenShift are limited to cloud services and RedHat derivatives which is an undesirable limitation. Fewer people will have experience with managing RedHat derivatives vs a less enterprise focused alternative such as Ubuntu. That leaves Docker Swarm and Kubernetes. The authors chose Kubernetes as the orchestration system due to it's greater popularity and wide community backing but note either would have likely been sufficient.

Having chosen to build on an orchestration platform the next choice is between the discussed home-automation platforms, HA or OpenHAB. Given the feature parity, the choice of OpenHAB or HA was made based on ease of use and popularity. *Setz et al.* [15] ranked Home Assistant as the top performer for user experience. GitHub's 2019 Octoverse report listed HA as the 10th biggest open-source project on its platform with 6,300 contributors [3], three years later in 2022 they placed it 2nd with 13,500 [4]. As a result Home Assistant was the chosen core user-facing system.

3.1   Architectural Choices

A major consideration in any IoT network is communication. Avoiding the introduction of unnecessary protocols to the setup reduces the technical overhead required to maintain it. Therefore, having them all use the same transport is preferable. One area where HA did not perform the best of the potential solutions is scalability. HA is a single-process application that cannot fully utilise multiple CPU cores. A publish-subscribe (Pub-Sub) message protocol between the sensors and HA can mitigate that. In the Pub-Sub model a broker allows clients to subscribe to topics they want to receive messages about. The broker tracks the subscriptions and relays messages to all clients subscribed to the topic. This moves the load associated with handling all concurrent clients away from the HA event bus and onto the broker, mitigating HA's concurrency limitation. Translation layers to read messages from protocols the system does not speak can also then relay these messages into the broker to enter them into the system without modifying the sensors directly. HA need not see or process the initial data before translation. Pub-Sub message brokers such as Mosquitto and ActiveMQ have been shown to scale from thousands to tens-of-thousands of messages per second [9]. The Pub-Sub protocol supported by Home Assistant is MQTT.

So far we have considered environment isolation, scalability, automated deployment, UI adaptability, and sensor support. Finally, there is also the task of easily creating and updating any experiment specific logic while minimising the need for specific technical knowledge. For this NodeRed was chosen. NodeRed is a low-code event driven graphical UI for connecting different APIs together. This allows for easily linking and updating logic for different sensors and events. NodeRed has a well used and tested third-party integration with Home-Assistant [10]. This is a major component of the system as it is what facilitates the ease of implementing and updating the experiment specific control logic for the lab. Therefore a well tested and supported integration was considered essential. According to HA's own analytics the NodeRed add-on is the most popular low-code add-on and is in use by 28% of installations totalling 40,175 at the time of writing [5].

3.2 The Implemented Architecture

The overall architecture is shown in Fig. 1. MQTT acts as the central hub of the system with Mosquitto as the chosen broker, chosen as it is the broker recommended by HA's MQTT integration. HA compatible MQTT layers with automated device-discovery have been implemented for thermal vision sensors, smart floor, and Ultra-Wide Band (UWB) positioning sensors. Home-Assistant has multiple camera and other integration's that facilitate recording. These were deliberately not used as it often involves lossy transcoding. The loss is undesirable as the images are quantitative measurements of pressure and temperature not RGB images; investigation and testing showed support for lossless video formats along the full recording to analysis pipeline was limited - which is one identified limitation of the system. Therefore, the recordings are stored temporarily on the sensors themselves. The UWB positioning system is designed for MQTT transport but does not report data in a way HA can ingest. A translation layer deployed on the Kubernetes cluster was added which subscribes to the UWB messages, translates, and re-publishes them with the device discovery information for HA. For the non-IP devices operating on 433MHz such as the

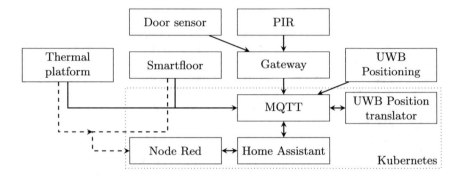

Fig. 1. The implemented architecture, the non-IP devices route through a gateway which translates them to MQTT. The thermal, smart floor, and UWB positioning platforms communicate directly with MQTT. A translator for the UWB system listens for its MQTT messages and translates them and re-publishes them to home assistant. Node red interacts with home assistant and also the thermal and smart floor systems for downloading recordings.

door and passive infrared sensors the gateway has been updated to publish the devices and their discovery information to HA.

System components were deployed on Kubernetes which successfully met the requirement of abstracting the network specifics. This includes automatic DNS resolution using their name, removing any need for them to know how to locate each other or deploy manual DNS entries to the university network. New versions of subsystems such as the UWB translator, which has undergone several iterations, have been deployed and in the process moved servers within the cluster, with Kubernetes automatically re-routing their traffic. Front-end services such as Home assistant, Node Red, and the MQTT broker are also automatically assigned domain names to permit the researchers to find them, abstracting away the network setup details from the users.

This setup was not entirely free of engineering effort, it required development to setup and connect Kubernetes to the DNS system for automatic registration and an ACME client for TLS certificate generation using the external-DNS and cert-manager projects. However, such setup is a one-off and ongoing maintenance effort is low compared to the work saved setting up individual services. Services are then managed using Kubernetes yaml configuration files. Deployment of updates on code changes can then be fully automated with CI/CD pipelines.

At the time of writing the system has been in place for several months. Uses to date include storing and retrieving data from one of the latest sensor revisions within the REAMIT project [14]. After the successful deployment and test of the

REAMIT sensor the system is now being used to manage and run the data collection for a multi-modal thermal fall detection project the authors are involved with. An example of an experiment specific dashboard, event driven controls, and data export is illustrated in the next section. A third usage is also planned for data collection utilising the multi-modal data and the integration with the UWB positioning system. As a final notable anecdote to illustrate the reduction in maintenance overhead; due to network restructuring at the universities new Belfast campus the exposed cluster services had to be transferred across networks and IP addresses. This process consisted of 1) Updating the IP address allocation in the Kubernetes pool, 2) triggering an automatic redeployment of the external services. All other actions such as DNS updates both internal and external were automatic.

4   Case Study: Multi-modal Thermal Fall Detection

The multi-modal thermal fall detection focuses on two of the sensors, the smart floor and the thermal sensor, which also monitors the low frequency audio dB level. They also possess the ability to do recordings of the 2D matrices they produce, registering a control switch to enable/disable their recording functionality. A dashboard panel was created for the data collection containing only the relevant sensors for the specific data collection. This reduces the clutter of the system and makes data collection more fluid. Creating the dashboard in HA only requires selecting the desired sensors after which is recommends a panel type and if approved adds it to the UI. No custom UI components have been needed to date. The lab space is shown in Fig. 2 and the first iteration of the dashboard is shown in Fig. 3.

NodeRed was also utilised to create a custom event driven flow for the data collection. First it uses it's NodeRed's HA integration to create a switch which is set to trigger the smart floor and thermal platform simultaneously. This makes it easier to use and minimises the amount of misalignment that can be present in the dataset by starting all the sensors at the same time. Secondly, as HA records the change in sensor state the difference in start time is documented to narrow it down further. This makes HA a reliable reference point for aligning multi-modal sensor timestamps. NodeRed flows triggered by the sensors stop-recording event of the thermal platform and the smart floor retrieve the finished recordings from the device and store them locally. In the case of the thermal platform this also extracts the dB level metric from home-assistant directly. Finally the flows send notifications through home assistant with links to download the resulting data files. These flows are shown in Fig. 4.

Fig. 2. Lab setup for thermal experiment. Thermal sensor is positioned on roof (grey sensor, middle ceiling) with the smart floor placed underneath covered by mats. Plastic bricks are used to construct different rooms such as the bedroom (back right). (Color figure online)

Fig. 3. Dashboard for the thermal fall detection experiment. It contains the smart floor and thermal sensor (grids, right), dB level, recording status, record button (top, left), and instructions for the experiment (far left).

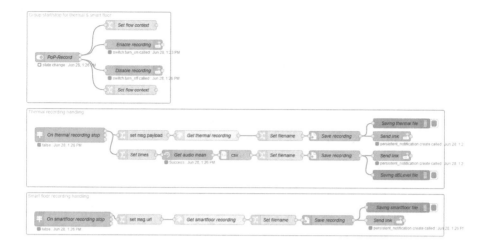

Fig. 4. NodeRed flows for the multi-modal thermal fall detection experiment. The top flow is the on/off button for the recording. The middle flow handles the data from the thermal sensor, both retrieving the recording and the audio data. The bottom flow retrieves the recording from the smart floor.

5   Conclusion

The system has met all the requirements, from easily customisable UI elements, through to making it easy to deploy additional components to the lab environment, using a core stack of fully open-source components. In the several months it has been operating, the system has not encountered any issues, with sensors continuing to be added as more of the previous smart-lab is brought online. The only issue identified to date has been the need to handle recording of the sensors producing 2D matrices separately from HA. Currently the MQTT broker is running stable at 3,000 messages/min (sent + received) with HA processing over 380 state changes/minute. Active sensors include two smart floor sensors (48×48, 48×32), 10 binary sensors, five UWB tags, and one 32×32 thermal sensor. Currently over 20 individual sensors are reporting over 50 entities through MQTT to HA. Collectively the MQTT broker, HA and NodeRed are utilising less than 10% of their available CPU resources leaving the system with significant room to grow. This illustrates that open source stacks in these areas have matured and are stable enough to be the base of smart lab environments and are capable of coping with the base workload.

Acknowledgments. Special thanks to Stuart Christy for his work in facilitating deployment of this architecture within the university network.

References

1. Amazon Web Services: AWS IoT (2023). https://aws.amazon.com/iot/
2. Cloud Native Computing Foundation: Kubernetes (2023). https://kubernetes.io/
3. GitHub: Octoverse report 2019 (2019). https://octoverse.github.com/2019/
4. GitHub: Octoverse report 2022 (2022). https://octoverse.github.com/2022/state-of-open-source
5. Home Assistant: Home assistant analytics, June 2023. https://analytics.home-assistant.io/add-ons
6. Home Assistant Core Team and Community: Home assistant (2023). https://github.com/home-assistant
7. Microsoft Azure: Azure IoT (2023). https://azure.microsoft.com/en-gb/solutions/iot/
8. Mijuskovic, A., Ullah, I., Bemthuis, R., Meratnia, N., Havinga, P.: Comparing apples and oranges in IoT context: a deep dive into methods for comparing IoT platforms. IEEE Internet Things J. **8**(3), 1797–1816 (2021). https://doi.org/10.1109/JIOT.2020.3016921
9. Mishra, B., Mishra, B., Kertesz, A.: Stress-testing MQTT brokers: a comparative analysis of performance measurements. Energies. **14**(18), 5817 (2021). https://doi.org/10.3390/en14185817, https://www.mdpi.com/1996-1073/14/18/5817
10. Nijhof, F.: Node red addon, home assistant community Add-ons (2023). https://github.com/hassio-addons/addon-node-red
11. OpenHAB foundation and Community: Openhab (2023). https://github.com/openhab
12. Rafferty, J., Synnott, J., Ennis, A., Nugent, C., McChesney, I., Cleland, I.: SensorCentral: a research oriented, device agnostic, sensor data platform. In: Ochoa, S.F., Singh, P., Bravo, J. (eds.) UCAmI 2017. LNCS, vol. 10586, pp. 97–108. Springer, Cham (2017). https://doi.org/10.1007/978-3-319-67585-5_11
13. Rafferty, J., et al.: A scalable, research oriented, generic, sensor data platform. IEEE Access **6**, 45473–45484 (2018). https://doi.org/10.1109/ACCESS.2018.2852656
14. Ramanathan, R., et al.: Motivations and challenges for food companies in using IoT sensors for reducing food waste: some insights and a road map for the future. Sustainability. **15**(2), 1665 (2023). https://doi.org/10.3390/su15021665, https://www.mdpi.com/2071-1050/15/2/1665
15. Setz, B., Graef, S., Ivanova, D., Tiessen, A., Aiello, M.: A comparison of open-source home automation systems. IEEE Access **9**, 167332–167352 (2021). https://doi.org/10.1109/ACCESS.2021.3136025
16. SlashData: The state of cloud native development q1 2021 | key insights for the cloud native computing foundation (2021). https://www.cncf.io/wp-content/uploads/2021/12/Q1-2021-State-of-Cloud-Native-development-FINAL.pdf
17. Synergy Research Group: Q1 cloud spending grows by over $10 billion from 2022; the big three account for 65% of the total. https://www.srgresearch.com/articles/q1-cloud-spending-grows-by-over-10-billion-from-2022-the-big-three-account-for-65-of-the-total
18. de la Torre, C.: Containerized Docker Application Lifecycle with Microsoft Platform and Tools. Microsoft Developer Division, NET and Visual Studio product teams, One Microsoft Way, Redmond, Washington 98052–6399 (2022)

Secure, Transparent, Supply Chain Supporting Agri-Food Provenance and Sustainability

Thomas Martin[1]([⊠]) [iD], Tom Cowderoy[1], Joseph Rafferty[1], Trevor Cadden[1], Ronan McIvor[1], and Lorraine Brennan[2]

[1] Ulster University, Belfast, UK
t.martin@ulster.ac.uk
[2] Ulster College Dublin, Dublin, Ireland

Abstract. The protein-I project is a cross Ireland initiative that takes a food systems approach to enhancing the sustainability of protein production across the island of Ireland, as part of the protein-I project, this study aims to produce a smart agricultural supply chain solution. There is a need for the requirements of the system to be tailored for the specific use case, agriculture on the island of Ireland is different from agriculture in other parts of the world. The average size of a farm in the Republic of Ireland is 32.4 ha [8], compare this to Australia where the average is 4,331 ha [7], therefore farm practices in Australia will be different. To develop a solution that is catered to the needs of the Agri-food sector on the island of Ireland this study gathered requirements from a series of workshops and interviews with stakeholders. Several use cases were presented by the stakeholders to identify the deficiencies in the current supply chain, and to address these challenges, a custom solution has been designed and implemented.

Keywords: Secure supply chains · blockchain · distributed ledger technology · carbon calculation · sustainable living · smart agriculture · Internet of Things · Industry 5.0 · protein-I project

1 Introduction

The modern food production process, the steps involved to produce food for consumers contain a variety of complex supply chains, the 'farm to fork' process has supply chains that encompass the entire planet, in 2018 Europe was estimated to import 14 million tonnes of soya each year [1]. Modern farming practices rely on these complex supply chains to maintain high yields and to ensure food availability, but these supply chains are not visible to the consumer, around 1/4 of the soya imported to Europe comes from Brazil. Soyabean production is a contributing factor to the deforestation of the amazon rainforest, and while most consumers want to protect the environment, they are not given information that would allow them to make informed decisions on their purchases, the information is lost/hidden as the product is transported between the different supply chain actors. The main problem caused by the complexity of the Agri-food supply chain is reduced provenance/traceability, because of the many supply chain actors involved and the aggregation of products in various warehouses information/data is lost. In most cases the traceability of farm products is paper based, forms such as grain passports are used to confirm a product's origin. The problem with the current solution is that it is easy to manipulate, the data isn't available to the consumer, and it only achieves low-level traceability [14].

2018 Strawberry contamination - In September 2018 boxes of strawberries from Queensland Australia were contaminated with sewing needles, the needles had been hidden inside the strawberries. After a couple of hospitalizations, the strawberries from affected brands were pulled from shelves. The investigation ended with no convictions due to lack of evidence, and while the police did eventually track the source down to one packing facility, by that time the damage had already been done. This is a case of agri-food terrorism which posed a serious risk to consumers and cost farmers thousands, one farm had to burn off 500,000 plants even though they were unrelated to the incident. It may be impossible to prevent future cases of agri-food terrorism but improving traceability in agri-food supply chains is necessary to reduce the financial impact, protect lives and convict the persons responsible.

2013 Horse meat scandal - In 2012 the food Safety Authority of Ireland was the first to report the presence of horse meat in products containing beef. In the following year more cases of horse meat being sold as beef arose across the UK and Ireland, further investigation found that equine DNA was present in meat products across all of Europe. The problem raised ethical questions and health concerns, a drug called phenylbutazone is used treat horses and was found to be present in the tainted meat products, the drug is banned in food products due to its dangerous ability to suppress white blood cells. The presence of pig DNA was also found to be present in many products claiming to be beef, selling beef contaminated with pork is unethical due to the various religious groups that forbid its consumption. The horse meat scandal damaged consumer trust in the European agri-food sector and is a clear example of why the current agri-food supply chain needs to be more traceable and transparent.

Greenwashing - As the world is becoming more aware of the climate crisis and the need to improve environmental protection many farmers are switching to sustainable farming methods and attempting to reduce their greenhouse gas emissions. The issue farmers are facing is that their efforts are going unrecognized, the agri-food sector has a problem with 'greenwashing', this is when a business lies or misleads the market by creating a false image of environmental sustainability. Sustainable farming methods have lower yields and require more work to manage, this makes them less competitive when compared to traditional methods, the upside is that the low emission/environmentally friendly products are more desirable and can be sold at a premium. A survey performed by YouGov found that 67% of the UK public are worried about climate change [2] and 57% are willing pay more for climate friendly produce [3], younger consumers are more likely to care about the carbon footprint of their food, this suggests that in the coming years the percentage of people who are willing to pay a premium for 'green' produce will increase. The increasing desire for sustainable produce will only encourage more business to lie about their carbon footprint, the current agri-food supply chain's lack of transparency means it doesn't have the ability to properly manage or fight greenwashing, this will reduce consumer trust in carbon values and hurt farmers/business which have embraced sustainable practices.

An efficient, traceable digital solution could be used to securely record the movement of agri-food products throughout the supply chain. The security of data could be ensured through the use of distributed ledger technologies (DLT). DLTs such as blockchain have already been implemented in a variety of other domains to provide secure supply chain functionality, however, these solutions have some deficiencies which reduce their applicability within the agricultural domain.

The protein-I Project is a cross Ireland initiative that brings together six research organizations. The project takes a food-systems approach to enhancing the sustainability of protein production across the island of Ireland. The main goal of the project is to diversify protein sources on the Ireland of Ireland by encouraging farmers to grow protein crops. As part of the protein-I project this study aims to produce a smart agricultural supply chain solution to be deployed across the twin jurisdictions on the island of Ireland. The smart supply chain solution should achieve the following objectives:

- Encourage shorter supply chains on the island of Ireland.
- Integrate Carbon calculation across the whole supply chain.
- Provide consumers with product information, allowing them to make informed decisions.
- Improve marketability of produce from the island of Ireland.

2 Related Work

To date, a number of solutions have been developed to provide assurance within supply chains. Several of the most relevant and advanced are discussed within this section.

Salah et al. have proposed a DLT based supply chain solution for traceability in Soybean production [4], their framework focused on the use of Ethereum smart contracts to create a smart system that can ensure the quality of the product delivered to the customer. The smart contracts are used to track, trace, and perform transactions in the soybean supply chain, the execution of the smart contract is automatic, the smart contracts will be carried out by the mining nodes, and they will alert supply chain actors when violations occur. The proposed supply chain solution helps to keep the supply chain running optimally by alerting actors and performing transactions automatically. The solution encourages supply chain actors such as farmers to upload data and pictures of crop growth to an inter planetary file system (IPFS), the file is digitally signed, hashed, and then saved to the blockchain. The solution was designed for the soybean industry but was generic enough that it could be applied across the Agri-food sector, the solution tackles the industry's need for traceability but doesn't give the consumer any transparency, nor does it encourage sustainability.

Shahid et al. proposed a framework for digitally tracking agri-food products using blockchain smart contracts [6]. This solution is different from the previous proposed by Salah et al. because it implements two new systems, a trading and delivery system to secure transactions between supply chain actors, and a reputation system to monitor the credibility of supply chain actors. The reputation system is responsible for managing the credibility of supply chain actors, after a transaction takes place the supply chain actors are requested to register reviews of the interaction, the reviews are recorded in an IPFS, hashed, then stored on the blockchain. The reputation system ensures that trustworthy/immutable reviews of supply chain actors are maintained, the solution uses the reviews to give purchasers a trust value, where a high value indicates that the seller is reliable. The trading and delivery system is what tracks the movement of products and makes the solution auditable, the trading process requires the buyer to create an order which is sent to the seller, once the buyer receives the product the transaction is confirmed, in order to confirm the transaction both the buyer and seller pay a security amount to the contract. This solution attempts to make transactions trustworthy and achieves this thought the use of the reputation and trading systems, but it is probably unnecessary in the Irish agri-food sector, the additional cost of paying a security fee for each transition would likely discourage small farmers from participating in the network.

Verifish is an Irish technology company that works in the sea food sector, they worked alongside the Irish state body 'Bord lascaigh Mhara (BIM)' to develop a solution that could be used to increase traceability and sustainability of Irish seafood [5]. The project came about to help protect the local ecosystem and to support local fishermen. The salmon fishing industry in Ireland has been growing and the organically grown fish are now considered a premium product worldwide, to protect the industry from fraud BIM teamed up with Verifish to create a DLT based supply chain. The solution uses a private distributed ledge called Hyperledger which can only be accessed by users with valid cryptographic credentials. Hyperledger is an open-sourced project launched in 2015 with the purpose of advancing industry adoption of blockchain technologies, Hyperledger focused on industry applications and unlike Bitcoin or Ethereum it is not open to the public. Hyperledger was used by Verifish to provide a secure method of tracing the origin of Irish Seafood. The solution works by allowing supply chain actors such as fishermen and processors to input data though their software platform, "AquaEye", the software can be run on PCs and mobile devices enabling information on catches to be uploaded quicky. While AquaEye is a novel approach to sustainability in the sea-food industry it lacks features that would be desired in the Agri-food sector, such as Carbon calculation.

The discussed solutions all provide a novel approach to supply chain traceability; however, they have some deficiencies which reduce their applicability to the use case scenario presented by this study. These deficiencies are:

- Solutions are designed for different domains.
- Do not fulfill the previously mentioned requirements, such as encouraging shorter supply chains and enabling carbon calculation.
- Incorporation of user experience paradigms which are not developed with innate appreciation of the workflow of within a typical agricultural setting on the island of Ireland.
- Reliance on computationally intensive DLTS such as Blockchain.
- Inability to provide dynamic supply chain functionality with automated negotiation and healing.
- Inability to provide carbon calculation tailored to the transport and logistics within the island of Ireland.
- No method of providing transparency to the consumer.

3   A Solution to Provide Secure, Dynamic, and Traceable Agrifood Supply Chains

In order to address the deficiencies identified in Sect. 2, stakeholders were consulted, and a tailored digital solution has been designed and implemented. Subsection 3.1 provides a summary of requirements developed though consultation with a range of stakeholders including farm operators, consumers, retailers and supported by two national Agricultural development agencies. Subsection 3.2 presents the design of the solution developed though consideration of related work and the requirements gathered in Sect. 3.1.

3.1   Requirements Elicitation

There is a need for the requirements of the system to be tailored for the specific use case, agriculture on the island of Ireland is different from agriculture in other parts of the world. The average size of a farm in the republic of Ireland is 32.4 ha [8], compare this to Australia where the average is 4,331 ha [7], therefore farm practices in Australia will be different. The agriculture industry changes depending on a lot of factors such as climate, crops, traditions, policies, and population density. The problems faced in the agriculture sectors of other counties will be different than those faced on the island of Ireland, in order to develop a solution that catered to the need of the Irish Agrifood sector it is necessary to get requirements from key stakeholders in this domain. To gather these requirements a series of workshops were organized to gather requirements from participants, and following these workshops a number on interviews were held with stakeholders ranging from small farms producing specialty crops such as flax, and large producers which package and supply supermarkets with fresh produce. The project initially took a holistic approach to the agri-food sector, but the project would later change to focus on the arable sector. The project held over 20 interviews with stakeholders, the interviews were conducted in a semi-structured manner to ensure flexibility, the interviews and workshops were used to develop a list of requirements for different stakeholder groups.

The solution should: a). Offer low friction interfaces to streamline data entry while considering current agricultural processes and workflows. B) eliminate duplicate data entry, as would historically be in place. C) provide data which can enable seamless management of stock/inventory d) Increase marketability of products viva by proving origin and quality E). Provide traceability (country of origin, age etc.) 6. Show carbon footprint of product. G) provide low-cost overheads, as compared to existing and alternative solutions H) Provide consumers with a manifest of the history, and provenance, of the product. In order to address these requirements, a custom solution has been designed and implemented. This solution is presented in the following Subsect. 3.2.

3.2   Solution Design

The developed solution is a dual homed distributed ledger-based service to enable secure and transparent tracking of the origin of agri-food products. The supply chain solution uses a technology designed and developed to be an alternative to traditional blockchains based on directed acyclic graphs (DAG). The developed supply chain solution uses a DAG to secure data. The downside of all blockchains, is that they are not databases, blockchains are immutable, distributed ledgers which allow users to store a cryptographic hash of data and a point to its location in a third-party system. The only way to query data in the DAG is via the data's unique message Id or index, so to solve this problem the developed supply chain solution uses a Mongo DB document store to keep a copy of the data for quick querying, files in the document store contain the data's message Id so that the data can be validated through the DAG. An API provides the frontend of the solution with a variety of REST endpoints, these endpoints can be used to retrieve, create, delete, edit, and update data. The frontend of the service is a Mobile application and a website, the mobile application is how supply chain actors will interact with the system, the app will allow users to create new batches of products/crops, view their inventory, create transactions, and produce QR codes that can be scanned to show the history of a product. The Website has dual functionality, the consumers will be taken to a webpage after scanning a QR code, this page will display all the information about the product, the origin, carbon footprint etc. The website will enable supply chain actors to view business analysis, they will also be able to view local businesses and what they produce, this is to encourage shorter supply chains (Fig. 1).

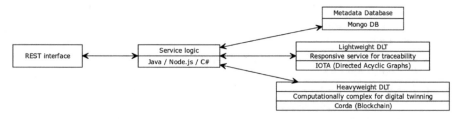

Fig. 1. The service architecture of the developed solution.

Distributed Ledger Technologies

The developed solution uses a lightweight blockchain alternative called the tangle, unlike traditional blockchains the tangle is not a linear chain of blocks but rather a directed acyclic graph (DAG). In a traditional blockchain the "miners" are used to validate transactions, but in the tangle, messages are validated via a parent/child relationship. The tangle is controlled by nodes, to add and read data from the tangle the solution uses a hornet node, the problem with the hornet node is that it doesn't store data permanently, the tangle was designed for IOT applications, so data is "pruned" after a while to prevent small mobile devices using up all available storage. To fix this problem and enable the solution to store data long-term we need a chronicle node, unlike other tangle nodes they store data permanently. The benefit of using the tangle over other traditional blockchain is that the tangle is free, lightweight but most importantly it has faster transaction speeds [18]. The downside is that the tangle's parent/child validation method is less secure than traditional blockchains. The tangle is used to store information on the history of a product, whenever a new transaction is registered the information is encrypted and stored in the payload of a tangle message, the tangle returns a message Id and then the solution saves a copy of the data with the message Id in the document store. The tangle messages create a virtual chain so that even if the data on the document store is deleted, it can be recovered and validated through the tangle.

Database

The solution uses a Mongo DB document store, the document store holds a copy of all information currently used by the solution, this includes a copy of data stored on the DAG to enable quick querying. MongoDB is a popular NoSQL database that stores data in BSON (Binary JSON) files, the schema-less model allows for good scalability and high performance in read and write operations [18].

REST Endpoints and Service Logic

The solution uses a.net core API, the API was built in visual studio using C#. The API provides the frontend of the solution with a variety of REST endpoints, these endpoints can be used to retrieve, create, delete, edit, and update data. The API uses a key to authenticate and control access to the service.

When registering a new transaction, the mobile application POSTs a custom signature object to the dedicated transaction REST endpoint, a signature object contains the shipment Id, new business Id, and current GPS coordinates of the user. The service logic uses the signature object to create a new transaction, during this process it uses the Google maps API service to calculate the distance the shipment has travelled. The shipment information contains the loading address, and the signature object has the current GPS coordinates, with the total distance travelled calculated the solution uses the vehicle details stored in the shipment information to estimate the amount of Carbon produced during transportation.

Smart Tags: QR codes

The solution uses QR codes to register transactions. The application will generate a QR code whenever a new shipment is created, the code represents the encoded shipment Id. The QR code can be scanned by the application which causes the shipment to be transferred into the inventory of the new business. The solution also uses QR codes to display a products information to consumers, the application can generate codes which represent an encoded webpage URL + product's message Id, this code can be scanned by a mobile device's camera which will then open a webpage displaying the history of a product.

Mobile Application

The mobile application is cross platform, created using React Native, the application interacts with the developed REST endpoints to provide functionality to the user. The app is used throughout the supply chain to log transactions, manage batches, and input data. Throughout application development the usability of the app was considered, a variety of UX techniques were used such as user personas and user journeys. It is important that the project takes Agri-workflows into consideration in order to develop a solution that is low friction for farmers. Interviews with farmers and on farm visits were used to shape the look of the application, it was important to farmers that the application be simple and quick to use (Fig. 2).

Fig. 2. Mobile application and QR code.

Sensor Integration

A smart environment uses IoT devices and data collection to enhance functionality and reduce repetitive tasks. Modern farms already implement a variety of IoT sensors, many farms across the island of Ireland use humidity and temperature sensors in grain stores, some use soil and moisture sensors in fields, and a few use subdermal sensors in livestock. The data collected from these sensors is fragmented across multiple computer systems and applications, by automatically uploading the data to the blockchain we can create smart contracts that provide additional functionality and reduce the amount of repetitive data inputs across the supply chain.

4   Agri-food Use Cases

Through consultation with key stakeholders, this study has identified a number of use cases for a secure supply chain solution in agrifood, one of these use cases is in the oats supply chain. To understand the current supply chain activities occurring in the agri-food sector, a process map of porridge oats has been drawn, this is outlined in Fig. 3 below. This process map has been outlined through secondary research and interviews with key stakeholders, such as farmers. Throughout all the stages data is having to be gathered and inputted. The current systems used to track and store information on food provenance is limited [8, 9].

4.1   Challenges

There are currently several challenges faced by supply chain actors. Firstly, much of the data is not shared between businesses due to interfaces not corresponding with each other, this is due to the various centralized and stand-alone information systems within each organization [13], this creates "information islands" [9], and is particularly noticeable

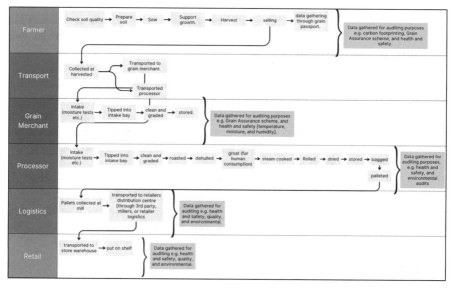

Fig. 3. Oats supply chain process map [9–12].

in the grain industry. Businesses not sharing data is problematic because it decreases efficiency and transparency.

Due to the physical nature of the documentation, there is a higher chance of the paperwork not being submissible. This may be through getting wet, damaged, or lost. Particularly challenging in the grain industry as there is a large volume of paperwork gathered for each load. Furthermore, for assurance schemes and certifications data is being entered after the activity is carried out, with some entering data the day before an audit, decreasing credibility. These challenges that emerge were also highlighted by Zhang *et al.* (2020) [9] and Francisco and Swanson (2018) [14] stating that it hinders and limits traceability.

Finally, interviews have highlighted the lack of market incentive for environmental sustainability, data collection, and sharing, due to limited financial benefit, this is because much of the data does not go passed the farm gate, it is not being provided to the end consumer. Sustainability is of vital importance with Heard and Bogdan's 2021 poll highlighting that 73% of people think sustainable food products are important [15]. However, it is challenging for consumers to understand sustainability and be informed about the environmental impact of their food [16]. There have been various scandals in the agri-food sectors in relation to counterfeit products and product quality, highlighting the importance of quality management [17].

4.2 How Does the Solution Address the Challenges?

The problems associated with data fragmentation will be fixed by bringing supply chain actors onto the one system and away from paper-based solutions, the developed smart supply chain enables product information to be passed between businesses intact, and without the need for reinputting data. When product is aggregated, such as when grain is siloed with grain from other suppliers, unlike the current system traceability isn't limited or hindered, the developed system keeps track of the origins of the products during mixing, so information is not lost.

To improve transparency in the supply chain, businesses have access to all data entered by previous supply chain actors, this enables business further down the supply chain to see the origin, the storage conditions, and the farming practices used to grow the products they purchase. The increased transparency along with the integration of a farm carbon calculator will make sustainability claims more trustworthy, grain stockholders will no longer be able to sell grain as "green" without proof nor be able to lie about a product's origin because it will be immutably stored on the DLT.

Finally, the products farmers grow will have increased marketability, with consumers willing to pay more for environmentally friendly produce [3], the ability to prove the origins, carbon footprint, and sustainability of produce, and then display that to the consumer through a QR code should increase the desirability of the product.

5   Future Work

The solution intends to have second heavyweight DLT to work alongside the lightweight DAG (see Fig. 1), this DLT which will be a traditional blockchain will be implemented to provide simulation and intelligent orchestration of the supply chain via smart contracts and chain code. This intelligent orchestration will be able to provide value added services such as the ability to dynamically calculate the most optimal supply chain configurations and provide recommendations to supply chain actors nominating substitution suppliers when needed. This will be especially useful to all buyers along the supply chain in times of shortage or sourcing issues.

The current solution is to be piloted on test sites at the end of the year (2023). Piloting the solution will help identify any challenges associated with adoption of the solution, and any issues with the technology, which will lead to future development. The data collected during this pilot will be used in a consumer trial, the trial will identify if the addition of a QR code on a product effects consumer purchasing behavior, then this data will be analyzed to get better understanding of any financial benefits of adopting the solution.

6   Conclusion

This study identified problems in the agri-food supply chain through the use of workshops and interviews with stakeholders, the current agri-food supply chain is full of separate unintegrated technologies, this is problematic because it is causing data fragmentation. Data fragmentation in the supply chain enables malicious actors to lie about and manipulate the origins of a product, an example of this is the 2013 horse meat scandal which plagued all of Europe. To ensure a products origin the supply chain needs to be traceable and auditable, but inefficiencies in the current supply chain means that this isn't always possible [8], events like the 2018 strawberry contamination in Australia, when needles hidden in strawberries caused injures show us that traceability is necessary. Now climate change has brought about a desire for sustainable products [3], traceability isn't enough, consumers don't fully trust existing quality labels, and claims of greenwashing show that there is a need for transparency. The solution this study developed helps tackle the problems identified in the current agri-food supply chain, the developed solution addresses these problems through the use of distributed ledger technologies and the facilitation of transparency through QR codes. The piloting of the solution at the end of 2023 will help identify any challenges associated with adoption of the solution, and any issues with the technology.

References

1. United States is Europe's main soya beans supplier with imports up by 112% (2019) European Commission. https://ec.europa.eu/commission/presscorner/detail/en/IP_19_161. Accessed 22 July 2023
2. Morris, J. (2022) Most people are worried about climate change – but what are they willing to do about it?, YouGov. https://yougov.co.uk/topics/politics/articles-reports/2022/11/22/most-people-are-worried-about-climate-change-what-. Accessed 22 July 2023
3. Nguyen, H. and Dsouza, R. (2021) Global: Consumer willingness to pay for environmentally friendly products, YouGov. https://yougov.co.uk/topics/consumer/articles-reports/2021/04/29/global-willingness-pay-for-sustainability. Accessed 22 July 2023
4. Salah, K., Nizamuddin, N., Jayaraman, R., Omar, M.: blockchain-based soybean traceability in agricultural supply chain. IEEE Access (2019)
5. Shahid, A., et al.: Blockchain-based agri-food supply chain: a complete solution. IEEE Access **8**, 69230–69243 (2020). https://doi.org/10.1109/access.2020.2986257
6. Fleming, F.: Verifish blockchain technology goes live with Aldi, Verifish. https://veri.fish/verifish-blockchain-technology-goes-live-with-aldi/. Accessed 22 July 2023
7. Agricultural Commodities, Australia, 2015–16 (no date) Australian Bureau of Statistics. https://www.abs.gov.au/AUSSTATS/abs%40.nsf/Lookup/7121.0Main+Features12015-16. Accessed 22 July 2023
8. Caro, M.P., Ali, M.S., Vecchio, M., Giaffreda, R.: Blockchain-based traceability in Agri-Food supply chain management: a practical implementation. In: IoT Vertical and Topical Summit on Agriculture - Tuscany (IOT Tuscany), pp. 1–4 (2018)
9. Zhang, X., et al.: Blockchain-based safety management system for the Grain Supply Chain. IEEE Access **8**, 36398–36410 (2020)
10. Decker, E.A., Rose, D.J., Stewart, D.: Processing of oats and the impact of processing operations on nutrition and health benefits. Br. J. Nutr. **112** (2014)
11. Tesco: Eat Happy project: Oaty Porridge (2016). https://www.eathappyproject.com/online-field-trips/oaty-porridge/. Accessed 24 Jan 2023
12. Flahavans: Our Oats. Flahavans (2023). https://www.flahavans.ie/inside-flahavans/our-oats/. Accessed 24 Jan 2023
13. Saberi, S., Kouhizadeh, M., Sarkis, J., Shen, L.: Blockchain technology and its relationships to sustainable supply chain management. Int. J. Prod. Res. **57**(7), 2117–2135 (2019)
14. Francisco, K., Swanson, D.: The supply chain has no clothes: technology adoption of blockchain for supply chain transparency. Logistics **2**(1) (2018)
15. Heard, H., Bogdan, A.: Healthy and sustainable diets: consumer poll. Food Standards Agency (2021)
16. Van Bussel, L.M., Kuijsten, A., Mars, M., van 't Veer, P.: Consumers' perceptions on food-related sustainability: a systematic review. J. Clean. Prod. **341**, 130904 (2022). https://doi.org/10.1016/j.jclepro.2022.130904

17. Chen, S., Shi, R., Ren, Z., Yan, J., Shi, Y., Zhang, J.: A blockchain-based supply chain Quality Management Framework. In: 2017 IEEE 14th International Conference on e-Business Engineering (ICEBE) (2017, preprint). https://doi.org/10.1109/icebe.2017.34
18. Jose, B., Abraham, S.: Performance analysis of NoSQL and relational databases with MongoDB and MySQL. Mater. Today Proc. **24**, 2036–2043 (2020). https://doi.org/10.1016/j.matpr.2020.03.634

Dataset Generation and Study of Deepfake Techniques

Sergio A. Falcón-López, Antonio Robles-Gómez$^{(\boxtimes)}$ (ID), Llanos Tobarra(ID), and Rafael Pastor-Vargas(ID)

Universidad Nacional de Educación a Distancia, ETSI Informática, C/Juan del Rosal, 16, 28040 Madrid, Spain
`arobles@scc.uned.es`

Abstract. The consumption of multimedia content on the Internet has nowadays been expanded exponentially. These trends have contributed to fake news can become a very high influence in the current society. The latest techniques to influence the spread of digital false information are based on methods of generating images and videos, known as *Deepfakes*. This way, our research work analyzes the most widely used Deepfake content generation methods, as well as explore different conventional and advanced tools for Deepfake detection. A specific dataset has also been built that includes both fake and real multimedia contents. This dataset will allow us to verify whether the used image and video forgery detection techniques can detect manipulated multimedia content.

Keywords: Deepfake · Dataset Generation · Detection Techniques · Multimedia Manipulation

1 Introduction

Any news can nowadays travel around the world in a few minutes. If the news that is transmitted also contains false information, the consequences could reach the point of manipulating the opinion of society [6]. For instance, society could watch a fake video in which an election candidate is performing immoral acts or even a photo expressing symbols or ideals against a campaign. Not only the detection of this information is necessary, but it also has to be effective in time, since the generation of false evidence can even violate the integrity of people.

In addition to this, the consumption of information on the Internet is almost unlimited, mainly due to the increment of mobile devices usage around the world. These devices make possible to create a large amount of daily multimedia content at an incredible speed of propagation. Message applications, social networks, and news websites allow false content to spread easily. For this reason, it is necessary to continue investigating to detect false information as quickly and rigorously as possible.

In the world of cybersecurity, criminals never rest, and they are analyzing new technologies with which to satisfy their interests regardless of the consummation of some crimes in the process. Current applications are sometimes not capable of

J. Bravo and G. Urzáiz (Eds.): UCAmI 2023, LNNS 842, pp. 197–206, 2023.
https://doi.org/10.1007/978-3-031-48642-5_19

detecting these cutting-edge manipulations, since criminals also rely on forensic techniques. Therefore, it is important to continue with continuous research that allows, through the use of new techniques and technologies, to detect and contain the achievement of crimes.

Therefore, the main objective of this work is to study of Deepfake techniques for image and video manipulation. The use of tools capable of generating Deepfakes will be studied, as well as Deepfake-oriented detection techniques are also tested. A specialized Deepfake dataset has also been generated, including both fake and real multimedia content. This dataset will allow us to study deeply the manipulation of multimedia content.

This paper is organized as follows. In Sect. 2, the principal Deepfake generation and detection techniques are studied. After that, Sect. 3 details the repository created and the used tools for its generation. Finally, Sect. 4 ends with some conclusions and possible future work.

2   Deepfake Studies

2.1   Generation Techniques

The main techniques known for the generation of Deepfakes are based on solutions implemented in a basic way through GANs (Generative Adversarial Networks), and several variations. The research community especially distinguishes four different methods for the generation of fakes in humans, based on the main types of existing facial manipulations. According to [21], the following methods are identified: *Entire Face Synthesis*, *Identity Swap/Faceswap*, *Attribute Manipulation*, and *Expression Swap*.

The *Entire Face Synthesis* technique consists of creating non-existent complete faces. An approach can be found in [13]. There is another work [14], where faces of people who do not exist [10] are generated. The *Identity Swap/Faceswap* consists of replacing the face of a person in a video with the face of another person. The *Attribute Manipulation* technique is also known as face editing or facial retouching. An image in which a face appears is used as a source, and the manipulation consists of modifying some of the attributes of the face, such as the color of the hair and skin, generating an appearance of the face that is more aged or even more rejuvenated, and adding elements to the face such as glasses or other accessories. One example can be found in [12].

Finally, the *Expression Swap* also known as *face reenactment*, consists of modifying the person's facial expression. The most popular approaches are based on Face2Face [20], which uses computer graphics to get the first few frames of a video. From these frames, a temporary facial identity is obtained, and from there, all subsequent facial expressions are tracked. Another of the most prominent approaches, such as Neural Textures [19], is based on neural textures, carrying out a rendering approach that uses the original video data to learn a neural texture of the target person, including a rendering network.

Different tools capable of generating Deepfakes have been used in this work to cover the four generation techniques mentioned above: Entire Face Synthesis through the Dall-E2 [2] project; Identity Swap/FaceSwap via DeepFaceLab [3]; Attribute Manipulation with the FaceApp [1] mobile application; and Expression Swap using Avatarify [4].

2.2   Detection Techniques

As a second part of the review, a set of Deepfake detection tools are studied. First, we have explored the tools that use conventional anomaly detection techniques in images and videos. Specifically, the application *Forensically* [5] will allow us to apply different techniques based on blind methods since, when analyzing Deepfakes, we do not have data collections related to the image or video to analyze. Results of the set of techniques applied with this tool must be evaluated by a human agent, so they are left to the criteria and level of expertise of the forensic analyst. For the analysis of images and videos, the same process must be carried out by taking into consideration that, with the videos we have analyzed individually the frames that compose them. The process consists of using the different utilities offered by the tool for each content.

According to this, the *MantraNet* application is capable of detecting several manipulation techniques and even unfamiliar ones. As a result, this tool returns a map of manipulation probabilities for each pixel in an image to be checked. The code of the notebook that is provided in [7] repository has been adapted to our purposes [18]. It is not available for videos.

The *Image Forgery Detection with CNNs* tool, which also uses Deep Learning, detects tampering based on the use of copy-move, delete and splice techniques. The script provided by the tool has been adapted to read all images found in the directory at the same time. This script returns 1 if the image has been manipulated and 0 if the image is not considered as manipulated. The adapted script has been run with the images from our repository, using for each run a different model, available in the application repository. The details about the parameters are described in [16]. Since this application is not designed for video detection, the model with the best statistics was selected and applied to all video frames in the repository.

Two plug-ins designed for the forensic analysis tool, *Autopsy*, have also been used. One of the plugins is used to detect manipulated images and another one is used to detect videos. The plugin can be downloaded and installed by following the instructions in the repository [17]. Once the application is installed and the plugins are placed in the corresponding directory, Autopsy has been loaded to launch the plugins against our repository. To use the image plugin, we activated the module called "Detect photo manipulations". To launch the module to detect Deepfake in videos, we activate the module "Detect Deepfake videos". Once the fingerprinting is completed, this tool returns the probability of tampering for images. For Videos, the tool returns whether they are Deepfake or not. As an example, Fig. 1 shows a set of images with their probability of being manipulated, among other options.

The *MesoNet* tool is based on the technique shown in [11]. The Python code located in [8] has been used for our purposes. Two already trained models can

Fig. 1. A set of images with their probability of being manipulated in the Autopsy tool.

be found in the repository, both of which we are to use for testing: Meso-4 and MesoInception4. For image detection in MesoNet, its code has been modified so that it only analyzes images, and the displayed result has been customized. The required adaptations have also been made to analyze only the videos in the repository.

Finally, the *Deepware Scanner* application is an online tool that is based on various Deepfake video detection projects. This application implements various Deepfake detection methods for videos. All we have to do is go to the Deepware Scanner section and, then, upload the video we desire to analyze. Next, a report will indicate if it is Deepfake and the probability of detection of the different models used. An example is shown in Fig. 2.

3   Deepfake Dataset

3.1   Discovery Tools

A Deepfake consists of and its generation, the discovery of techniques capable of detecting anomalies in both images and video begins, which are often used in the forensic analysis environment. The first open tool that has been found and used for the generation of videos or images with Deep Learning has been DeepFaceLab [3].

One of the applications that has been used to generate images has been the FaceApp application on an Android phone [4]. The images that come in the demo have been used to generate fake images, since it is not allowed to test own images for free of charge.

New advances in the creation of images from descriptive texts have been explored, as occurs in the Dall-E2 project. This solution is based on an online tool, which can currently be accessed from [2]. An example of this tool can be observed in Fig. 3.

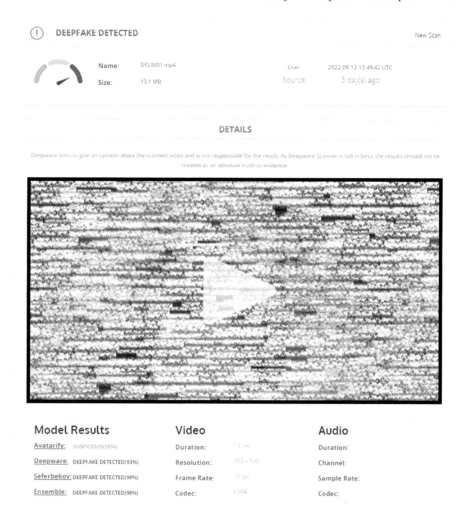

Fig. 2. Use of Deepware Scanner for an image to detect the Deepfake probability.

When accessing this website [10], the image of the face of a person who "does not exist" will directly appear, and that image has been generated through the implementation of the generation technique described in [14], using a StyleGan. This implementation has been used to download Deepfake images, generated entirely synthetically. For this, we have only accessed the web and downloaded the generated image.

Different fake video-generating apps generally use face swapping. So, it was decided to use Avatarify Desktop, which implements the expression transfer technique by obtaining facial expressions captured through a camera and transferring them to a face photo in real time. We can download it from the following project: Avatarify Desktop [1].

Fig. 3. Interface of the Dall-E2 tool by including some examples.

As there are several techniques for generating false multimedia content, the applications mentioned above have been used to create images and videos with different methods so that the techniques for detecting anomalies in images and videos can then be tested. For this reason, it has been decided to create a repository [18], in which there are images and videos generated with DeepFaceLab, images created with FaceApp, Dall-E2 and Thispersondoesnotexist.com, and finally videos generated with Avatarify. In order to discriminate whether the detection tools are capable of distinguishing between real images or videos, real multimedia content has been added from repositories used in other research on Deepfakes.

3.2   The Generated Dataset

In order to analyze the different Deepfake detection tools, a dataset has been created that includes images and videos generated by different applications in order to test the detection capacity of these tools [18].

For the generation of images, the applications *FaceApp*, *Dall-E2* and the images generated from project *thispersondoesnotexist* [10] and *DeepFaceLab* have been used. It has been decided to generate the content with several applications because each one generates a deepfake with different methods and produces images with different results and patterns. For example, DeepFaceLab performs Faceswap, and what it does is exchange the face generated in the original content, which suggests that the splice detection tools (slicing) could have a high probability of detecting the fake, while the images generated by Dall-E2 are completely synthetic, and it could happen that these detection techniques do not work for these cases. With the *FaceApp* application, they have been used as input images tagged as originals from the Celeb-DF repository,[15]. With *Thispersondoesnotexist* and *Dall-E2* completely synthetic images have been obtained. Lastly, we have used the frames generated by the application *DeepFaceLab*, which it uses to generate deepfake videos.

For the generation of videos, the application *DeepFaceLab* has been used, whose source and destination videos, labeled as real, have been selected from the dataset Celeb-DF, [15], Finally, *Avatarify* has been used to generate deepfake videos through the *expression swap* method. A selection of those images labeled as real from the dataset, Celeb-DF [15], have been used as source images. Next, using a video camera, the expression of a real face has been transferred to the image, and the process has been recorded to generate a final video captured with OBS Studio [9].

Table 1. Dataset features for generated Deepfake images. The repository is located in [18].

Prefix	Tool	Description	Example
FA	FaceApp	4 images from the demo images of the application itself	FA0001.jpg
FARC	FaceApp	11 images modified from real images of the Celeb-DF dataset [15]	FARC0001.jpg
D	Dall-E2	A total of 20 images: 10 images generated by Dall-E2 (in PNG format); and 10 images converted to JPG format	D0001.jpg
TPNE	Thispersondoesnotexist	9 web-generated images [10], which implements [14]	TPNE0001.jpg
DFL	DeepFaceLab	A total of 16 images: 8 frames taken from Deepfake videos generated with face swapping (in PNG format); and 8 converted to JPG format	DFL00006.png

Table 2. Dataset features for generated Deepfake videos. The repository is located in [18].

Prefix	Tool	Description	Example
DFL	DeepFaceLab	A total of 4 videos: 3 original videos from the Celeb-DF repository [15]; and a video generated from DeepFaceLab demo videos	DFL0001.mp4
AV	Avatarify	4 videos based on real images from the Celeb-DF repository [15]	AV0001.mp4

The generated multimedia content has been uploaded to the dataset and has been structured in two directories:

Table 3. Dataset features for real images. The repository is located in [18].

Prefix	Tool	Description	Example
RC	Celeb-DF	11 images from the "img_align_celeba.zip" in the Celeb-DF repository [15]	RC0001.jpg
RVC	Celeb-DF	6 images from videos of the Celeb-DF repository [15]: Celeb-real directory in"Celeb-DF.zip"	RCV00025.jpg
RDFL	DeepFaceLab	A total of 4: 2 images extracted from DeepFaceLa (in PNG format); and 2 images converted to JPG format	RDFL00009.jpg

Table 4. Dataset features for real videos. The repository is located in [18].

Prefix	Tool	Description	Example
C	Celeb-DF	3videos of the Celeb-DF repository [15]: Celeb-real directory in "Celeb-DF.zip"	C0001.mp4

– */images*: containing two sub-directories, "fake" containing the Deepfakes images generated, and "real", where the real images are located and pulled from the repository Celeb-DF [15].
– */videos*: Within this directory, there are two sub-directories called "fake" in which the generated Deepfakes videos are found, and "real", in which the real videos downloaded from the repository Celeb-DF [15] have been added.

Tables 1 and 2 specify the details of the generated fake files contained in the repository. The first one represents the generated images and the second one the generated videos. Tables 3 and 4 show the real images and videos included in our dataset. In each of the different tables, the nomenclature and content of the repository are described in order to easily identify the origin of the images and videos.

All the tables have the same structure. The first column describes the prefix used to name each image or video file. The second column indicates which tool has been used to generate them or, in the case of real images and videos, the repository where they come from. The third column indicates the number of images generated and some comments about their generation. Finally, there is an example column of the nomenclature used to name each image or video.

4   Conclusions and Further Works

Without the use of computational resources, it is now very difficult or almost impossible to detect the manipulation of multimedia content. Internet has also generated a new trend in which Deepfake contents are already part of our daily digital lives. This fact has led to the appearance of multiple mobile applications and tools for Deepfake generation, which are available to any consumer of technology. An additional problem is the possibility of immediate sharing and propagation of the fake multimedia content created. The integrity of human beings, the security of a country, or media manipulation to change the intention to vote in elections, among others could be violated.

The detection of these manipulations is absolutely necessary to be able to give a quick response to society and to deny or mitigate those attacks. Therefore, this work has first studied different generation techniques of Deepfakes, in order to raise awareness about the relevance of this research topic. Several conventional and advanced tools for Deepfake detection have been also explored that could be employed to mitigate the Deepfake problem on the Internet. A specific Deepfake dataset has also been created for further validation purposes with these detection techniques.

As future work, it is proposed to generate a more robust dataset with enough content for each type of existing generation technique and, additionally, study of conventional techniques and specific anomaly detection techniques.

Acknowledgements. Authors would like to acknowledge the support of the 2022–2023 SUMA-CITeL research project (096-043077), the 2023–2024 LearnIoTOnCloud research project (2023-PUNED-0018), the CiberGID UNED innovation group with the CiberScratch 2.0 project, as well as the I4Labs UNED research group with the 2022–2024 In4Labs research project (TED2021-131535B-I00), officially recognized by the Ministry of Science and Innovation. The authors also acknowledge the support of the E-Madrid-CM Network of Excellence (S2018/TCS-4307) from the Madrid Regional Government; and the SNOLA Network of Excellence (RED2018-102725-T) from the Spanish Ministry of Science, Innovation and Universities.

References

1. Avatarify. https://github.com/alievk/avatarify-desktop. Accessed 23 Oct 2023
2. Dall-E2. https://openai.com/dall-e-2/. Accessed 23 Oct 2023
3. DeepFaceLab. https://github.com/iperov/DeepFaceLab. Accessed 23 Oct 2023
4. FaceApp. https://play.google.com/store/apps/details?id=io.faceapp&gl=ES&pli=1. Accessed 23 Oct 2023
5. Forensically. https://29a.ch/photo-forensics. Accessed 23 Oct 2023
6. INCIBE: Deepfakes (In Spanish). https://www.incibe.es/aprendeciberseguridad/deepfakes. Accessed 23 Oct 2023
7. MantraNet. https://github.com/ISICV/ManTraNet. Accessed 23 Oct 2023
8. MesoNet. https://github.com/DariusAf/MesoNet. Accessed 23 Oct 2023
9. OBS Studio. https://obsproject.com/. Accessed 23 Oct 2023
10. This Person Does Not Exist. https://this-person-does-not-exist.com/. Accessed 23 Oct 2023
11. Afchar, D., Nozick, V., Yamagishi, J., Echizen, I.: MesoNet: a compact facial video forgery detection network. In: 2018 IEEE International Workshop on Information Forensics and Security (WIFS). IEEE (2018). https://doi.org/10.1109/wifs.2018.8630761, https://doi.org/10.1109%2Fwifs.2018.8630761

12. Choi, Y., Choi, M., Kim, M., Ha, J.W., Kim, S., Choo, J.: StarGAN: unified generative adversarial networks for multi-domain image-to-image translation (2017). https://doi.org/10.48550/ARXIV.1711.09020
13. Gonzalez-Sosa, E., Fierrez, J., Vera-Rodriguez, R., Alonso-Fernandez, F.: Facial soft biometrics for recognition in the wild: recent works, annotation, and COTS evaluation. IEEE Trans. Inf. Forensics Secur. **13**(8), 2001–2014 (2018). https://doi.org/10.1109/tifs.2018.2807791
14. Karras, T., Laine, S., Aittala, M., Hellsten, J., Lehtinen, J., Aila, T.: Analyzing and improving the image quality of styleGAN (2019). https://doi.org/10.48550/ARXIV.1912.04958
15. Li, Y., Yang, X., Sun, P., Qi, H., Lyu, S.: Celeb-DF: a large-scale challenging dataset for DeepFake forensics (2019). https://doi.org/10.48550/ARXIV.1909.12962, https://arxiv.org/abs/1909.12962
16. Rao, Y., Ni, J.: A deep learning approach to detection of splicing and copy-move forgeries in images. In: 2016 IEEE International Workshop on Information Forensics and Security (WIFS), pp. 1–6 (2016). https://doi.org/10.1109/WIFS.2016.7823911
17. Ferreiras, S.: Photo and video manipulations detector. https://github.com/saraferreirascf/Photo-and-video-manipulations-detector. Accessed 23 Oct 2023
18. Sergio A. Falcón: Deepfake Repository. https://github.com/oigres5/DeepfakeTFM_UNED. Accessed 23 Oct 2023
19. Thies, J., Zollhöfer, M., Nießner, M.: Deferred neural rendering: image synthesis using neural textures (2019). https://doi.org/10.48550/ARXIV.1904.12356, https://arxiv.org/abs/1904.12356
20. Thies, J., Zollhöfer, M., Stamminger, M., Theobalt, C., Nießner, M.: Face2face: real-time face capture and reenactment of RGB videos (2020). https://doi.org/10.48550/ARXIV.2007.14808, https://arxiv.org/abs/2007.14808
21. Tolosana, R., Vera-Rodriguez, R., Fierrez, J., Morales, A., Ortega-Garcia, J.: DeepFakes and beyond: a survey of face manipulation and fake detection (2020). https://doi.org/10.48550/ARXIV.2001.00179, https://arxiv.org/abs/2001.00179

Understanding the Association of Driving Safety and Visual Behaviors Collected Through Smart Sensing Technology

Ernesto M. Vera-Uribe$^{(\boxtimes)}$ [ID], Josué S. Armenta [ID], and Marcela D. Rodríguez [ID]

Engineering Faculty, Universidad Autónoma de Baja California, 21100 Mexicali, BC, México
{ernesto.vera,josue.armenta,marcerod}@uabc.edu.mx

Abstract. Driving safety is a concern primarily in young and older drivers. Even though most traffic accidents happen due to human factors, studies do not consider visual behavior variables to model driving safety. In this study, we used the Intelligent Multimodal Monitoring System to Infer Points of Visual Attention (SiMIPAV) to collect data from 10 young drivers to analyze the feasibility of using visual behaviors in assessing driving safety. To this end, participants' driving performance was evaluated using on-road driving tests and cognitive measures. We see the potential to generate models that predict cognitive maturity to drive from visual behavior properties. Specifically, the frequency of on-road visual behavior could be used to determine scanning skills.

Keywords: driving performance · wearable sensors · correlation analysis

1 Introduction

Driving is a complex activity that requires several skills like lane keeping, speed control, distance monitoring, and environmental scanning [1]. These abilities are related to drivers' cognitive skills, such as decision-making, mental status, executive functioning, attention, memory, and visuospatial skills [1]. According to the Insurance Institute for Highway Safety (IIHS), older adults and young adults (ages 16–24) are more likely to be involved in a fatal crash than any other age group [2]. Young adults are a high-risk group for reasons like distracted driving [3], inexperience [4], and their cognitive skills not fully developed [5]. The most reliable method to measure the driving safety is evaluating it under naturalistic conditions [6]. Another approach is to apply cognitive measurement instruments such as the TMT-A, TMT-B, Digit Span, and Reaction Time, among others [7]. The problem is that there is no consensus on which has the best predictive capacity, so the recommendation is to use a combination of tests assessed by a clinical specialist [7]. On the other hand, inattention and deficits in visuospatial perception are associated with poor driving performance [8]; for instance, performing tasks with high visual-manual demand, such as eating or using the telephone, substantially increases the risk of accidents, affecting young and elderly more than middle-aged drivers [9]. With the above, we hypothesized that *visual behavior is associated with poor*

J. Bravo and G. Urzáiz (Eds.): UCAmI 2023, LNNS 842, pp. 207–213, 2023.
https://doi.org/10.1007/978-3-031-48642-5_20

driving performance. We collected data from 10 young drivers using SiMIPAV, an intelligent sensing system validated to collect drivers' visual behaviors [10]. In addition, we administered neuropsychological tests to assess driving-related cognitive skills [5]. This study determines the feasibility of producing prediction models of driving performance.

2   Method

2.1   Study Procedure

We conducted a study with 10 participants in the age range of 16–29 years. They were active drivers with valid driving licenses. Each participant was compensated $300 pesos (around 19 USD dlls). They or their parents (for drivers under 18 years old) provided oral and written consent. First, we gathered data on driving experience and cognitive skills by administering paper-based neuropsychological tests [5]. Then, we measured participants' simple and peripheral reaction times using computerized versions of these tests. Finally, they drove in their cars for approximately 25 min (see Fig. 1b) while using SiMIPAV (Fig. 1a) to collect their visual behaviors [10]. Also, a driving instructor from the municipal transit department accompanied them to evaluate their on-road driving performance using eDOS, a valid scoring system [11].

Fig. 1. Driving setting: a) SiMIPAV's devices: (1) Yost Labs 3-SpaceTM (2), Google Pixel cell phone, (3) tablet for video recording the sessions; b) driving route followed by participants.

2.2   Cognitive Skills Measures

Most of the tests were paper-based (see Table 1), e.g., the Montreal Cognitive Assessment (MoCA) [12], Trail Making Test (TMT), parts A and B [13], Digit Symbol Substitution (DSS) test [14], Digit Span Forward (DS-F) and Backward (DS-B) tests [15] and Clock Drawing Test (CDT) [16]. To assess the simple reaction time (SRT), we implemented a computerized tool that presents a visual stimulus (a red triangle) on a monitor (see Fig. 2a). Participants were required to say "EMA" when detecting it within 5 s to be valid [17]. We implemented the Peripheral Reaction Time (PRT) tool [17] to present 2 triangles while using a driving simulator (carla.org) (see Fig. 2b).

2.3 On-Road Driving Assessment

We implemented an assessment tool based on the Driver Observation Schedule (eDOS) to get an objective measure of driving safety [11]. It runs on any mobile device with Android version 6 or later. As presented in Fig. 2b, the eDOS considers the level of complexity of the environment in which maneuvers took place, such as the intersection type and the speed zone. Thus, the maneuver: "Right turn at a nondirectional intersection on the residential street" has a lower complexity and, therefore, a lower score than "Right left across traffic at nondirectional intersection on a major road." Besides the maneuver, 13 different errors may be registered [11].

Table 1. Neuropsychological and on-road measurements used to collect data.

Measurement	Brief description of the data collected	Unit	M (SD)
MOCA [12]	Attention and concentration, executive functions, memory, language, visuospatial abilities, conceptual reasoning, calculation and orientation	Score (0–30)	25.5 (4)
TMT-A [13]	Processing speed	Seconds (s)	28.78 (5.05)
TMT-B [13]	Ability to switch attention	Seconds (s)	76.22 (36)
DSS [14]	Motor speed, attention, and visuo-perceptual functions	Correct responses	55.5 (6.31)
DS-F, DS-B [15]	Working memory	Correct responses	11.9 (3.25)
CDT [16]	Impairment	Score (0–10)	8.5 (2.46)
SRT [18]	36 measurements were taken with the SRT tool	Mean in s	1.78 (0.75)
PRT [17]	6 measures were with the PRT tool	Mean in s	2.09 (0.6)
eDOS [11]	Driving safety based on maneuvers-environ. Errors	Score > 0	11.8 (6.46)
Experience	Years (y) of driving experience	Mean in y	2.9 (1.64)

(*continued*)

Table 1. (*continued*)

Measurement	Brief description of the data collected	Unit	M (SD)
D-on, D-off, F-on, F-off [10]	Glance duration to on-road (D-on), off-road (D-off)	Mean in s	3.04 (0.98), 3.24 (1.39)
	Glance frequency to on-road (F-on), off-road (F-off)	Quantity of glances	234.2(77.19), 187.1 (90.45)

As depicted in Fig. 2a, the eDOS tool consists of three modules. The *Trip Data Collector* provides the User Interfaces to register participants' maneuvers and errors (see Fig. 2b). It supports two modes of use: vehicle mode and simulator mode. The main difference between these is that the vehicle mode activates the *GPS Processing Unit* to track its location and estimate driving speed. In the simulator mode, they are estimated by the driving simulator. The *Score Estimator* module calculates the weighted eDOS and road complexity scores to be stored in the *Proxy Storage*.

2.4   Visual Behavior Measures

Glance duration and quantity of glances are relevant properties to understanding visual behaviors [8] and, therefore, were used as measures for testing our hypothesis. Glance duration, or eye fixation, is when the eyes stop moving to stay at a specific focal point [8]. Glance duration has been used to determine how a focal point demands processing time and paying attention [10]. The quantity of glances refers to the number of times drivers take their eyes off a focal point and back again. These two properties of visual behaviors were measured for on-road and off-road using the SiMIPAV system [10]. It tracks drivers' head orientation, which is a good proxy for estimating eye gaze [19]. It consists of two different sensors to compensate for their limitations. One is a wearable sensor on the head, a Yost Labs 3-Space™ Mini Bluetooth device (yostlabs.com), which integrates inertial sensors and advanced filtering algorithms to estimate movements. The second sensor is a mobile phone camera. A Facial Recognition Application (FRA) was implemented using the CameraView v2.6.1 software to capture video stream frames and the SDK ML Kit Firebase to estimate the Euler angles of head movements. These CV open-access libraries are limited in estimating Euler angles in pitch movement. This limitation was compensated with the inertial sensor mounted on the drivers' heads, which provides reliable estimations on pitch and roll but not on the yaw movement (Euler Z) since the vehicle turns to left and right affect the head orientation's estimate on the vertical Z-axis (yaw). Both sensors capture data at 50 Hz, which are classified into on-road and off-road visual points of attention by the k-NN algorithm with an accuracy of 98%, as explained in [10]. The k-NN was implemented using Android Studio v3.5 with the API 19 (Kit Kat) and deployed on a Google Nexus tablet.

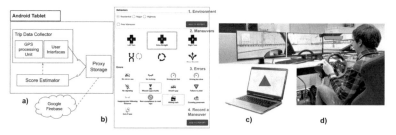

Fig. 2. Data collection tools: eDOS architecture (a) and user interface (b); SRT (c) and PRT (d).

3 Results

Descriptive statistics for all measures are presented in Table 1, and Fig. 3 shows some of the associations that were statistically significant at p set at .01 (99%), .05 (95%) or .1 (90%), using two-tailed tests. To this purpose, we estimated the Pearson correlation coefficients (r). As expected, we found associations between variables of the same measurement category, for instance, PRT is associated with PRT-s (r = 0.83; $p < .01$) and PRT-r (r = 0.81; $p < .01$), and SRT with PRT-l (r = 0.76; $p < .05$). Regarding visual behaviors, F-on and D-off (r = −0.56; $p < .1$) and F-off and D-on (−0.73; $p < .05$) are correlated. These results are in line with our previous results [10].

Related to neuropsychological measures, we found a correlation between DSS and TMT-B (r = −0.67; $p < .05$) as both assess similar attentional dimensions, and MoCA with CDT (r = 0.75; p < .05), which can be due to CDT version is part of the MoCA test, and that both are used to detect cognitive impairment [16]. On the other hand, MoCA and SRT (r = 0.58; $p < .1$) and Mem and TMT-A (r = −0.58; $p < .1$) are associated. Finally, MoCA is the only test correlated with age (r = 0.55; $p < .1$). This result may be due to cognitive skills being fully developed between 22 and 27 years [5].

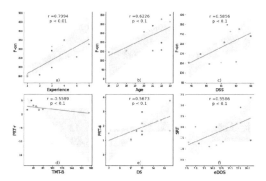

Fig. 3. Associations found between measures that were statistically significant.

Regarding visual behavior measurements, F-on has a strong positive correlation with driving experience (r = 0.79; p < .01) (see Fig. 3a) and a moderate positive correlation with age (r = 0.62; $p < .1$) (see Fig. 3b) and with the DSS (r = 0.58; $p < .1$) (see

Fig. 3c). The above indicates that the maturity of young drivers in visuospatial and attentional skills related to scanning (DSS test) impacts on the visual behavior (F-on) related with distractions or scanning outside the on-road zone. Peripheral reaction time are moderately correlated with two cognitive tests, i.e., PRT-r with TMT-B ($r = -0.55$; $p < .1$) (see Fig. 3d), and PRT-s with DS ($r = 0.56$; $p < .1$) (see Fig. 3e). This makes sense since TMT-B assess divided attention, also evaluated with PRT. Finally, eDOS has a moderate correlation with SRT ($r = 0.55$; $p < .1$) (see Fig. 3f).

4   Conclusions

Our results confirm that age is associated with general cognitive performance (MoCA) and that driving experience is related to the frequency of gazing (F-on), which is correlated with the DSS test (scanning skills). Therefore, our preliminary conclusion is that the frequency of glances (F-on) can be one of the essential variables for predicting attention skills critical for safe driving. On the other hand, the peripheral reaction time (PRT) measurements are determinants of evaluating divided attention capacity. This leads us to explore the generation of models that consider visual scanning and the peripheral reaction time to predict attention skills necessary to drive safely. A limitation of this study is the lack of age variability; thus, including more drivers between 16 and 18 may help confirm or identify new correlations.

Acknowledgement. We thank to CONAHCYT for the scholarship granted to the first author.

References

1. Grundler, W., Strasburger, H.: Visual attention outperforms visual-perceptual parameters required by law as an indicator of on-road driving performance. PLoS ONE **15**, e0236147 (2020)
2. Fatality Facts 2021 Older people. https://www.iihs.org/topics/fatality-statistics/detail/older-people. Accessed 24 July 2023
3. Young, K.L., Lenné, M.G.: Driver engagement in distracting activities and the strategies used to minimise risk. Saf. Sci. **48**, 326–332 (2010)
4. Sheykhfard, A., Qin, X., Shaaban, K., Koppel, S.: An exploration of the role of driving experience on self-reported and real-world aberrant driving behaviors. Accid. Anal. Prev. **178** (2022)
5. Ledger, S., Bennett, J.M., Chekaluk, E., Batchelor, J.: Cognitive function and driving: Important for young and old alike. Transp. Res. Part F **60**, 262–273 (2019)
6. Kavouras, C., Economou, A., Liozidou, A., Kiosseoglou, G., Yannis, G., Kosmidis, M.H.: Off-road assessment of cognitive fitness to drive. Appl. Neurops. Adult. (2020)
7. Wolfe, P.L., Lehockey, K.A.: Neuropsychological assessment of driving capacity. Arch. Clin. Neuropsychol. **31**, 517–529 (2016)
8. Sun, Q.C., Xia, J.C., He, J., Foster, J., Falkmer, T., Lee, H.: Towards unpacking older drivers' visual-motor coordination: a gaze-based integrated driving assessment. Accid. Anal. Prev. **113**, 85–96 (2018)
9. Guo, F., et al.: The effects of age on crash risk associated with driver distraction. Int. J. Epidemiol. **46**, 258–265 (2017)

10. Armenta, J.S., Andrade, A.G., Rodriguez, M.D.: An intelligent multi-sourced sensing system to study driver's visual behaviors. IEEE Sens. J. **21**, 12295–12305 (2021)
11. Chen, Y.T., Gélinas, I., Mazer, B.: Development of a weighted scoring system for the electronic driving observation schedule (eDOS). MethodsX **7**, 101099 (2020)
12. Pike, N.A., Poulsen, M.K., Woo, M.A.: Validity of the montreal cognitive assessment screener in adolescents and young adults with and without congenital heart disease. Nurs. Res. **66**, 222–230 (2017)
13. Reitan, R.M.: Trail Making Test Manual for Administration and Scoring. Reitan Neuropsychology Laboratory. Scientific Research Publishing (1992)
14. Jaeger, J.: Digit symbol substitution test. J. Clin. Psychopharm. **38**, 513–519 (2018)
15. Diamond, A.: Executive functions (2013).www.annualreviews.org. https://doi.org/10.1146/annurev-psych-113011-143750
16. Freund, B., Gravenstein, S., Ferris, R., Burke, B.L., Shaheen, E.: Drawing clocks and driving cars: use of brief tests of cognition to screen driving competency in older adults. J. Gen. Intern. Med. **20**, 240–244 (2005)
17. Stinchcombe, A., Gagnon, S., Zhang, J., Montembeault, P., Bedard, M.: Fluctuating attentional demand in a simulated driving assessment: the roles of age and driving complexity. Traffic Inj. Prev. **12**, 576–587 (2011)
18. Simple Reaction Time Task. https://scienceofbehaviorchange.org/measures/simple-reaction-time-task/. Accessed 25 July 2023
19. Metz, B., Schömig, N., Krüger, H.P.: Attention during visual secondary tasks in driving: adaptation to the demands of the driving task. Transp. Res. Part F **14**, 369–380 (2011)

Data Science

Clustering ABI Patients for a Customized Rehabilitation Process

Alejandro Moya[1] ⓘ, Luis Zhinin-Vera[1] ⓘ, Elena Navarro[1](✉) ⓘ, Javier Jaen[2] ⓘ, and José Machado[3] ⓘ

[1] LoUISE Research Group, Computing Systems Department, University of Castilla-La Mancha, 02071 Albacete, Spain
{Alejandro.Moya,Luis.Zhinin,Elena.Navarro}@uclm.es
[2] Instituto Universitario Mixto de Tecnología de Informática, Universitat Politècnica de València, Valencia, Spain
fjaen@upv.es
[3] Centro Algoritmi/LASI, University of Minho, Braga, Portugal
jmac@di.uminho.pt

Abstract. Acquired Brain Injury (ABI) is a medical condition resulting from injury or disease that affects the functioning of the brain. The incidence of ABI has increased in recent years, highlighting the need for a comprehensive approach to treatment and rehabilitation to improve patients' quality of life. Developing appropriate therapies for these patients is a challenging task because of the wide diversity of effects and severity they may suffer. This problem exacerbates the complexity of designing the rehabilitation activities, which is a time-consuming and complicated task that may cause poor patient recovery, if such activities are poorly designed. In order to overcome this problem, it is common practice to create groups of patients with similar complaints and deficits and to design rehabilitation activities that may be reused internally by such groups, facilitating comparative analyses. Usually, such grouping is conducted by specialists who may neglect to detect commonalities due to the huge amount of information to be processed. In this work, a clustering of ABI patients is performed following a systematic methodology, from preprocessing the data to applying appropriate clustering algorithms, in order to guarantee an adequate clustering of ABI patients.

Keywords: Acquired Brain injury · clustering · rehabilitation

1 Introduction

Over the past decade, the number of people living with disabilities has increased. Several factors have contributed to this increase, both globally and within countries. First, medical and technological advances have improved the detection and diagnosis of disabilities, leading to an increase in the identification and registration of people with disabilities. Furthermore, improved access to health services and medical care has contributed to the survival of people with severe medical conditions or acquired disabilities, which has increased the number of people living with disabilities. The ageing of the population

© The Author(s), under exclusive license to Springer Nature Switzerland AG 2023
J. Bravo and G. Urzáiz (Eds.): UCAmI 2023, LNNS 842, pp. 217–228, 2023.
https://doi.org/10.1007/978-3-031-48642-5_21

is another factor influencing the increase in the number of people with disabilities. As life expectancy increases, so does the likelihood of developing age-related disabilities such as visual impairment, hearing impairment, or reduced mobility. The Convention on the Rights of Persons with Disabilities (CRPD) in 2022 [20] emphasizes the need for protection and care for people with disabilities, as their fundamental rights are often denied. Within this group, people with Acquired Brain Injuries (ABI) are particularly relevant.

ABI is caused by a bump, blow, or jolt to the head or a penetrating head injury. It disrupts the normal functioning of the brain. Several factors, including brain tumours and blood vessel degeneration, can also cause ABI [13]. The number of people affected by ABI has increased significantly in recent years, making it a major societal concern that explains why some researchers have called it the "silent epidemic" [7]. ABI can cause mild, temporary problems or more severe neurological impairment, particularly in children whose brains are still developing [2, 16].

One of the main challenges in rehabilitating people with ABI is the wide range of deficits they can experience, depending on the specific area of the brain that has been damaged [1]. These deficits can be categorised into three main groups [9]: (1) physical/motor (10 different deficits), (2) cognitive (36 different deficits), and (3) behavioural/emotional (5 different deficits). Both the diversity and severity of deficits that ABI patients may suffer make the development of rehabilitation activities a challenging effort. As noted in [10], there is an inherent variability among ABI patients that results in a wide range of characteristics that need to be considered when designing rehabilitation activities for them. Therefore, there is a clear need to customize such rehabilitation activities according to patients' needs.

A proposal for automating such customization is presented in [10]. Here, the design and implementation of a genetic algorithm to automatically generate physical/cognitive rehabilitation activities considering patients' physical/cognitive characteristics are presented. A fundamental aspect of this algorithm is the experts' knowledge about which features are better rehabilitating which deficits and which ones are worse. This knowledge is specified as a *contribution map* that relates features used for defining rehabilitation activities with deficits. This contribution map is currently hard coded in the algorithm without any chance of customizing it according to the patient' outcomes. Ideally, it should be defined for every patient. However, this would be an endless task due to both the number of patients and the inherent complexity of such a task. Therefore, an interesting alternative is to group patients according to their damage level so that their expected outcomes are similar. Then, different contribution maps may be defined for each of these groups, facilitating their evolution. In order to carry out such grouping, a clustering algorithm is presented and applied in this paper.

This paper describes the non-straightforward process to be followed to achieve successful clustering, carried out with actual data of patients extracted from the Federal Interagency Traumatic Brain Injury Research (FITBIR) Informatics System [4] (Study: Transforming Research and Clinical Knowledge in Traumatic Brain Injury (TRACK-TBI) – Adult [5]). In this respect, we believe this work is of interest for technologists and practitioners working in the field of ABI, as it illustrates the numerous challenges to be addressed when clustering actual data from ABI patients.

The paper has been structured as follows: Sect. 2 presents the work related to the clustering of ABI patients. Section 3 presents the clustering proposal that has been conducted, from the collection of a dataset of ABI patients to the pre-processing, clustering, and post-processing of the results obtained. Finally, Sect. 4 outlines the conclusions and future work.

2 Related Work

The use of clustering techniques is common in the medical world. The utilization of clustering techniques in the field of medicine offers numerous benefits. For instance, they have been applied for creating clusters of areas or villages impacted by COVID-19 in order to generate mapping results to identify which affected regions require vaccination urgently [19]. However, they have mainly applied for the classification of patients. As stated in [17], individuals with a particular disease often share a significant set of symptoms. Additionally, according to [15], clinical practice guidelines, which provide appropriate and validated treatment recommendations, are typically established for specific patient groups. Therefore, it is crucial to identify patients who exhibit similar features and/or diseases and group them together in clinically meaningful clusters. This enables the assignment of the most suitable therapies, enhances the accuracy of diagnoses, and contributes to an overall improvement in patient care.

A literature review on how to perform clustering on ABI patients was presented in [11]. This study showed the different techniques, from pre-processing the data to performing and evaluating the cluster itself, that should be followed to perform good clustering. Furthermore, the study evaluated how the clustering was performed on the different studies. As it was presented in that study, most of the articles analysed do not indicate how the pre-processing of the data is carried out, or if they do, they tend not to use the appropriate techniques, such as dimensionality reduction or data normalisation. Regarding the choice of the clustering algorithm to be used, all the papers tend to use the most traditional clustering algorithms, which are easier to carry out and interpret but may not be the most appropriate for the data being used. Furthermore, when it comes to reporting the execution parameters used in the clustering algorithms, many do not report them, or rely on human knowledge, or use the information provided by the execution of one clustering algorithm to execute another clustering algorithm, which may lead to obtaining groups of patients with ABI that are not entirely appropriate. In general, it was observed that clustering is carried out by personnel who are not experts in clustering and who do not follow all the steps, and the methodology adequate to guarantee the success of the groups obtained. Therefore, this paper shows how the clustering of patients should be carried out, from the pre-processing of the data to the obtaining of the groups and their quality assessment.

3 Clustering for ABI Patients

To carry out a clustering of any type of information, there are a series of steps to follow to ensure that the clustering has been carried out correctly and that the groups obtained offer relevant information for the user analysing it. The work of [14] indicates a series of linear steps to follow when performing clustering. These steps (see Fig. 1) are: (1) obtaining a dataset; (2) pre-processing the dataset; (3) executing the corresponding clustering algorithms; and (4) post-processing the results obtained. In the first step, it is essential that the raw data have been properly collected, and that we are clear about the meaning of each dimension of data and the possible values that each of these dimensions can take. As for the second step, the raw data obtained, in most cases, cannot be used directly by the clustering algorithms, it is necessary to perform a series of conversions and transformations on the raw data to adapt it to the requirements of the clustering algorithm(s) to be used. In the third step, it is essential to choose the clustering algorithm or algorithms to be used, depending on the shape of the data or the type of data, some algorithms will be more suitable than others. As for the fourth step, the analysis of the information obtained after clustering is fundamental. In this step, the groups of clusters obtained are analysed, with different graphs or statistical calculations in order to determine the quality of the groups, as well as their validity. Figure 1 shows the steps described by [14] and the choices that have been made for this work on ABI patient clustering. The following subsections detail each one of these steps.

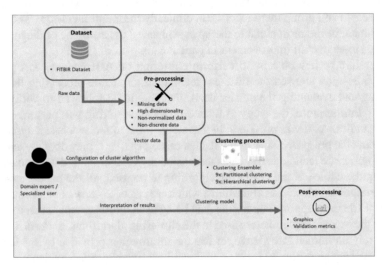

Fig. 1. Methodology applied for the clustering of patients with ABI. Based on [11]

3.1 Dataset

It is worth highlighting that Traumatic Brain Injury (TBI) is used to specify that the injury to the brain has been caused by an external force or trauma to the head, such as a fall, vehicle accident, or sports injury. In the literature, it is used frequently to distinguish between people with ABI and people with TBI. For this work, a dataset published, by Federal Interagency Traumatic Brain Injury Research (FITBIR) Informatics System [4] has been used. FITBIR was developed to enable the sharing of research data related to Traumatic Brain Injury (TBI). It has several modules that allow for raw data per patient as well as summaries and interpretations of the information. The main objective of FITBIR is to accelerate research in the area of TBI by allowing researchers who have access to the platform to obtain any data/information they need from any study and to perform re-analysis of the information, comparison of results, etc.

In this paper, the study entitled "Study: Transforming Research and Clinical Knowledge in Traumatic Brain Injury (TRACK-TBI) - Adults" [5] was chosen. TRACK-TBI is a public-private partnership of researchers, philanthropists, and industry leaders who aim to accelerate clinical research in TBI. Its primary goal is to create a comprehensive, high-quality TBI database that integrates clinical, imaging, proteomic, genomic, and outcome biomarkers. In addition, it will provide analytic tools and resources to improve the diagnosis and prognosis of TBI, as well as evaluate outcomes and compare the effectiveness and costs of tests, treatments, and services. The project collected and analysed detailed clinical data from 3000 subjects from 11 centres, covering the full spectrum of TBI.

Among all the existing information in the study, we have focused on those tests to assess the patients most commonly used in the literature, according to [11]. And, among all of them, after analysing the amount of information per patient existing in each of them, in this first approach, we focused on the Glasgow Coma Scale (GCS) test [21] which is used to objectively assess the degree of impairment of consciousness. This test assesses three aspects of responsiveness: (1) ocular, (2) motor, and (3) verbal.

3.2 Pre-processing

The pre-processing step is one of the most important stages in clustering. The data must be perfectly prepared to guarantee the success of the clustering algorithm, thus avoiding any possible noise that may arise. In the work of [11], a number of the most commonly used preprocessing techniques are indicated and should be taken into consideration:

- Missing data. At this stage, it is a matter of deciding which technique to apply for the treatment of missing data. Poor treatment of missing data can lead to errors or false results [22]. We proceeded to remove those attributes from the dataset where all values were null for all records existing in the dataset. In addition, for those records where there was some missing data in some attribute, due to the context of the problem, the technique of deleting the entire record was chosen because filling in those missing values may result in using completely erroneous information.
- Outliers. It is necessary to make an adequate treatment of those data that may be outliers in order to avoid a lack of grouping of patients. In the case of data obtained from FITIBIR, the platform indicates the range of values that each attribute can take. Therefore, the existence of outliers is only reduced to those attributes that take numerical values. After using different information visualisation methods, such as histograms, no significant outliers were detected.
- High dimensionality. A high dimensionality of the information may lead to the use of irrelevant information for the study, and therefore, to inappropriate groupings. In the present study, there are a total of 36 attributes, of which only 3 were chosen for the first preliminary tests, therefore, the use of any technique that reduces dimensionality at the attribute level was ruled out. The fact that only three attributes were chosen

corresponds to the three attributes that refer to the raw results obtained by the GCS test, relating to eye, motor, and verbal responses. The remaining attributes may be interesting to take into account in future studies, since they refer to attributes of a demographic, temporal, or other nature. Regarding the possible high dimensionality related to the number of records, after visualising the dataset, it was found that there was more than one record of the same patient for a specific day. In order to avoid this problem, it was decided to group all the records of the same patient and of the same day into one, applying an average to the numerical variables.

- Non-normalized data. All the data for the different attributes must be placed on the same scale, in order to ensure that all the attributes have the same importance in the execution of any clustering algorithm. There are several normalisation techniques, among all of them, in this work we used the "z-score" technique, which normalises by subtracting the original value minus the mean of the whole data and dividing it by the standard deviation of the data.
- Non-discrete data. Discretisation of the information is necessary in most clustering algorithms. In the case of the dataset used, it stores the information on certain attributes using descriptive text. This cannot be used in a clustering algorithm, therefore, the transformation of this text to a numerical value was carried out, this transformation is specified in the FITBIR documentation. In addition, after this transformation, it was observed that certain values collected were missing, but were listed as a "category". Therefore, this information was transformed into a "missing value" and the missing data processing techniques indicated above were applied.

After applying all the above pre-processing techniques, the data cloud is represented in Fig. 2.

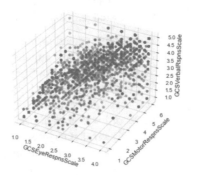

Fig. 2. Visualization of pre-processed data.

3.3 Clustering Process

There are a variety of clustering algorithms that can be used for the clustering of ABI patients. Each clustering algorithm has its strengths and weaknesses, which means that the choice of one algorithm over another is not a trivial task and has to be decided on a judgemental basis. According to the study in [11], the most widely used types of clustering algorithms in the literature are hierarchy-based clustering algorithms and partition-based clustering algorithms. As discussed, the choice of these types of algorithms is made mainly because of their simplicity of execution, and because they are the most frequently used algorithms. This way of choosing the clustering algorithm may be problematic because the algorithm selected may not be the most appropriate for the type of data being used. Furthermore, it should be noted that, given a set of data, different clustering algorithms, or the same clustering algorithm with different input parameters may result in totally different groupings.

In order to mitigate this problem, there is a growing tendency, known as *Cluster Ensemble* [18], to use the execution of different clustering algorithms with different configurations in order to obtain a final solution from all of them. In this way, a more robust and high-quality solution is achieved, reducing the weaknesses that algorithms may exhibit when used in isolation. This explains why this technique was exploited in this work.

According to the literature review on Clustering Ensemble presented in [23], there are four ways to carry out the execution of clustering algorithms. These different ways are: (1) using different clustering algorithms; (2) using the same clustering algorithm and using different initialisation parameters; (3) using different subsets of data for the different clustering algorithms; and (4) using different subsets of attributes for the different clustering algorithms. For this work, we have opted for the first and second option, because the data size does not allow an effective use of options three and four. The two algorithms most frequently used in the literature for clustering patients with ABI, K-means and agglomerative hierarchical, were used in this work, which was run several times. The initialisation parameter that was changed from run to run is the number of clusters desired, using values between [2, 10] because, after analysing the dataset, this is the expected range for the number of clusters. The main goal was to obtain robust and high-quality solutions taking advantage of the strengths of these clustering algorithms and reducing the limitations that both algorithms have if they are executed individually.

After the execution of the different clustering algorithms specified in the ensemble, it is necessary to obtain a final solution that encompasses the results obtained by the clustering algorithms executed individually, which is called a consensus function. There are different consensus functions, among which the following stand out:

- Co-occurrence matrix: the co-occurrence matrix of all the individual solutions obtained in the ensemble is calculated, and a hierarchical clustering of the co-occurrence is performed, treating it as a similarity matrix.
- Majority vote: based on the work of [6], this algorithm assigns clusters to the same group if they are clustered together at least 50% of the time. It joins clusters greedily, with evidence that at least one pair of elements from two different clusters is placed together most of the time.
- Mixture model: creates the clusters according to probability maximisation based on the clustering solutions observed in the whole set.

For this work, we have chosen to run the three consensus functions so that the different results obtained can be compared according to the consensus function chosen.

3.4   Post-processing

In the last step, the analysis of the results obtained is carried out, with an interpretation of the clusters formed and a mathematical analysis of the quality of the clusters obtained. As mentioned above, three different consensus functions have been used to obtain a final solution. These solutions are illustrated in Fig. 3.

As can be seen, the Majority Vote and Mixture Model solutions are very similar, giving a total of three clusters. In contrast, for co-Occurrence Matrix, a total of five clusters are obtained. For Majority Vote, it can be observed that it forms more precise groups when it comes to distinguishing whether a patient is in "good", "fair", or "bad" condition. Three well-spaced clusters are formed in the data space. In contrast, in the groups formed by the Mixture Model, a large group of ABI patients with a "fair" situation was created, and then the other two groups corresponded to extremes, corresponding to patients who are in very "good" or very "bad" conditions. In the case of co-Occurrence Matrix, the interpretation is more complex, since a total of five clusters have been formed. Distinctions are made in this solution, in the clusters based on of the visual response attribute.

Continuing with the interpretation of the groups formed in Majority Vote and Mixture Model, the first cluster (Majority Vote - Red; Mixture Model - Blue), is a cluster formed mainly by patients with very impaired motor and verbal responses, with visual responses not being relevant for this group. The second cluster (Majority Vote - Blue; Mixture Model - Green), corresponds to patients with very little impairment of their motor, visual, and verbal responses, with most patients in this group having a high score on the GCS test. Finally, the last cluster is formed by patients with moderate motor and verbal response impairments.

In order to evaluate how effective a solution is, there are a series of metrics that mathematically/statistically evaluate its goodness. Consideration must be given to the fact that, although mathematically or statistically good, this does not mean that the clusters formed make sense in the context where they are applied. Therefore, the knowledge of an expert will always be necessary to determine whether the clusters formed are valid in the applied context.

Table 1 shows the metrics run for the solutions obtained by the three consensus functions. A description of the metrics used in this work can be found in [3]. It should be noted that the metrics used correspond to internal validation metrics [8], i.e., without considering any external information, the clustering structure's viability is assessed by analysing the clustered data itself.

Looking at the result of the Silhouette Coefficient metric, we can see that the Majority Vote solution is ranked with the highest value, which means that the solution obtained has created clusters whose distance between them is significant, that is, they are clearly differentiated. The Co-occurrence Matrix solution has obtained the lowest value for this metric, as can be seen. On the other hand, if we check the R Squared metric, it can be observed that the co-occurrence Matrix obtains the highest score that is, it creates clusters that are more clearly different from each other. This shows us that the choice of one metric over another is not trivial and always has to be made with the knowledge of the context in which it is applied, with some metrics being more interesting than others. For this work, the best solution is the one offered by co-occurrence Matrix, since it

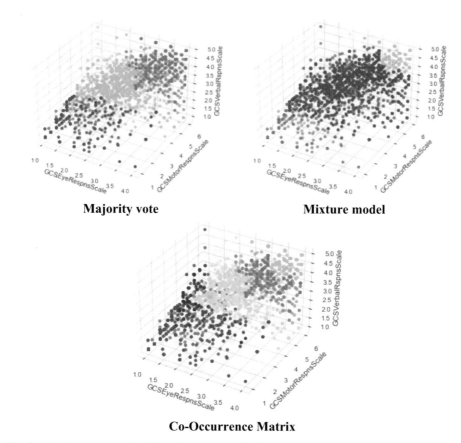

Majority vote   **Mixture model**

Co-Occurrence Matrix

Fig. 3. Final clustering of ABI patients using: (1) Majority vote, (2) Mixture model and (3) Co-Occurrence Matrix

offers a much more informative solution than the others, as shown in the previous Table. Moreover, the visualisation of the results shows that the clustering carried out using all the attributes offers a more accurate and useful classification for the experts.

4   Conclusions and Future Work

As can be seen, clustering patients with ABI is not a trivial task. Following the methodology described in this article ensures that the clustering algorithm has been executed correctly. In the first stages, it is necessary to have a suitable dataset with sufficient information so that, clustering can be performed in a correct and informative way. Then, a pre-processing of the data is conducted in order to guarantee that the data has been transformed, to meet the needs normally required by all clustering algorithms: the non-existence of outliers, the non-existence of missing data, normalisation, etc. For the clustering process stage, it is difficult to choose an algorithm according to the data, which has motivated us in this work to propose the design of an ensemble clustering. Our proposal

Table 1. Results of internal validation metrics

	Majority vote	Mixture model	Co-occurrence matrix
McClain-Rao Index	0.15410	0.15050	0.10746
Wemmert-Gancarski	0.0	0.0	0.0
C-Index	0.01295	0.03419	0.01009
Det_Ratio Index	22.72373	16.72992	57.31150
Ksq_DetW Index	95015924854.61777	129057182112.53427	84765035040.83104
Log_Det_Ratio Index	26892.55806	24256.08074	34857.59664
Point biserial	−1.81990	−1.68335	−1.69778
R squared	0.64672	0.65687	0.72182
Root mean square	0.47050	0.46370	0.41755
Silhouette Coefficient	0.70141	0.67279	0.60934
Trace_W index	6027.01708	7075.13191	4171.40956

includes the two traditional clustering algorithms most commonly used in the grouping of patients with ABI: hierarchy-based and partition-based clustering. These algorithms may not be the most suitable for the grouping of patients with ABI in some situations, but the collective performance of the ensembles renders them appropriate, as it mitigates the constraints that they experience. Finally, we have carried out the post-processing of the results, where it is necessary to take into account the context of the problem being addressed in order to be able to decide whether the groups formed are informative, and, in addition, the execution of objective metrics that evaluate the quality of the structure of the results obtained.

There are numerous future works that can be proposed. It would be interesting to increase the number of attributes to be used in clustering to obtain more informative clusters. These additional attributes could come from other attributes gathered from the same test, in this case the GCS, from other tests carried out at the same time, or from other personal information about the patient, such as patient demographic data. Another area of future work is related to the final results obtained in the clustering process. Clustering is an unsupervised learning technique where the data are not "labelled", which means that these techniques perform a grouping without any prior knowledge of the problem. As discussed in the post-processing stage, it is necessary for an expert to determine whether the clusters formed make sense with respect to the context in which the clustering is carried out. It may be the case that such clusters are good at the internal validation level, but that they are not informative when applied in the context of the problem. One way to solve this problem lies in the application of semi-supervised clustering techniques, called interactive clustering algorithms [14]. It is a type of clustering algorithm that lies between supervised and unsupervised learning, in which unlabelled data exists, but in the process of clustering, the expert can introduce knowledge, causing changes in the way different clusters are formed. In this respect, other clustering algorithms, such as Group-Based Trajectory Modelling (GBTM) [12], will be considered in a future work.

Acknowledgements. This paper is part of the R+D+i projects PID2019-108915RB-I00 and PID2022-140907OB-I00, and the grant PRE2020-094056 funded by MCIN/AEI/10.13039/501100011033. It has also been funded by the University of Castilla-La Mancha (2022-GRIN-34436) and by 'ERDF A way to make Europe, the PhD scholarship 2019-PREDUCLM-10772 and co-financed by the FSE Operational Programme 2014-2020 of Castilla-La Mancha through Axis 3.

References

1. ADACE CLM: ADACE - Association of ABI of Castilla - La Mancha. https://www.adaceclm.org/
2. Choudhry, O.J., Prestigiacomo, C.J., Gala, N., Slasky, S., Sifri, Z.C.: Delayed neurological deterioration after mild head injury: cause, temporal course, and outcomes. Neurosurgery **73**(5), 753–760 (2013)
3. Desgraupes, B.: Clustering indices (2017)
4. FITBIR: Federal Interagency Traumatic Brain Injury Research (FITBIR). https://fitbir.nih.gov/
5. FITBIR: Transforming Re-search and Clinical Knowledge in Traumatic Brain Injury (TRACK-TBI) – Adult. https://fitbir.nih.gov/portal/study/viewStudyAction!view.action?studyId=FITBIR-STUDY0000267
6. Fred, A.: Finding consistent clusters in data partitions. In: Kittler, J., Roli, F. (eds.) MCS 2001. LNCS, vol. 2096, pp. 309–318. Springer, Heidelberg (2001). https://doi.org/10.1007/3-540-48219-9_31
7. Lenrow, D.A.: Physical medicine and rehabilitation: an update for internists. Med. Clin. North Am. **104**(2), xvii–xviii (2020)
8. Liu, Y., Li, Z., Xiong, H., Gao, X., Wu, J.: Understanding of internal clustering validation measures. In: 2010 IEEE International Conference on Data Mining, pp. 911–916. IEEE (2010)
9. Montero, F., López-Jaquero, V., Navarro, E., Sánchez, E.: Computer-aided relearning activity patterns for people with acquired brain injury. Comput. Educ. **57**(1), 1149–1159 (2011)
10. Moya, A., Navarro, E., Jaén, J., López-Jaquero, V., Capilla, R.: Exploiting variability in the design of genetic algorithms to generate telerehabilitation activities. Appl. Soft Comput., 108441 (2022)
11. Moya, A., Pretel, E., Navarro, E., Jaén, J.: A systematic literature review of clustering techniques for patients with traumatic brain injury. Artif. Intell. Rev. (2023)
12. Nagin, D.S., Odgers, C.L.: Group-based trajectory modeling in clinical research. Annu. Rev. Clin. Psychol. **6**(1), 109–138 (2010)
13. Network, T.A.: Definition of ABI. http://www.abinetwork.ca/definition
14. Neubauer, T.R., Peres, S.M., Fantinato, M., Lu, X., Reijers, H.A.: Interactive clustering: a scoping review. Artif. Intell. Rev. **54**(4), 2765–2826 (2021)
15. Parimbelli, E., Marini, S., Sacchi, L., Bellazzi, R.: Patient similarity for precision medicine: a systematic review (2018). https://doi.org/10.1016/j.jbi.2018.06.001
16. Podell, J., et al.: Rapid prediction of secondary neurologic decline after traumatic brain injury: a data analytic approach. Sci. Rep. **13**(1), 403 (2023)
17. Saxena, A., et al.: A review of clustering techniques and developments. Neurocomputing **267**, 664–681 (2017)
18. Strehl, A., Ghosh, J.: Cluster ensembles-a knowledge reuse framework for combining multiple partitions. J. Mach. Learn. Res. **3**, 583–617 (2002)

19. Trisuciana, F.M., Witarsyah, D., Sutoyo, E., Machado, J.M.: Clustering of COVID-19 vaccination recipients in DKI Jakarta using the K-medoids algorithm. In: 2022 International Conference Advancement in Data Science, E-learning and Information Systems (ICADEIS), pp. 01–07. IEEE (2022)
20. UN: Convention on the rights of persons with disabilities (2022)
21. Wilson, J.T., Pettigrew, L.E., Teasdale, G.M.: Structured interviews for the Glasgow outcome scale and the extended Glasgow outcome scale: guidelines for their use. J. Neurotrauma **15**(8), 573–585 (1998)
22. Wu, J., Song, C.-H., Kong, J.M., Lee, W.D.: Extended mean field annealing for clustering incomplete data. In: 2007 International Symposium on Information Technology Convergence (ISITC 2007), pp. 8–12. IEEE (2007)
23. Wu, X., Ma, T., Cao, J., Tian, Y., Alabdulkarim, A.: A comparative study of clustering ensemble algorithms. Comput. Electr. Eng. **68**, 603–615 (2018)

Comparative Study of Large Language Models as Emotion and Sentiment Analysis Systems: A Case-Specific Analysis of GPT vs. IBM Watson

David Carneros-Prado[1]([⊠]), Laura Villa[1], Esperanza Johnson[2],
Cosmin C. Dobrescu[1], Alfonso Barragán[1],
and Beatriz García-Martínez[1,3]

[1] Departament of Information Technologies and System, University of Castilla-La Mancha, Paseo de la Universidad 4, 13071 Ciudad Real, Spain
{David.Carneros,Laura.Villa,Cosmin.Dobrescu,Beatriz.GMartinez}@uclm.es
[2] Game School - Department of Game Development, Høgskolen i Innlandet, Holsetgata 31, 2318 Hamar, Norway
esperanza.johnson@inn.no
[3] Social Sciences and Information Technology Faculty, University of Castilla-La Mancha, Av. Real Fábrica de Sedas s/n, 45600 Talavera de la Reina, Spain

Abstract. Sentiment analysis and emotion-detection techniques have wide applications in diverse fields. Various systems such as IBM Watson NLU have been developed for this purpose. Separately, large language models (LLMs) like GPT-3.5 have shown promise for diverse natural language processing (NLP) applications. This study investigates whether an LLM without explicit training could perform sentiment and emotion classification comparably to customized systems. For this purpose, a comparative analysis was conducted between GPT-3.5 and IBM Watson's sentiment analysis, and emotion classification, using a dataset of 30,000 tweets related to the Covid-19 pandemic. The results revealed the versatility of LLMs, suggesting their potential transferability to diverse NLP tasks beyond their original training objective when properly prompted. GPT-3.5, despite not being explicitly trained for these tasks, achieves competitive performance with IBM Watson's emotion classification capabilities when provided with a suitable prompting context. Precisely, GPT-3.5 demonstrates surprising adaptability to detect nuanced sentiments, such as irony, compared to Watson's rigid emotion model. However, GPT-3.5 also struggles to fit textual expressions into the prescribed emotion classifications. Overall, this motivates expanded research into leveraging large pre-trained language models for affective computing applications by means of thoughtful prompt and evaluation design.

Keywords: Large Language Model · IBM Watson · Sentiment Analysis · Emotion Recognition · GPT

J. Bravo and G. Urzáiz (Eds.): UCAmI 2023, LNNS 842, pp. 229–239, 2023.
https://doi.org/10.1007/978-3-031-48642-5_22

1   Introduction

Advances in computational capabilities and novel deep-learning techniques have fueled the growth of sentiment and emotion analysis technologies. These innovations, as part of affective computing [13], have opened new doors in terms of creating systems that can recognize emotions and improve human-system interactions. Sentiment and emotion analysis tools have been used in a broad spectrum of fields. From a critical role in social media analysis, helping marketers better tune their strategies for customer satisfaction [1,9], to aiding in understanding students' emotional states to optimize teaching and learning environments [15], their contributions are truly substantial.

Various techniques have been adopted for sentiment analysis and emotion recognition in diverse contexts. These include facial image analysis [10], voice-tone examination [3], and gait analysis [2]. However, within the sphere of Natural Language Processing (NLP), sentiment analysis has taken a central stage and has become one of the most utilized methods in the last years [4,8,11,16].

Sentiment analysis, defined as extracting valuable semantic information from text to understand a writer's attitude, uses NLP techniques to identify positive, negative, or neutral sentiments [12]. Its scope is not limited to categorizing the text as positive or negative. This goes beyond the analysis of evaluative dimensions, such as agreement or disagreement, and judgments of good or bad [11].

Different NLP systems have been recently developed. For instance, IBM Watson, initially designed to answer questions, has now branched out to offer a variety of cloud-based services, including sentiment and emotion classification, as part of its Natural Language Understanding service. The service categorizes sentiments into positive, negative, and neutral based on valence (i.e., agreeableness) and identifies emotions following a modified version of Ekman's six basic emotions model [6].

Furthermore, pretrained language models (PLMs) with transformers have already demonstrated their effectiveness in tackling NLP tasks. However, when these models are scaled up, their capabilities increase significantly, and unlocking abilities are not present in their smaller versions. This led to the emergence of Large Language Models (LLMs) [17]. The Generative Pretrained Transformer (GPT) series by OpenAI [5] is a prominent example of such models that garnered much attention after its public release via ChatGPT.

The aim of this study is to compare IBM Watson with a Large Language Model (LLM), specifically GPT-3.5, for sentiment analysis and emotion classification, through a direct comparison of their performance. We chose a public dataset of tweets from the Covid-19 pandemic period, as it captures the thoughts, feelings, and emotions expressed on social media during the lockdown [7]. To make a fair comparison, we will instruct the LLM through its "prompt" to provide results akin to IBM Watson. Our goal is to explore whether an LLM that is not specifically trained for sentiment and emotion analysis tasks can match the performance of a system specifically built for these tasks.

The remainder of this paper is organized as follows. Section 2 outlines the methods used in this study, including data processing, sentiment recognition, and emotion recognition. Section 3 presents the results of the comparison. Section 4 discusses the findings and Sect. 5 highlights the main conclusions of this research.

2   Materials and Methods

The Materials and Methods section outlines the processes involved in data acquisition and processing, techniques employed for sentiment analysis and emotion detection, and the evaluation methods used to assess the effectiveness of the implemented models.

2.1   Data Processing

The selected dataset, referred to as *Coronavirus (covid19) Tweets*[1], comprises 4.9 million tweets collected between the dates of 29-03-2020 and 15-04-2020, using the hashtags *#coronavirus, #coronavirusoutbreak, #coronavirusPandemic, #covid19, #covid_19, #epitwitter, #ihavecorona, #StayHomeStaySafe,* and *#TestTraceIsolate*. Given the substantial size of this dataset, a data cleaning process was implemented to retain only the most relevant tweets.

The data cleaning process, depicted in Fig. 1, began by filtering English tweets to prevent language bias in the results. Retweets and quoted tweets were eliminated. The remaining dataset was further refined by selecting only those tweets with at least 100 retweets or likes. Tweets containing fewer than five words; those with more hashtags, mentions, URLs, or emojis than words; and tweets beginning with more than three repeated characters were subsequently removed.

From the processed dataset, 30,000 tweets were randomly selected. These patients underwent a text processing procedure that involved converting the text to lowercase, removing links and special characters, while retaining the "#" symbol. Figure 2 presents a word cloud of the 30,000 randomly selected tweets used in this study. The sample size of 30,000 tweets was determined based on the usage limits of the Watson and GPT-3.5 APIs to access their models.

2.2   Sentiment Analysis and Emotion Recognition

IBM Watson Natural Language Understanding provides a sentiment analysis service that returns a score between -1 and 1 to indicate whether the sentiment of a given text is negative, neutral, or positive. In addition, it offers an emotion recognition service that distinguishes between five different emotions (Anger, Disgust, Fear, Joy, and Sadness). A total of 30,000 selected tweets were processed through these services to determine their sentiment polarity and the predominant emotion they expressed.

For comparison, the OpenAI GPT 3.5 model was chosen as the representative Large Language Model. To conduct sentiment and emotion analyses, a task-specific context was first provided to the LLM. The selected context was as

---

[1] https://www.kaggle.com/datasets/smid80/coronavirus-covid19-tweets-early-april.

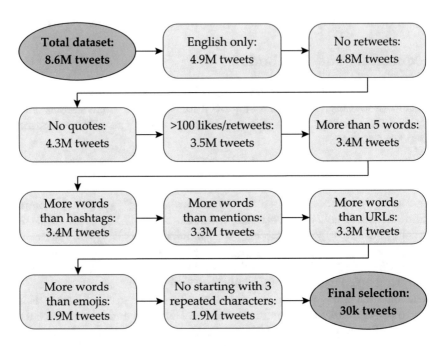

Fig. 1. Reduction process of an 8.6M tweet dataset to a final selection of 30k tweets, involving filtering for language, eliminating retweets, considering reach, word count, and removing repeated characters.

Fig. 2. A word cloud illustrating the most frequently occurring words within the 30,000 randomly selected tweets from the cleaned dataset.

follows: *"Please rate the sentiment between −1 (negative) and 1 (positive) in the above message as SA: [score], just the score, and the expressed emotion as EM: [emotion]. Only the following emotions are allowed: Joy, Sadness, Fear, Anger, and Disgust."*. The LLM was instructed to use the same emotion model as the IBM Watson NLU to enable direct comparison (Ekman's six basic emotions model [6] without surprise emotion). Consequently, each of the selected tweets was sent individually through the API in this context, resulting in sentiment and emotion outputs in the following format: *SA: [score] EM: [emotion]*.

2.3   Evaluation

The evaluation was conducted in two stages. First, the sentiment detection system was evaluated, followed by the emotion recognition system. For sentiment analysis, the Mean Squared Error (MSE) between the scores returned by GPT 3.5 and IBM Watson NLU was calculated. The main evaluation metric used was this MSE between the sentiment scores predicted by GPT-3.5 and IBM Watson NLU. Subsequently, the three possible classes - positive, neutral, and negative - were compared. In instances where significant discrepancies were noted, a manual analysis of the tweet in question was conducted to investigate the reasons for the divergence between both systems.

For emotion detection, a confusion matrix was generated between the labels to identify any misclassification among the different models. A manual check was performed on cases with the highest confusion rates with the aim of understanding the underlying reasons for such misclassifications. This was done through a peer-review process, where two reviewers independently categorized the sentiment of the tweet, followed by a consensus discussion to determine the final label.

To make a fair comparison, the two models were evaluated on the same dataset of 30,000 tweets related to Covid-19. No model was assumed to be superior as a gold standard reference.

3   Results

This section presents the results of a comparative study between IBM Watson NLU and GPT 3.5, focusing on their performance in sentiment analysis and emotion-detection tasks.

3.1   Sentiment Analysis

Upon obtaining the scores for both systems, their mean and standard deviation were calculated, yielding -0.21 ± 0.66 for IBM Watson NLU and -0.08 ± 0.59 for GPT-3.5. The MSE between the models, as the main evaluation metric, was computed as 0.35. Sentiments were classified as positive (greater than 0), negative (less than 0), or neutral (0). Of the total tweets analyzed, GPT-3.5 classified 44% as positive, 5% as neutral, and 51% as negative. In comparison, IBM Watson

NLU classified 31% as positive, 11% as neutral, and 58% as negative. Figure 3a shows the distributions of the three classes obtained from both systems. IBM Watson NLU classifies more entries as negative and neutral compared to GPT. On the other hand, GPT balances positive and negative classifications more evenly but identifies fewer neutral entries.

In Fig. 3b, the confusion matrix illustrates the classification overlap between the two systems, without assuming either one as the ground truth. A significant number of tweets were accurately classified as negative or positive using both IBM Watson NLU and GPT-3.5. However, the neutral class exhibited some discrepancies. The majority of tweets categorized as neutral by IBM Watson NLU were labeled as negative by GPT-3.5, indicating differences in neutrality perceptions between the two systems.

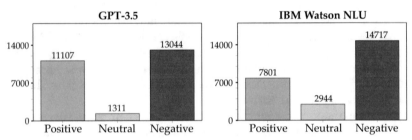

(a) Histogram displaying the distribution of the three sentiment classes (negative, neutral, and positive), as identified by both systems.

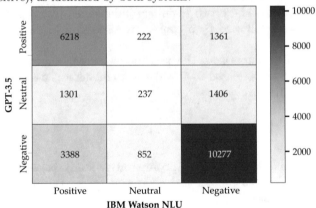

(b) Confusion matrix comparing sentiment classifications between the analyzed systems.

Fig. 3. Confusion matrix comparing sentiment classifications by GPT-3.5 to those by IBM Watson NLU.

Interestingly, certain tweets were classified as positive by one system and negative by the other, or vice versa. To investigate these discrepancies, the tweets

@user2 @user3 @user4 @user5 many many doctors in the world have released results of #hydroxychloroquine working as part of a cocktail of drugs for #covid19 but yeah we should listen to this physical therapist dude who has better expertise #covidiot

(a) Example of a tweet rated positive by IBM Watson NLU (score: 0.44) and negative by GPT-3.5 (score: -0.8).

@user

singapore quickly converted medical centres into public health preparedness clinics, making them the first line of defence for suspected cases of covid-19. phpcs also get priority access to vital equipment like masks. #foreigncorrespondent #coronavirus

(b) Example of a tweet classified as negative by IBM Watson NLU (score: -0.71) and positive by GPT-3.5 (score: 0.5).

Fig. 4. Illustrative examples of tweets with contrasting sentiment classifications. User-names and mentions in the tweets have been censored to maintain privacy.

and classifications were manually reviewed by two independent annotators in a peer-review process. Figure 4a presents an example of a tweet labeled as positive by IBM Watson NLU (0.44) and negative by GPT-3.5 (−0.8). This tweet exhibits an ironic tone that GPT-3.5 seems to detect; hence, its negative classification, whereas IBM Watson NLU appears to overlook the irony, leading to positive classification.

Another instance of contrasting classifications is presented in Fig. 4b, featuring a tweet categorized as negative by IBM Watson NLU (−0.71) and positive by GPT-3.5 (0.5). This tweet conveys a positive sentiment, praising Singapore's effective response to the pandemic, which was accurately captured by the GPT-3.5.

3.2   Emotion Recognition

Following emotion-detection, it was found that GPT classified 4538 tweets as "None," indicating that it did not detect any of the five emotions: "joy," "sadness," "fear," "anger," or "disgust." The tweets were removed to enable a more accurate comparison. Figure 5a illustrates the class distribution of emotion recognition for both systems. IBM Watson NLU was observed to classify a large number of tweets as "joy," followed by "sadness," while GPT's most frequently classified class is "joy," although in fewer instances than Watson, followed by "fear." It can be observed that GPT-3.5 achieved a more balanced classification across different emotions, unlike Watson, which presented a very unbalanced classification.

Figure 5b presents a confusion matrix between the classifications of both the systems, without assuming either one as the ground truth. The predictions of GPT and Watson revealed significant discrepancies, particularly in "joy", "sadness", and "anger", suggesting that both models may interpret and classify emotions differently.

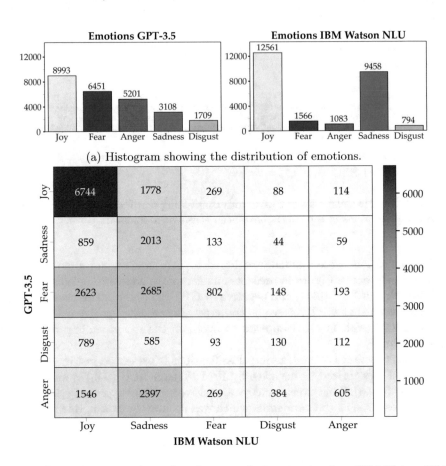

(a) Histogram showing the distribution of emotions.

Fig. 5. Illustrative comparison of sentiment analysis outcomes from IBM Watson NLU and GPT-3.5.

Figure 6 shows two examples of tweets classified differently. First, Fig. 6a features a tweet classified as "joy" by Watson and "anger" by GPT, which shows frustration with the coronavirus situation and decisions made by the governor. Second, in Fig. (6b), a tweet classified as "joy" by Watson and "sadness" by GPT is shown, which expresses sadness/anger owing to the increased wage gap caused by the pandemic.

@user

That said @user1 continues to try putting band aid on hot spot while rest of state bleeds.
Stay at home for south fl while rest of florida goes to beaches, pastor held large church services, etc.
problem is bigger than just south fl! #coronavirus #coronavirusflorida

(a) Example of a tweet rated as joy by IBM Watson NLU and anger by GPT-3.5.

@user

today is #equalpayday and we can't forget the fact that women each lose an average of 10,000+
to the wage gap each year the equivalent of several months of rent. amidst the #covid19 pandemic,
too many families are scrambling to be able to pay rent tomorrow. source npwf

(b) Example of a tweet classified as joy by IBM Watson NLU and sadness by GPT-3.5.

Fig. 6. Illustrative examples of tweets with contrasting emotions classifications. Usernames and mentions in the tweets have been censored to maintain privacy.

4   Discussions

The comparative analysis between IBM Watson NLU and GPT-3.5 for sentiment analysis and emotion classification reveals several key insights. Overall, the natural language processing capabilities of GPT-3.5 has allowed it to perform surprisingly well with regards to sentiment and emotion classification without specific training, often outperforming custom-built IBM Watson NLU services. This demonstrates the versatility and potential transferability of large pre-trained language models such as GPT-3.5 to diverse NLP applications when provided with suitable prompting. Despite the small sample size, these results still provide valuable insights into the transfer capabilities of large language models. However, caution must be taken in generalizing the findings more broadly. Further research with larger samples is warranted to confirm the observed trends and obtain more robust and generalizable conclusions. Neither model was assumed to be the gold standard. Both presented different strengths and weaknesses based on the comparative analysis.

However, providing the appropriate prompting context poses a challenge to large language models such as GPT-3.5. As evidenced by the classification of several tweets as "None" for emotions, GPT-3.5 struggles to fit some textual expressions cleanly into a discrete emotion classification approach, even when explicitly instructed to do so. This highlights the importance of prompt engineering when applying LLMs to constrained classification tasks.

On the other hand, a major limitation of IBM Watson is its reliance on Ekman's basic emotions model, which lacks nuance and fails to detect more complex aspects, such as irony, that requires understanding context and subtle cues (as seen in Fig. 4a or Fig. 6a). In contrast, the contextual prompting approach with GPT-3.5 enabled it to better grasp ironic sentiment despite not being explicitly trained for sentiment analysis.

In addition, the evaluation methodology in this study using coarse positive/negative/neutral sentiment classes and a limited 5-emotion scheme has limitations. More nuanced evaluation frameworks, such as Russell's circumplex model [14], that accounts for emotion dimensions, such as arousal and valence, could provide deeper insights into the sentiment and emotion-detection capabilities of LLMs. Further research with enhanced prompting strategies and evaluation methodologies is warranted to fully unlock the capabilities of large language models in affective computing tasks.

5   Conclusions

This comparative study demonstrates the promising capabilities of large language models such as GPT-3.5, for sentiment analysis and emotion classification without specialized training. With suitable prompting, GPT-3.5 performed competitively with IBM Watson's customized services. However, challenges remain regarding prompt engineering and rigorous evaluation. The key conclusion is that GPT-3.5 showed a surprising adaptability to these affective computing tasks, overcoming limitations such as detecting irony that rigid emotion models miss. However, difficulties constraining GPT-3.5's open-ended language generation to fixed classification frameworks have been observed. Overall, this motivates further research into leveraging large pre-trained language models for diverse NLP applications through thoughtful prompt design and evaluation strategies. The results revealed the versatility of LLMs, suggesting their potential transferability to various practical contexts, beyond their original training objectives. However, the small sample size limits the generalizability of the results. Further studies with larger sample sizes are needed to corroborate these preliminary findings.

Funding Information. This research was funded by the MINISTERIO DE CIENCIA, INNOVACIÓN Y UNIVERSIDADES, grant number project: PDC20200-133457-I00 (sSITH: Self-recharging Sensorized Insoles for continuous long-Term Human gait monitoring); the JUNTA DE COMUNIDADES DE CASTILLA-LA MANCHA, grant number project: SBPLY/21/180501/000160 (SHARA3); and the 2022-PRED-20651 predoctoral contract by UNIVERSITY OF CASTILLA-LA MANCHA.

References

1. Al Ajrawi, S., et al.: WITHDRAWN: Evaluating Business Yelp's Star Ratings Using Sentiment Analysis. Elsevier (2021). isbn: 2214–7853
2. Altamirano-Flores, Y.V., et al.: Emotion recognition from human gait using machine learning algorithms. In: Proceedings of the International Conference on Ubiquitous Computing & Ambient Intelligence (UCAmI 2022), pp. 77–88. Springer (2022)
3. Alu, D., Zoltan, E., Stoica, I.C.: Voice based emotion recognition with convolutional neural networks for companion robots. Sci. Technol. **20**(3), 222–240 (2017)
4. Batbaatar, E., Li, M., Ryu, K.H.: Semantic-emotion neural network for emotion recognition from text. IEEE Access **7**, 111866–111878 (2019)
5. Brown, T., et al.: Language models are few-shot learners. Adv. Neural. Inf. Process. Syst. **33**, 1877–1901 (2020)
6. Ekman, P.: An argument for basic emotions. Cogn. Emotion **6**(3–4), 169–200 (1992)

7. Garcia, K., Berton, L.: Topic detection and sentiment analysis in Twitter content related to COVID–19 from Brazil and the USA. Appl. Soft Comput. **101**, 107057 (2021)
8. Guo, J.: Deep learning approach to text analysis for human emotion detection from big data. J. Intell. Syst. **31**(1), 113–126 (2022)
9. Jang, H.-J., et al.: Deep sentiment analysis: mining the causality between personality-value-attitude for analyzing business ads in social media. Expert Syst. Appl. **40**(18), 7492–503 (2013)
10. Jayalekshmi, J., Mathew, T.: Facial expression recognition and emotion classification system for sentiment analysis. In: 2017 International Conference on Networks & Advances in Computational Technologies (NetACT), pp. 1–8. IEEE (2017). isbn: 1-5090-6590-3
11. Nandwani, P., Verma, R.: A review on sentiment analysis and emotion detection from text. Soc. Network Anal. Mining **11**(1), 81 (2021)
12. Onyenwe, I., et al.: The Impact of political party/candidate on the election results from a sentiment analysis perspective using# anambradecides 2017 tweets. Soc. Netw. Anal. Min. **10**, 1–17 (2020)
13. Picard, R.W.: Affective Computing-Mit Media Laboratory Perceptual Computing Section Technical Report No. 321. In: Cambridge, MA 2139, p. 92 (1995)
14. Russell, J.A.: A circumplex model of affect. J. Personality Soc. Psychol. **39**(6), 1161 (1980)
15. Sangeetha, K., Prabha, D.: Sentiment analysis of student feedback using multi-head attention fusion model of word and context embedding for LSTM. J. Ambient. Intell. Humaniz. Comput. **12**, 4117–4126 (2021)
16. Tesfagergish, S.G., Kapočcūutė-Dzikienė, J., Damaševičius, R.: Zero-shot emotion detection for semi-supervised sentiment analysis using sentence transformers and ensemble learning. Appl. Sci. **12**(17), 8662 (2022)
17. Zhao, W.X., et al.: A survey of large language models. In: arXiv preprint arXiv:2303.18223 (2023). arXiv: 2303.18223

Meta-learning. An Approach Applied to Clinical Data

Sandra Amador[1], Kelly Turbay[1], Alicia Montoro[2], Macarena Espinilla[2], Higinio Mora[1], and David Gil[1(✉)]

[1] Department of Computer Technology and Computation, University of Alicante, Alicante, Spain
david.gil@ua.es
[2] University of Jaen, Jaén, Spain

Abstract. Currently, there is a large amount of digitized data. In the field of health, an important part of the data is obtained from electronic health records, related to the health of people.

This stream of data has led to a multitude of research efforts in the healthcare domain, focusing on various aspects such as disease prediction, clinical risk assessment, mortality analysis, and more. These predictions assist healthcare providers in early identification of potential risks, leading to better patient care.

The objective of this work is to propose a meta-learning model applied to clinical data in which the different algorithms included can be compared and the benefits of each of them appreciated. Meta-learning involves algorithms learning about other algorithms through experience. However, the application of meta-learning to diagnosis poses considerable practical challenges.

A case study is used to test the model and assess its effectiveness. In this study, the application of meta-learning showed very promising initial results, adapting the algorithm to the database used depending on the disease to be predicted.

Keywords: meta-learning · clinical data · artificial intelligence

1 Introduction

At present, with the development of technology, there are large volumes of data where much of this data is collected on healthcare. As the development of technology advances, we find ourselves with a greater number of data. Therefore, for many decades, the link between technology and the field of health is of great importance. In recent years, there have been numerous investigations related to technology and health due to the need to exploit the large amount of information obtained. Much of the research is carried out for the prediction, classification or detection of diseases, clinical risk and mortality, among others. With these kinds of predictions, healthcare providers are helped to identify potential risk at an early stage, thus improving patient care.

While much of the clinical data comes from people, and the planet has reached 8 billion people by 2022 according to [1], when considering a particular disease the number of patients is very small compared to that figure. In many cases, researchers are faced with the difficulty that the number of sample data available is extremely limited. Taking these difficulties into account, it leads us to the importance of designing accurate prediction models that allow working with limited samples of patients, making efficient use of the available knowledge. Meta-learning is the process of learning to learn. Within the psychological sciences, this idea originated in 1949 [2]. In the field of artificial intelligence, it refers to learning algorithms that learn about other algorithms using experience [3,4]. A set of algorithms is capable of solving a task. This is where meta-learning comes into play. In this example, meta-learning would determine which algorithms to use and, in turn, how to combine them to get the best possible results. Meta-learning aims to dynamically search as the number of tasks increases, the best learning strategy [5].

Within meta-learning, there is the possibility of applying different algorithms and combining them with deep learning. Deep neural networks, in general, given a set of input data, so that the desired target outputs match the network outputs, adjust the neural network weights [4]. There are numerous studies that apply meta-indexing with deep neural networks such as. Among them, the application for initial weights of the network [6] and the weight update rule [7–10] stand out. The aim of this work is to propose a meta-learning model applied to clinical data in which the different algorithms included can be compared and the advantages of each of them can be appreciated. A case study is used to test the model and evaluate its effectiveness. Specifically, we will focus on ensemble methods, evaluating the types of ensemble stacking, bagging and boosting. To do so, we explore the application of several algorithms, such as DecisionTreeClassifier, SVM, LogisticRegression, KNeighborsClassifier, AdaBoost, Gradient Boosting and RandomForest, on two databases widely used in the scientific community: Early Detection of Diabetes and Obesity Levels. The source code can be consulted at [11].

In the world of machine learning, ensemble models have emerged as powerful tools to improve prediction accuracy and overall performance. The proposed architecture in this study harnesses the potential of meta-learning models, specifically bagging, boosting, and stacking models, to create a robust and reliable system. Ensemble models leverage the collective knowledge of multiple classifiers, rather than relying on a single model, to make more accurate predictions and overcome individual model limitations. Ensemble models leverage the collective knowledge of multiple classifiers, rather than relying on a single model, to make more accurate predictions and overcome individual model limitations. [12]. To do so, we will perform a series of experiments, evaluating different combinations of hyperparameters to identify the optimal models that best fit our datasets.

We compare two different approaches, one employing each ensemble technique separately with different codes, and the other using a loop to iterate through each model.

This paper discusses two datasets that follow a specific architecture for data preprocessing and modeling. The datasets undergo cleaning, which involves removing duplicates, filling in missing values, and eliminating columns with high correlation. The cleaned data is then balanced using the SMOTE technique. Categorical data is associated with numerical values using the StandardScaler class for normalization. The data is divided into a 70/30 ratio for training and evaluation to prevent overfitting. The trained model is then tested on new data to evaluate its accuracy. The case studies section provides detailed information on the parameters and hyperparameters used in the models and ensembles. Tables present the types of assemblies, estimators, and their respective ranges for hyperparameters and parameters. Overall, this process yields a final model with a certain level of accuracy determined through validation.

We present the results obtained in the experiments, which reveal high accuracy rates and solid performance for most classes in both datasets. In particular, the Stacking Classifier model stands out for its excellent results. Additionally, we evaluate each model with several metrics, such as confusion matrix, ROC curve, precision, recall, and F1 score, to provide a comprehensive assessment of its performance. Overall, our results highlight the effectiveness of ensemble techniques in achieving accurate classification in various machine learning applications. The remainder of this paper is structured as follows: in Sect. 2, we show possible architecture models in the form of meta-learning approaches. Thereafter, in Sect. 3, we explore various ensembles techniques by conducting a complete series of experiments, finishing the paper with the conclusions.

2   Architecture

The meta-learning models to be used in this proposed architecture are ensembles, namely bagging, boosting, and stacking models. Ensemble models utilize the collective decisions of multiple models to enhance performance on a broader scale. Instead of relying on a single classifier, ensemble models consist of a group of classifiers known as an ensemble of classifiers. Each classifier within the ensemble is trained independently, often using different subsets of the data or employing diverse algorithms. Once trained, their individual predictions are merged using methods like voting or averaging to generate the final classification for unseen instances [12–14]. In summary, ensemble models use the combined knowledge of multiple models, allowing them to make more accurate predictions and improve overall performance (see Fig. 1).

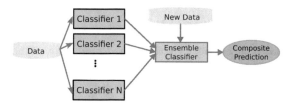

Fig. 1. Esquema general de un ensamble de clasificadores

In a bagging approach, accuracy improvement is achieved by reducing the variance of individual classifiers using the composite model. It demonstrates increased resilience against the impact of noisy data and overfitting. Models can be run in parallel to optimize computational efficiency [12,14]. (See Fig. 2).

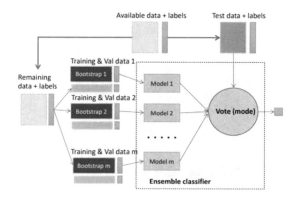

Fig. 2. Ensemble methods Bagging

In boosting, weights are also assigned to each training instance. A series of classifiers is learned iteratively. After each classifier is learned, the weights are updated to allow the next classifier to focus more on the instances that were misclassified by the previous one. The final boosted classifier combines the votes of each individual classifier, where the weight of each vote is determined by the classifier's accuracy [14]. (See Fig. 3).

Fig. 3. Ensemble methods Boosting

Stacking is a technique that learns how to optimally combine the predictions of multiple high-performing machine learning models. Stacking, as an ensemble method, has distinct characteristics that set it apart from other ensemble techniques such as bagging and boosting. These differences lie in the diversity of models, the training data, and the approach to combining predictions [14]. The key distinctions between stacking and other ensemble methods like bagging and boosting are that, in contrast to bagging, the models employed in stacking are usually diverse and trained on the same data set, and in contrast to boosting, a single model is utilized to determine the optimal way to combine the predictions from the contributing models. This approach enhances the ensemble's overall performance and predictive capabilities [12] (see Fig. 4).

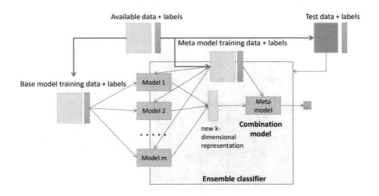

Fig. 4. Ensemble methods Stacking

First, the data are cleaned by removing repeated entries, filling in missing values and removing columns with high correlation. Next, the cleaned data are balanced using the SMOTE technique. Next, a normalization process is applied to associate categorical data with numerical values using the StandardScaler class. For training, the data is split in a 70/30 ratio, with 70% being used for training and 30% for evaluation. This helps to avoid overfitting. The trained model is then tested with new data to assess its accuracy. Parameters and hyperparameters were used in the models and data set. See details in https://github.com/kellyturbay/EnsamblesDiabetesObesity.

3   Case Study

In this section, we will explore various ensemble techniques, including stacking, bagging and boosting. We will use algorithms such as DecisionTreeClassifier, SVM, LogisticRegression, KNeighborsClassifier, AdaBoost, Gradient Boosting and RandomForest on two databases widely used by researchers: Early Detection of Diabetes and Obesity Levels. See Table 1 and 2.

To achieve this goal, we will conduct a series of experiments in which we will adjust different key parameters. We will perform an exhaustive search to find the best parameters and evaluate various combinations of hyperparameters to identify the optimal model that best fits our data sets. In addition, we will perform two different approaches: in the first one, we will apply each ensemble technique separately using different codes, while in the second one we will iterate in a loop each model.

Table 1. Description of variables in the dataset Obesity levels

Variable	Description	Data type	Value Range	Considered
Gender	Individual gender	Categorical	Male/Female	Yes
Age	Age of individual	Integer	14–61	Yes
Height	Individual height	Float	–	Not
Weight	Individual weight	Float	1.61–1.98	Not
family history with overweight	Family history of excess weight	Categorical	Yes/No	Yes
FAVC	Frequent consumption of high calorie foods	Categorical	Yes/No	Yes
FCVC	Vegetable and fruit consumption	Float	1–3	Yes
NCP	Number of main meals a day	Integer	1–4	Yes
CAEC	Consumption of food between meals	Categorical	no /Sometimes /Frequently /Always	Yes
SMOKE	Smoking habit	Categorical	Yes/No	Yes
CH2O	Water consumption in liters per day	Integer	1–3	Yes
SCC	Consumption of carbonated drinks	Categorical	Yes/No	Yes
FAF	Weekly physical activity	Integer	0–3	Yes
TUE	Time spent using electronic devices in hours	Integer	0–2/3–5 /More than 5	Yes
CALC	Alcohol consumption	Categorical	no /Sometimes /Frequently /Always	Yes
MTRANS	Means of transportation used	Categorical	Automobile /Motorbike /Bike /Public Transportation /Walking	Yes
NObeyesdad	Classification of levels of obesity	Categorical	Insufficient /Normal /Overweight LevelI /Overweight LevelII /Obesity TypeI /Obesity TypeII /Obesity TypeIII	Yes

Table 2. Description of variables in the early detection of diabetes data set

Variable	Description	Data type	Value Range	Considered
Age	Age of patient	Integer	16 – 90	Yes
Gender	Gender of patient	Categorical	Male/Female	Yes
Polyuria	Presence of polyuria	Categorical	Yes/No	Yes
Polydipsia	Presence of polydipsia	Categorical	Yes/No	Yes
sudden weight loss	Sudden weight loss	Categorical	Yes/No	Not
weakness	Presence of weakness	Categorical	Yes/No	Yes
Polyphagia	Presence of polyphagia	Categorical	Yes/No	Not
Genital thrush	Presence of genital thrush	Categorical	Yes/No	Yes
visual blurring	Presence of visual blur	Categorical	Yes/No	Not
Itching	Presence of itching	Categorical	Yes/No	Not
Irritability	Presence of irritability	Categorical	Yes/No	Yes
delayed healing	Late healing	Categorical	Yes/No	Yes
partial paresis	Presence of partial paresis	Categorical	Yes/No	Not
muscle stiffness	Presence of muscle stiffness	Categorical	Yes/No	Not
Alopecia	Presence of alopecia	Categorical	Yes/No	Yes
Obesity	Presence of obesity	Categorical	Yes/No	Yes
class	Variable that indicates whether or not the individual has diabetes	Categorical	Positive /Negative	Yes

3.1   Levels of Obesity

The dataset used in this study was obtained from the UCI Machine Learning repository, specifically from the following link:https://archive.ics.uci.edu/dataset/544/estimation+of+obesity+levels+based+on+eating+habits+and+physical+condition

It is composed of 2111 records and 17 attributes that include information on eating habits, physical condition and other relevant variables. The dataset was subjected to a preprocessing process that included duplicate data detection and removal, data transformations to ensure correct formatting, and manual feature extraction and selection of relevant features. No missing values or outliers were found in the dataset, so no data imputation was performed. In addition, no balancing of the target variable was performed to classify obesity levels. After applying transformations, we obtained a dataset that summarizes about 2110 records in 15 column. Subsequently, the data were prepared and two different encodings were applied to evaluate each result obtained. The Fig. 5 show the precision, fitting and training time, and the accuracy and execution time comparison of each model, respectively.

The text discusses the results of evaluating different models for classification performance. The StackingClassifier model performs well with less training time, but the RandomForestClassifier model achieves the highest accuracy of 80.41%. Additional metrics such as confusion matrix, ROC curve, and classification metrics like accuracy, recall, and F1 score are used to evaluate each

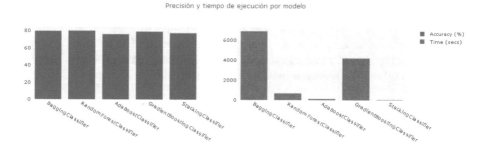

Precisión y tiempo de ejecución por modelo

Fig. 5. Accuracy and run time per model on the obesity level data set

model. The RandomForestClassifier model shows high classification performance for different classes, with recovery rates ranging from 65% to 100% and F1 values ranging from 60 to 99%. The model can correctly classify instances into different classes, although some classes have higher accuracy than others. Overall, the RandomForestClassifier model is reliable for classifying instances in the dataset. See Fig. 5 and for more details see https://github.com/kellyturbay/EnsamblesDiabetesObesity.

The classification error of the Stacking model was 3.0%, indicating that it is performing well on the training set. The performance of each individual model was evaluated using metrics such as precision, recall, and f1 value. Ver Fig. 6.

```
==== Sumario de la clasificación ====
            precision   recall  f1-score   support

         0      1.00      0.99      0.99        93
         1      0.99      0.90      0.94        78
         2      0.98      0.97      0.98       104
         3      0.98      0.99      0.98        99
         4      1.00      1.00      1.00        90
         5      0.96      0.99      0.97        86
         6      0.92      0.98      0.95        83

  accuracy                          0.97       633
 macro avg      0.97      0.97      0.97       633
weighted avg    0.98      0.97      0.97       633

Accuracy -> 97.47%
```

(a) Obesity Levels data set classification report

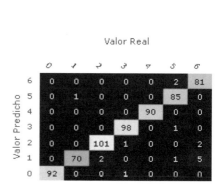

Matriz de Confusión

(b) Evaluation by confusion matrix in the Obesity dataset

Fig. 6. Performance evaluation of the Stacking method on the Obesity Levels dataset

The overall accuracy of the Stacking model was 97.47%, with accuracies for each class ranging from 92 to 100%. The recall metric showed that the model was able to correctly retrieve most instances of each class, with values ranging from 90% to 100%. The F1 value metric also indicated high performance across all classes. Overall, the Stacking model achieved optimal accuracy for the obesity levels dataset. Additionally, it was observed that varying the number of cross-validation partitions did not necessarily result in lower training error. https://github.com/kellyturbay/EnsamblesDiabetesObesity.

3.2   Early Detection of Diabetes

The dataset used in this study was obtained from the UCI Machine Learning repository [16]. It is composed of 519 rows with 17 columns. Prior to performing the analysis, various preprocessing processes were applied to the data. This included duplicate data detection and removal, as well as transformations to ensure that the data were in the correct types. Manual feature extraction and selection of the most relevant features was also performed. In addition, target class balancing was performed to ensure a balanced distribution between cases of people with and without diabetes. Importantly, no missing values or outliers were identified in the data set, so no data imputation was necessary. After applying transformations, we have obtained a dataset summarizing about 510 records in 13 columns.

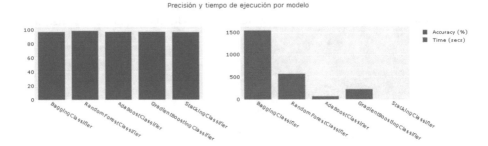

Fig. 7. Accuracy and run time per model in the diabetes early detection data set

Based on the above results (See Fig. 7), it is observed that the accuracy of all models in the validation set is remarkably high, exceeding 90%. These findings indicate that all models perform satisfactorily as classifiers on this data set. The StackingClassifier stands out for its good results and shorter training time. However, the RandomForestClassifier is determined to be the best classifier, achieving an accuracy of 98.96% in 576.85 s. Evaluation metrics such as confusion matrix, precision, recall, and f1-score are also presented for the RandomForestClassifier, which shows high performance in classifying positive instances with an overall accuracy of 98.96%.

Next, using the stacking technique, the parameters found to perform best for each base model in the diabetes early detection dataset are shown. In evaluating the performance of the Stacking technique, a classification error of 2.1%

was obtained. This indicates that the model is performing satisfactorily, as it misclassifies only 2.1% of the instances in the training set. For more details see https://github.com/kellyturbay/EnsamblesDiabetesObesity.

In addition, each model has been evaluated using additional metrics, which include generating the confusion matrix and calculating metrics such as precision, recall and f1 value for each class, as well as the overall model accuracy. See Fig. 8.

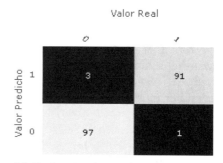

```
==== Sumario de la clasificación ====
             precision   recall  f1-score   support

          0       0.97     0.99      0.98        98
          1       0.99     0.97      0.98        94

   accuracy                         0.98       192
  macro avg       0.98     0.98      0.98       192
weighted avg      0.98     0.98      0.98       192

Accuracy -> 97.92%
```

(a) Diabetes Dataset Classification Report

(b) Evaluation by confusion matrix in the early detection dataset of Diabetes

Fig. 8. Performance evaluation of the Stacking method in the Diabetes early detection dataset

The Stacking model applied to the diabetes level data set shows an overall accuracy of 97.92%, indicating that it is able to correctly classify approximately 98% of the test data. The individual accuracy for each class varies between 97% and 99%, demonstrating good performance across all classes. The recall metric shows that the model is able to correctly recover most of the positive cases in each class, with values ranging from 97% to 99%. In addition, outstanding performance is observed in the f1-score metric for all classes. In conclusion, the Stacking model is effective in predicting the level of diabetes in the data set, with acceptable overall accuracy. It is also highlighted that performing 10 partitions in the cross-validation generates a lower training error for each estimator. For more details see https://github.com/kellyturbay/EnsamblesDiabetesObesity.

4 Conclusions

In conclusion, this project focuses on the application of several ensemble models to the dataset obtained from the UCI Machine Learning repository website. The results obtained are expected to contribute to the development of efficient tools for the detection and prevention of obesity, based on the information collected in the reference article and the available dataset. Furthermore, these sophisticated generic models that have been developed could well be adapted to other more complex datasets. This issue will be dealt with extensively in future lines.

By applying the meta-learning approach and using ensemble techniques such as Stacking, Bagging and Boosting, it has been shown that it is an effective strategy that improves the accuracy and efficiency in the classification of clinical data. By combining different algorithms, it takes advantage of the collective results of the base models, and thus, it is possible to reach better conclusions by overcoming the limitations of each one individually. This ability to learn and compare different algorithms to fit specific data sets proves to be a powerful tool in medical decision making. Regarding future work, other lines of research are proposed that can extend and enrich this study, such as experimenting with the application of these models to more complex and specific datasets with a greater variety of medical characteristics. Likewise, it would be enriching to analyze the performance of ensemble models in different populations and clinical contexts, allowing generalization to different scenarios. It is also proposed to use other machine learning strategies to establish parameters and hyperparameters, such as genetic algorithms or gradient descent techniques. Additionally, it is possible to experiment with the use of other machine learning techniques and advanced algorithms, such as convolutional neural networks or recurrent networks to address more complex and multidimensional diseases in the field of health. Finally, tools could be developed to visualize the results, such as generating real-time charts that allow medical professions to easily interpret and make decisions.

Acknowledgements. This result has been supported through the Spanish Government by the project PID2021-127275OB-I00, FEDER "Una manera de hacer Europa".

References

1. United Nations: Population— United Nations (2022). https://www.un.org/en/global-issues/population
2. Harlow, H.F.: The formation of learning sets. Psychol. Rev. **56**(1), 51 (1949)
3. Vilalta, R., Drissi, Y.: A perspective view and survey of meta-learning. Artif. Intell. Rev. **18**, 77–95 (2002)
4. Wang, J.X.: Meta-learning in natural and artificial intelligence. Curr. Opin. Behav. Sci. **38**, 90–95 (2021)
5. Thrun, S., Pratt, L.: Learning to learn: introduction and overview. In: Thrun, S., Pratt, L. (eds.) Learning to learn. Springer, Boston, pp. 3–17 (1998). https://doi.org/10.1007/978-1-4615-5529-2_1
6. Finn, C., Abbeel, P., Levine, S.: Model-agnostic meta-learning for fast adaptation of deep networks. In: International Conference on Machine Learning, pp. 1126–1135. PMLR (2017)
7. Ravi, S., Larochelle, H.: Optimization as a model for few-shot learning. In: International Conference on Learning Representations (2016)

8. Vinyals, O., Blundell, C., Lillicrap, T., Wierstra, D., et al.: Matching networks for one shot learning. In: Advances in Neural Information Processing Systems, vol. 29 (2016)
9. Andrychowicz, M., et al.: Learning to learn by gradient descent by gradient descent. In: Advances in Neural Information Processing Systems, vol. 29 (2016)
10. Snell, J., Swersky, K., Zemel, R.: Prototypical networks for few-shot learning. In: Advances in Neural Information Processing Systems, vol. 30 (2017)
11. Turbay, K.: EnsamblesDiabetesObesity (2023). https://github.com/kellyturbay/EnsamblesDiabetesObesity
12. López, J.A.: Bloque iii. técnicas de análisis ensembles. In: *2023 UA Bloque III. Técnicas de análisis Ensembles*. Universidad de Alicente (2023)
13. Sollich, P., Krogh, A.: Learning with ensembles: how overfitting can be useful. In: Advances in Neural Information Processing Systems, vol. 8 (1995)
14. Pedregosa, F., et al.: Scikit-learn: machine learning in Python. J Mach. Learn. Res. **12**, 2825–2830 (2011). https://scikit-learn.org/stable/modules/ensemble.html
15. "Estimation of obesity levels based on eating habits and physical condition", UCI Machine Learning Repository (2019). https://doi.org/10.24432/C5H31Z
16. "Early stage diabetes risk prediction dataset", UCI Machine Learning Repository (2020). https://doi.org/10.24432/C5VG8H

Bridging the Gap: Enhancing Geospatial Analysis with Natural Language and Scenario Generation Language

Jonathan Frez[1]([✉]) and Nelson Baloian[2]

[1] Universidad Diego Portales Vergara 435, Santiago, Chile
jonathan.frez@mail.udp.cl
[2] Department of Computer Science, Universidad de Chile, Beauchef 851, Santiago, Chile
nbaloian@dcc.uchile.cl

Abstract. Scenario Generation Language (SGL) is a powerful tool that simplifies geospatial analysis and decision-making processes, removing the requirement for users to have expertise in GIS or SQL. However, users still need to understand the SGL grammar. This paper introduces a novel approach that utilizes GPT (Generative Pre-trained Transformer) - LLM (Large Language Model) to generate SGL statements directly from natural language questions. By leveraging the capabilities of GPT-LLM, this approach bridges the gap between user intent and technical query construction, enhancing the usability and accessibility of SGL. It enables decision-makers to interact with geospatial data using familiar natural language queries, without the need for in-depth knowledge of SGL or complex geospatial querying techniques. The integration of natural language processing with SGL empowers users to effortlessly generate accurate and syntactically correct statements, streamlining the analysis process and facilitating scenario exploration. Experimental results indicate that directly utilizing GPT-LLM for geospatial analysis may not yield satisfactory results. However, the approach presented in this paper demonstrates its effectiveness in simplifying geospatial analysis and supporting informed decision-making.

Keywords: Scenario Generation Language (SGL) · geospatial analysis · decision-making · GPT (Generative Pre-trained Transformer) · LLM (Large Language Model)

1 Introduction

Decision Support Systems (DSS) are interactive digital platforms designed to assist decision-makers in tackling unstructured challenges by utilizing data and models. The decision-making process typically involves problem definition, objective identification, scenario creation, alternative scrutiny, choice selection, and sensitivity analysis. DSS addresses ill-structured problems, where definitions may lack clarity, and data sources can be diverse and contradictory. Potential solutions are numerous, without a single "best" or "correct" answer. Despite the

J. Bravo and G. Urzáiz (Eds.): UCAmI 2023, LNNS 842, pp. 252–263, 2023.
https://doi.org/10.1007/978-3-031-48642-5_24

wide-ranging features of DSS described in academic literature, there is consensus on these key traits.

In this complex process, human judgment plays a crucial role in generating alternatives, reassessing objectives, and refining decisions. Since decision problems lack precise inputs, various scenarios are generated to explore "what if" questions. Computers support this by aggregating data, producing decision alternatives, evaluating outcomes against objectives, visualizing results, and facilitating communication.

Spatial Decision Support Systems (SDSS) are a specialized category of Decision Support Systems that incorporate geographical or spatial data. By merging Geographic Information Systems (GIS) capabilities with traditional DSS components, SDSS becomes a comprehensive tool for spatially-oriented decision-making. Spatial data involves location-specific information such as coordinates, area, or elevation. SDSS effectively manages and manipulates this data, allowing users to visualize patterns, trends, and relationships that might not be apparent in non-spatial data.

SDSS finds applications in various domains where location-specific data is crucial. These include Urban Planning and Development, Natural Resource Management, Public Health, Emergency Response, Agriculture, and more. SDSS enhances decision-making by enabling visualization, manipulation, and analysis of spatial data, allowing decision-makers to comprehend complex spatial relationships and make well-informed choices. It often involves generating heat maps or suitability maps that visually represent the degree of compliance with predefined conditions. However, conducting such analyses requires expertise in handling geographic data and producing visualizations.

This work introduces a framework to address technical complexities in [4,5], aiming to assist decision-makers in generating and collaboratively analyzing suitability maps. The core element of this framework is the Scenario Generating Language (SGL) definition, similar to SQL, but specifically designed for specifying suitability map requirements with machine-interpretable sentences. Unlike traditional systems using relational databases, this framework provides a graphical interface, making it more intuitive and user-friendly, eliminating the need for users to know the language syntax beforehand.

However, this approach's drawback is that it limits the language's flexibility and expressiveness, restricting its potential. This research delves into exploring the use of natural language for specifying map requirements, potentially enhancing the language's capabilities and user experience.

2   Related Work

2.1   Using GPT-LLM to Generate Structured Output from Natural Language Instructions

Currently, the utilization of natural language to generate geographical decision-making scenarios remains a relatively unexplored area. The main reason for this is the lack of existing technology capable of transforming natural language instructions into analytical processes or sets of query and calculation operations. However, in the past year, with the advent of GPT-LLM, we have witnessed the emergence of various natural language transformers for different types of generation. In particular, we are interested on text2text transformers that generates an structured text given a natural language expression.

For instance, in [1], ProGen is introduced as a language model that can generate protein sequences with reliable functions across large protein families. This process is akin to generating grammatically and semantically accurate natural language sentences on diverse subjects. This remarkable development demonstrates the potential of using an GPT-LLM to generate text strings that adhere to standardized grammatical conventions.

In [2], the authors introduce MathPrompter, which is capable of generating algebraic expressions based on math problems expressed in natural language. Furthermore, they leverage the power of GPT-LLMs to develop Python programs that can execute and solve the math problems originally stated in natural language, using the generated mathematical expressions. The importance of this strategy for our work is that it demonstrates algebra's role as an intermediate language between the GPT-LLM and code execution. By leveraging algebra, we can directly express and solve problems using natural language, while also generating executable code. This approach highlights the versatility of algebra as a bridge between the language model and the practical implementation of solutions.

2.2   The Scenario Generation Language

The SGL is a tool developed to create suitability maps and compare outcomes based on different hypotheses. It aims to generate scenarios using expert knowledge, empirical data, and existing environmental models. The language allows users to specify scenarios based on specific queries related to three main objects of analysis: 1) People, 2) Transportation demand, and 3) Burglary risk. These objects must be modeled based on the Suitability Object definition, and the hypotheses needed to answer the queries are supported using Data Objects as support data.

SGL queries consist of three main components: 1. Object of Analysis: This component defines what is being analyzed in the scenario. For example, to generate a suitability map for people, the user should use the following sentence:

```
suitable @people
```

2. Hypotheses: This component allows experts to express their knowledge using multiple hypotheses. For instance, when looking for persons, one hypothesis may be

"people use to be in cinemas with a mass of 20%" or "people are in schools or workplaces with a mass of 30%," with additional filters like considering cinemas with more than 200 people capacity:

```
hypothesis {@cinema}20\% {@school, @workplace}30\%
where @cinema.capacity > 200
```

3. Model Interactions: This component represents real-world interactions between the elements in the Data Object set. For example, it can express that a high density of people is not expected in a lake or sea, but rather in a sports stadium during an event. These interactions are represented using values in an interval. For example:

```
Model@stadium{day{13:100},day{17:100}} @lake{0%}
```

means a 100% value for stadiums at 1 pm and 5 pm, and 0% for lake areas.

A full Scenario definition may look like this:

```
suitable @persons hypothesis {@cinema}20% {@school,@workplace}30%
{@shops}20% where @shops.capacity > 20 Model @stadium{100} @lake{0%}
@school{day{8:100}, day{14:100}}
```

The SGL language is designed to be extended to allow users to specify other types of maps (e.g., belief or plausibility maps) and incorporate additional filters like defining polygons for map visualization and specifying temporal attributes for data used in generating the map. This flexibility makes it a versatile tool for scenario generation and analysis.

As part of the prompt provided to the GPT-LLM, it includes the formal grammar definition of SGL as presented in Fig. 1:

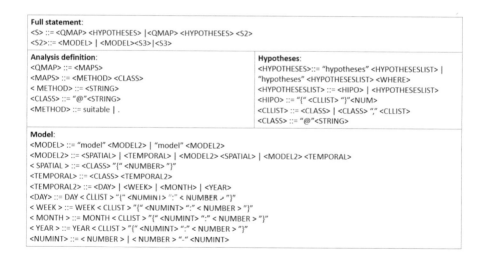

Fig. 1. SGL formal Grammar

3   Implementing a Natural Language Interface for SGL

As the reader may imagine, there are multiple potential points of failure in scenario generation. While most of these can be managed, there is one particular challenge: the GPT-LLM's ability to consistently generate correct expressions in SGL is not always guaranteed. To address this, the concept of "reflection" comes into play, leveraging a property of certain GPT-LLM. Reflection [3] refers to the GPT-LLM's capacity to engage in self-analysis by verbally reflecting on task feedback signals. This self-reflective process involves maintaining a dedicated episodic memory buffer to enhance decision-making in subsequent trials. In our case, we have adapted the syntax analyzer of SGL to offer verbal feedback to the LLM whenever a syntax error is detected. This integration of reflection enables the GPT-LLM to learn from its mistakes and improve its ability to generate accurate SGL expressions over time (See Fig. 2).

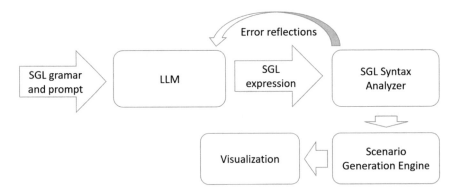

Fig. 2. Diagram of scenario generation process

The scenario generation process initiates by informing the LLM that it will be operating with a specific language, which may be unfamiliar or not present in its training set. This initial prompt serves as an extended version of the following text, starting with an explanation of the language components, using examples and ending with the formal grammar description:

The Prompt:
SGL is a Scenario Generation Language, it purpose is to make spatial decision scenarios based on queries like the followings:

- 1. Which are the areas with most concentration of people if we assume that there will be people at cinemas, workplaces and around schools

- 2. What is the point of high demand for public transportation if we assume that people will require it near shopping areas, schools, and hospitals?
- 3 others examples

The main components of an SGL query are the following:

- 1. Suitable object
- 2. Hypotheses
- 3. Model Interactions

An SGL sentence will look like: <Suitable object> hypotheses <Hypotheses> model <Model Interactions>

When the Object of Analysis component is to generate the suitability map for people, the user should use the following sentence:

"suitable @people"

In the Hypothesis component, the expert can express his/her knowledge using multiple hypotheses. In the following example, the expert is looking for persons; then one hypothesis may be"people use to be in cinemas with a mass of 20% or schools and workplaces with 30%":

"hypothesis @cinema20% @school, @workplace30%

The model characteristics statement is designed to represent real-world interactions between the elements in the Data Object set. For example, a high density of people is not expected in a lake or sea. On the contrary, we expect a high density of people inside a sports stadium at an event. These kind of interactions complements the behavior statement by adding environmental rules. This rule can be expressed using values in an interval. For example, if we are generating a scenario for persons, we add in the model statement a 0% value for lake areas and 100% value for stadiums at 1 pm and 5 pm. This value can decrease or increase the certainty level in the indicated area:

"Model @stadiumday13:100,day17:100 @lake0"

A full Scenario definition will then look like the following:

"suitable @persons hypothesis @cinema20% @school,@workplace30% @shops20% Model @stadium100 @lake0 @schoolday8:100, day14:100"

The SGL grammar is the following:
<S> ::= <QMAP> <HYPOTHESES>|<QMAP> <HYPOTHESES> <S2>
<S2> ::= <MODEL> | <MODEL><S3>|<S3>

Analysis definition:
.... Grammar here ...
Hypotheses
.... Grammar here ...
Model Interactions
.... Grammar here ...

User Input:
At the conclusion of the prompt, several semantic rules are established. For instance, a list of permissible objects or classes that can be utilized is provided. Finally, the user's input is to be incorporated in the following manner: "I am seeking an SGL statement that precisely addresses the subsequent question. Please provide only the SGL statement in English, devoid of any additional explanations: «user input»."

4   Preliminary Results

We have extensively experimented with various attempts and prompt adaptations to ensure the generation of syntactically correct statements. However, the generative aspect of the LLM introduces an inherent creativity factor, essential for producing plausible scenario formulations, but potentially challenging for adhering strictly to grammar rules. To overcome this challenge, we implemented the reflection process described earlier. Through this process, we modified the SGL's syntax analyzer to provide error messages in a more expressive manner. These error messages are incorporated at the outset of the prompt, as outlined in the previous section, before re-querying the system. As a result, for SGL, we observe that it takes fewer than 10 iterations to generate a syntactically correct statement in response to a given input.

To better illustrate the input types utilized and the results achievable, let us present some illustrative examples:

Each SGL statement is ultimately handed over to the interpreter, which diligently analyzes and executes the required tasks to collect and scrutinize data concerning the hypotheses. The interpreter evaluates the probabilistic models associated with the chosen classes, such as the movement of individuals around a point of interest, and carefully assesses the interactions between elements as described in the modeling. Finally, the interpreter generates a suitability map, which can be easily visualized using a simple viewer. For example, the following visualization (Fig. 3) correspond to the evaluation of the first SGL statement in Table 1:

Table 1. Example SGL statements generated from natural language inputs

NL	im looking for crimes that might occur in the vicinity of schools. I have received information suggesting a potential association between crimes and locations such as ATM machines and bus stops.
SGL	suitable @crimes hypotheses {@atm, @bus_stops}100% {@schools}50%
NL	I'm looking to determine a location for a shop. Ideally, I am seeking a place that experiences high foot traffic and is in close to other shops. Considerations such as a bustling mall or a location near popular restaurants and cinemas would be suitable.
SGL	suitable @consumingPlaces hypothesis {@people}50% {@shops}50% model @mall{100} @restaurant {day{14:100}, day{19:100}} @cinema{100}
NL	im looking for locations to places ev charging stations, parkings slots will do, also universities, restaurants and bars.
SGL	suitable @evchargingstations hypothesis {@parking}50\% Model @university{100} @restaurant{50} @bar{50}
NL	I know that public transportation demand is strongly related to amenity places like cafe,fast_food,food_court, pub,restaurant,college, library, school, atm, bank, clinic, hospital, pharmacy, and cinema . i want a map of possible transportation demand.
SGL	suitable @transportation_demand hypotheses {@cafe, @fast_food, @food_court, @pub, @restaurant, @college, @library, @school, @atm, @bank, @clinic, @hospital, @pharmacy, @cinema}100% model @cafe{90} @fast_food{90} @food_court{90} @pub{90} @restaurant{90} @college{100} @library{100} @school{100} @atm{70} @bank{70} @clinic{60} @hospital{100} @pharmacy{70} @cinema{100}.

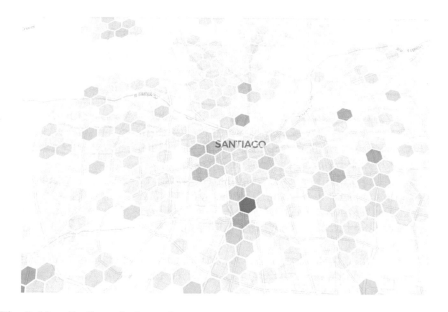

Fig. 3. Visualization of crimes that can occur near a school given a natural language input

5   Generalizing the Process

To achieve results that are currently only possible through specific query-like languages using natural language, the method must be generalized by leveraging a LLM. This involves adapting the concept of "reflection" and integrating it into the "compiling" process of the query-like language.

The initial step involves elaborating a comprehensive prompt that explains the language syntax. This entails providing verbal explanations of the various syntax options, offering specific and detailed usage examples, and providing explanations for each example. Additionally, it is essential to specify the formal grammar of the language. We call this prompt "language prompt" (See Fig. 4).

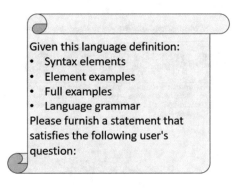

Given this language definition:
- Syntax elements
- Element examples
- Full examples
- Language grammar

Please furnish a statement that satisfies the following user's question:

Fig. 4. language prompt

For lesser-known languages, it is crucial to provide a comprehensive verbal explanation of the syntax elements. Additionally, defining the syntax rules and grammar is necessary. This step is essential for reducing the syntax errors and providing valuable input to the language model.

The next step involves concatenating (in order) the language prompt with the user prompt, and adding the following phrase: 'I want a language statement that answers the question, provide only the statement, all in English, without any additional discussion:' Once this full prompt is prepared, it is sent to the LLM for processing.

The LLM will return a language statement, witch could have lexical, syntactical or semantic errors. In these case, a feedback mechanism must be implemented to generate an improved statement.

The feedback mechanism is divided in two parts: An adapted syntax analyzer, and a reflection feedback mechanism:

Adapted Syntax Analyzer: A custom syntax analyzer designed specifically for the target language. This analyzer is responsible for validating the correctness of the generated expressions based on the defined grammar rules and providing a verbal explanation of any syntax errors encountered.

Reflection Feedback Mechanism: The utilized LLM may feature an episodic memory buffer dedicated to storing past inputs and their corresponding feedback signals. This memory enhances decision-making in subsequent trials. In case the LLM possesses this type of memory, a syntax error mechanism can be triggered, prompting an improvement of the generated language expression. See Fig. 5.

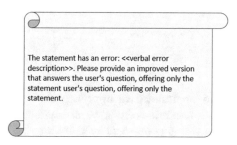

Fig. 5. Reflection Feedback Mechanism With episodic memory

If the LLM does not support an episodic memory, a similar prompt can be employed by concatenating the language prompt, user prompt, and a notification explaining that the previously generated statement contains an error. The user can then be asked to provide a new and corrected statement: "This statement '«previous statement»' has an error: «verbal error description». Please provide an improved version that answers the user's question, offering only the statement.". See Fig. 6.

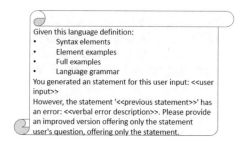

Fig. 6. Reflection Feedback Mechanism Without episodic memory

Through the feedback mechanism, the LLM can harness its episodic memory, learning from past mistakes and continuously improving its performance over time. By reflecting on the received feedback, the LLM gains valuable insights, enabling it to make more informed decisions while generating scenarios. Consequently, the LLM retains a deeper understanding of the language syntax, leading to the generation of more accurate and refined statements. This can be used to generate an iterative Refinement allowing the LLM to undergo iterative refinement, where it continuously generates statements, receives feedback, updates its episodic memory, and refines its generation process.

It's important to note that adapting this method to different languages and LLMs may require specific considerations depending on the characteristics of the language and LLM being used. Additionally, the quality and quantity of feedback data will also play a crucial role in the LLM's learning process and overall performance.

6   Conclusions

In this work, we present an innovative approach that leverages GPT-LLM to directly generate SGL statements from natural language questions. This approach effectively bridges the gap between user intent and technical query construction, significantly enhancing the usability and accessibility of SGL. The integration of natural language processing with SGL empowers users to interact with geospatial data using familiar natural language queries, eliminating the need for in-depth knowledge of SGL or complex geospatial querying techniques. Preliminary experimental results demonstrate the effectiveness of our approach in simplifying geospatial analysis and supporting informed decision-making.

According to our assessment, the most pivotal milestone in achieving our objective is to refine the syntax analyzer's output, enabling it to express errors verbally and provide more comprehensive explanations for syntactic issues. This strategic adjustment is crucial as LLMs possess a creativity parameter, vital for generating diverse outputs and embracing the concept of reflection (the LLM's ability to analyze its actions and learn from mistakes). However, this creative aspect may conflict with the necessity of adhering strictly to a predefined grammar. Thus, harnessing reflection as a syntactic corrector can continuously enhance the model's capacity to generate precise expressions over time, ultimately bolstering the reliability of our proposed approach.

As a future work, advancing towards semantic correctness in class selection and ensuring proper utilization of SGL modeling parameters is proposed. This advancement will enable the generation of dynamic scenarios, where the correct interpretation and meaningful representation of classes play a pivotal role. By incorporating semantic correction mechanisms, the model's capability to produce more accurate and contextually relevant scenarios will be significantly enhanced, offering a more sophisticated and adaptable approach for scenario generation.

References

1. Ali, M., et al.: Large language models generate functional protein sequences across diverse families. Nat. Biotechnol. **41**, 1–8 (2023)
2. Imani, S., Du, L., Shrivastava, H.: MathPrompter: mathematical reasoning using large language models. arXiv preprint arXiv:2303.05398 (2023)

3. Shinn, N., Cassano, F., Labash, B., Gopinath, A., Narasimhan, K., Yao, S.: *Refl exion: language agents with verbal reinforcement learning.* [arXiv preprint]. arXiv:2303.11366 (2023)
4. Frez, J., Baloian, N., Zurita, G., Pino, J.A.: Dealing with incomplete and uncertain context data in geographic information systems. In: Proceedings of the 2014 IEEE 18th International Conference on Computer Supported Cooperative Work in Design (CSCWD), pp. 129–134. IEEE explore, Hsinchu, Taiwan, May 21–23 (2014)
5. Frez, J., Baloian, N., Zurita, G.: Getting serious about integrating decision support mechanisms into geographic information systems. In: Ninth International Conference on Computer Science and Information Technologies Revised Selected Papers, pp. 1–11. IEEE explore, Yerevan, Armenia (2013)

DAKTILOS: An Interactive Platform for Teaching Mexican Sign Language (LSM)

Jesús Javier Gortarez-Pelayo[1] ⓘ, Ricardo Fernando Morfín-Chávez[2] ⓘ, and Irvin Hussein Lopez-Nava[2(✉)] ⓘ

[1] Faculty of Sciences, Universidad Autónoma de Baja California, 22860 Ensenada, BC, Mexico
jesus.gortarez@uabc.edu.mx
[2] Computer Science Department, Centro de Investigación Científica y de Educación Superior de Ensenada, 22860 Ensenada, BC, Mexico
{morfinrf,hussein}@cicese.edu.mx

Abstract. This paper presents an interactive platform for teaching Mexican Sign Language (LSM) based on Artificial Intelligence (AI) models named DAKTILOS. This platform was developed with recent Web technologies that allow integrating real-time hand tracking from 2D images natively. The platform was designed to recognize and score static and dynamic LSM alphabet signs made by users. Once the 21 hand-keypoints were recognized and extracted from the AI MediaPipe model, they were dynamically compared to the target manual configurations (letters) using AI FingerPose classifier, and scored to determine if it was performed correctly providing visual feedback. DAKTILOS allows exploring 27 letters indicating the correct configuration setting with a 3D hand. Preliminary tests with eight subjects were conducted to determine the functionality of the platform. Finally, this teaching tool can help to bridge the communication gap between the deaf and hearing communities.

Keywords: Mexican Sign Language · Sign Language Recognition · Learning platform · MediaPipe · FingerPose · Teaching tool

1 Introduction

In a world where communication is the cornerstone of human interaction, the significance of sign languages cannot be underestimated. For individuals who are deaf or hard of hearing, sign languages serve as lifelines, providing them with a means to express themselves, understand their surroundings, and connect with the world around them [2].

Sign languages, with their unique grammar, syntax, and vocabulary, represent complete and independent linguistic systems [3]. They have been crucial in breaking barriers for the deaf community and facilitating communication, enabling them to flourish in diverse environments.

Supported by Consejo Nacional de Ciencia y Tecnología.

In the realm of cognitive and linguistic development, studies have unveiled fascinating insights into the advantages of early exposure to sign languages. Research findings have indicated that deaf children exposed to sign languages from an early age exhibit enhanced language skills and cognitive abilities compared to their peers without access to sign languages [5].

This discovery underscores the pivotal role of sign languages as a foundational tool in shaping the cognitive potential of young deaf individuals. Numerous studies have highlighted the positive impact of incorporating sign languages in educational settings, demonstrating improved academic achievements and a better grasp of curriculum content among deaf students [4].

Furthermore, sign languages are more than just communication tools; they are the bedrock of cultural identity for the deaf community. Embracing sign languages means celebrating and preserving the unique cultural heritage of deaf individuals, allowing them to express their experiences, emotions, and stories in a language that resonates with their core being.

In particular, Mexican Sign Language (LSM) emerges as a distinct, and necessary, linguistic standard, offering a profound impact on the lives of Mexico's deaf community [6]. As a unique mode of communication, LSM not only provides deaf individuals with a means to interact and express themselves but also serves as an essential conduit for cultural preservation, social inclusion, and education [1]. Teaching LSM effectively requires innovative tools that fit different learning needs and encourage cultural understanding. Online and mobile platforms offer new ways of teaching LSM, breaking the constraints of physical space and enabling tailored, self-paced learning.

In this paper, we present an interactive web platform for teaching the basics of the LSM (DAKTILOS), using a 2D camera to track hands and providing real-time feedback. The following sections present (Sect. 2) the elements of the LSM that are taught through the platform, (Sect. 3) the technologies used for the development of the platform, (Sect. 4) the preliminary evaluation carried out and (Sect. 5) the next steps to continue the development of the LSM teaching tool.

2   LSM Alphabet

To simplify the teaching of the LSM, it can be divided into two components: (i) purely manual signs to represent the alphabet and spell words, and (ii) signs that require the movement of more than one hand, facial expressions, and/or body movements. The first component is called dactylology or fingerspelling, and is commonly used to spell proper names, words that do not yet have an associated sign, new or technical terms, and even words from other languages.

The LSM alphabet is composed of 27 letters, of which 21 are static, that is, the representation does not change in time, from beginning to end, and does not require any hand movement; meanwhile, the other 6 letters are dynamic, which means that a hand movement is needed to represent them. In both cases, the fingers remain immobile, and only in dynamic signs is there a change in position or orientation of the entire hand. Figure 1 shows the manual configuration of the LSM alphabet, indicating with arrows the movement of the hand for signs that are dynamic, e.g., the J sign requires a movement in a curve from back to front, and top-bottom-top.

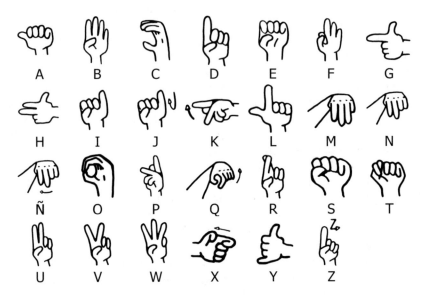

Fig. 1. LSM alphabet (signs are displayed as if the user is looking at them in a mirror).

3　Methods

The data flow of the DAKTILOS platform, starting with the input from the images captured with the camera to the visual feedback on the screen together with the automatic detection of the signs is shown in Fig. 2.

The proposed platform uses MediaPipe[1] and FingerPose[2] to track and detect manual configurations made by users. MediaPipe employs machine learning to infer 21 keypoints of the hand in real-time from 2D video, such as a webcam. FingerPose infers fingers position and orientation based on the keypoints detected.

The DAKTILOS platform utilizes a two-step process to recognize LSM signs. Firstly, keypoints of the hand are extracted in real-time using the Mediapipe library from 2D video capture by the camera. These keypoints are then processed (since they are not directly compatible with FingerPose) and passed to the FingerPose module, which infers the position and orientation of the fingers.

---

[1]　https://github.com/google/mediapipe.
[2]　https://github.com/andypotato/fingerpose.

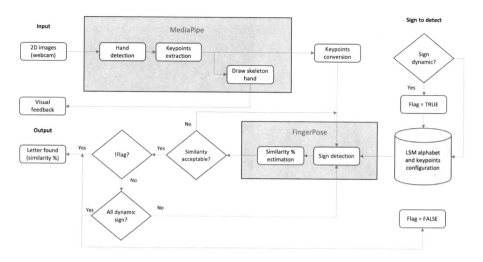

Fig. 2. Data flowchart of DAKTILOS.

Secondly, the platform uses templates that describe the curl and direction of each finger for every letter in the LSM alphabet (see Fig. 1), for dynamic letters, two hand positions are defined: the start position and the end position. By comparing the user's hand configurations with these templates, the platform estimates the similarity percentage for the identified sign.

Once the acceptance percentage threshold is reached, the platform determines whether the letter is static or dynamic. If the letter is static, the platform sends the identified letter along with its corresponding similarity percentage. However, if the letter is dynamic, the platform continues detecting the sign and calculating the similarity percentage based on the current phase of the sign.

Table 1. Classification of type of letters from LSM alphabet.

Static	Simple dynamic	Special dynamic
A, B, C, D, E, F, G, H, I, L, M, N, S, O, P, R, S, T, U, V, W, Y	J, K, Ñ, Q	X, Z

Classification is performed by separating the letter into three proposed types: static, simple dynamic, special dynamic, as shown in Table 1. Static signs are signs that involve stationary hand configurations. Simple dynamic letters are those in which the direction of the fingers change from their beginning to their end. On the other hand, special dynamics are those in which the direction of the fingers remains constant, but the position of the hand position change.

For simple and special dynamic letters, a flag is added to identify the phase of the sign. At all times, the platform provides a visual indicator of the similarity percentage, and in the case of dynamics its phase, as shown in Fig 3.

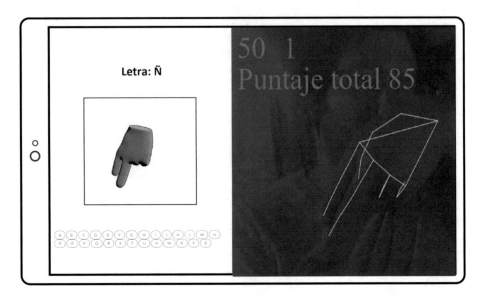

Fig. 3. Example of detection of letter Ñ. 50 is the current similarity of the phase, 1 is the current phase, 85 is the average of the score obtained in the 2 phases

4   Evaluation

A preliminary evaluation was carried out with 8 subjects (28 ± 12.9 years) with no prior knowledge of the LSM, to determine which letters might present more challenges during detection. The subjects were instructed to achieve a score of 100, which is the maximum. These results are shown in Fig. 4.

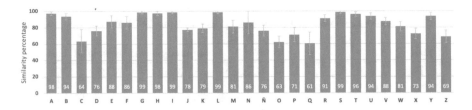

Fig. 4. Results of preliminary evaluation.

In summary, the similarity percentage of all letters among all subjects was 84.4 ± 11.4, and grouping the letters by type: static with 87.7 ± 11.3, simple dynamic 73.4 ± 8.6, and special dynamic 70.6 ±2.7. The detection of the dynamic letters, in general, had a major problem and is attributed to the hand tracking by MediaPipe, which caused difficulties in phase changes.

The letters with a detection greater than 90 were (11 in total): A, B, G, H, I, L, R, S, T, U, and Y; and with a detection of less than 70 were: C, O, and

Q; of which the first two are static letters. The reason for the low results for the letters C and O could be due to the finger flexion being very stiff compared to other configurations.

5   Conclusions

The teaching platform offers users the possibility to learn the 27 letters of the LSM alphabet by displaying and scoring their hand configuration. Sign detection was tested with a variety of subjects and accuracy was good. However, the platform faces difficulties in detecting the hand in unfavorable lighting conditions, occlusion by other hand segments, or with unclear backgrounds, which especially affects dynamic letters.

The current implementation does not include double letters (e.g., LL, RR, AA), which could be considered as special dynamic letters, although they are planned to be added in the future. In addition, the platform does not address the use of punctuation marks and umlauts, which limits the full teaching of the language.

DAKTILOS encourages practice and is designed so that users can also learn their name without having to search through all the letters, and is ready for learning the most commonly used words in daily life, along with a game of guessing words of a specific topic, e.g. countries, from dactylology.

References

1. Aldrete, M.C.: Gramática de la lengua de señas mexicana. El colegio de méxico (2008)
2. Bauman, H.D.L., Murray, J.J.: Deaf studies in the 21st century: "deaf-gain" and the future of human diversity. In: The Oxford Handbook of Deaf Studies, Language, and Education, vol. 2, no. 2, p. 210 (2010)
3. Emmorey, K.: Language, Cognition, and The Brain: Insights from Sign Language Research. Psychology Press (2001)
4. Marschark, M., Spencer, P.E.: The Oxford Handbook of Deaf Studies, Language, and Education, vol. 2. Oxford University Press (2010)
5. Petitto, L.A., Langdon, C., Stone, A., Andriola, D., Kartheiser, G., Cochran, C.: Visual sign phonology: insights into human reading and language from a natural soundless phonology. Wiley Interdisc. Rev. Cogn. Sci. **7**(6), 366–381 (2016)
6. Quinto-Pozos, D.G.: Contact between Mexican sign Language and American Sign Language in two Texas border areas. The University of Texas at Austin (2002)

Evaluation of Four Classification Algorithms Applied to Detection of Poisonous Mushrooms

Jacqueline S. Ortiz-Letechipia, Ricardo Villagrana-Bañuelos[✉],
Carlos E. Galván-Tejada, Jorge I. Galván-Tejada, and José M. Celaya-Padilla

Unidad Académica de Ingeniería Eléctrica, Universidad Autónoma de Zacatecas,
Jardín Juarez 147, Centro, 98000 Zacatecas, Zac, Mexico
{36172849,34150131,ericgalvan,gatejo,jose.c110padilla}@uaz.edu.mx

Abstract. The morphological classifications throughout time, have been created by qualitative manual methods, based on observations and classifying according to their characteristics; this method can have a wide range of errors and most of the time it occupies a considerable amount of time. In this work a morphological classification of two genera of mushrooms is performed, grouping the specimens in poisonous or edible, using four classification algorithms, random forest, linear regression, K-Nearest Neighbor and Support Vector Machines; with the aim of obtaining a more accurate, objective, efficient and fast classification. For the evaluation of the four classifiers, the area under the curve (AUC) was used, obtaining a maximum of 1 and 0.978 as minimum, giving an alternative solution to the traditional methods and in the future providing a greater taxonomic contribution.

Keywords: Machine Learning · Mushrooms · Classification

1 Introduction

The kingdom Fungi, commonly referred to as mushrooms, consists of eukaryotic, heterotrophic organisms that possess a diversity of structures, functions, growth forms and lifestyles [19]. It is classified within the domain Eukarya, according to the currently most widely accepted classification system [18]. Phylogenetically, this kingdom is more closely related to animals than to plants [6].

Mushrooms present a diversity of approximately 1.5 million members in the world [19] impacting all habitats, considering many cosmopolitan species belonging to the kingdom. These have been used throughout time, up to the present, being of great utility for human beings, in the medical field, as bioremediation, in the food industry, to mention a few. The consumption of mushrooms is presented, either as food or as a recreational activity, emphasizing the presence of mushrooms that cause serious diseases and intoxications and even becoming lethal, hence the importance of a good classification.

The genus *Agaricus* (Linnaeus, 1753) belongs to the kingdom Fungi, family Agaricaceae [26] in the clade Euagarics or Agaricales [17] in the phylum Basidiomycota. In the genus *Lepiota* (Pers.) Gray, within section Ovisporae (J.E. Lange)

J. Bravo and G. Urzáiz (Eds.): UCAmI 2023, LNNS 842, pp. 270–278, 2023.
https://doi.org/10.1007/978-3-031-48642-5_26

Kühner comprising species with ellipsoidal to ovoid, smooth and (with few exceptions) dextrinoid, non-metachromatic spores, with pileus coat formed by elongated hairs with or without hymeniform basal substrate (hymenotricodermis to trichodermis), and with fibulae [7]. They are saprotrophic organisms with a preference for calcareous soils [4], produce the toxin muscarinic muscarine, causing gastrointestinal micetism, and in complication can lead to death [22]. Saryoko et al. [1], tested several algorithms to obtain the best performance, namely by using neural network algorithm, logistic regression, support vector machine learning, Naïve Bayes, decision tree and random forest. The Neural Network algorithm occupied the highest performing accuracy value with an accuracy value of (92.98%). The DDN was optimized by using RMSprop, Adam SGD, Adagrad and Adadelta with 20 epoch learning rate comparison. The experiment showed that it produces better accuracy value, so they performed some experiments using Deep Neural Network (DNN) in terms of accuracy value. Deep Neural Network with non-transfer learning produced (99.38%)accuracy with Adagrad optimizer for mushroom agricultural plant classification. In the identification of poisonous and edible mushrooms, various classification and prediction models are employed. For example, Decision Tree, NaïveBayes and Support Vector Machine (SVM) [27]. The Decision Tree algorithm is the simplest algorithm to implement in [2] programming, it is the highest accuracy classification algorithm with the lowest level of error. The experiment is based on the dataset obtained from Audubon Society Field Guide to North American Mushrooms, available in the UCI repository [24] which includes specimens from the *Agaricus* and *Lepiota* mushroom families.

John Heland Jasper C. Ortega et al [20] used confusion matrix analysis and different classification algorithms for toxicity detection in mushrooms. The highest accuracy of all the techniques used was 88.2% corresponding to the Decision Tree model [23]. The following work focuses on the classification of *Agaricus* and *Lepiota*, two previously described mushroom genera; by means of Linear Regression, Support Vector Machine, Random Forests and K nearest neighbors (abbreviated as K-NN) in order to achieve a correct classification and prevent mushroom poisoning.

This paper is organized as follows, in the present section is presented an introduction to mushrooms description and current state of the art in automatic classification. Materials and methods are described in Sect. 2. Section 3 reports the results obtained from the methodology. The discussion and conclusions of this proposal are described in Sect. 4.

2   Materials and Methods

The methodology followed in this work for the classification of edible and poisonous mushrooms is shown in the following Fig. 1, section A) corresponds to the data acquisition, section B) data pre-processing, section C) demonstrates the implementation of the proposed classification algorithms evaluated and the last step section D) evaluates the results based on their accuracy, sensitivity, specificity and Area Under the Receiver Operating Characteristic Curve (AUC-ROC).

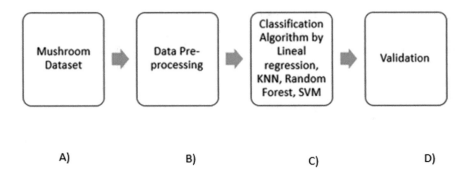

Fig. 1. Flow chart of the methodology for mushrooms classification

2.1 Data Description

The dataset used in this work came from the UCI Machine Learning repository, extracted from the Audubon Society Field Guide to North American Fungi (1981) [25]. The data correspond to 23 species of gilled fungi belonging to the genera *Agaricus* and *Lepiota*. Each species is identified as poisonous (1) or edible (0). The final dataset is comprised by two classes, Table 1 describes the features contained in the dataset with their respective values.

2.2 Data Preprocessing

To perform the evaluation of the classifiers proposed in this work, the variables with category outputs were converted to numerical values, and the stalk root feature was removed because more than 70% of the data were missing.

Additionally, in order to be able to perform the blind test, the data set was divided into two balanced subsets, corresponding to the blind test set (20%) and the training set (80%).

2.3 Classification Algorithms

This section describes the classification algorithms selected for evaluation.

Linear Regression. Regression is aimed at describing how the relationship between two variables X and Y is, in such a way that even predictions can be made about the values of variable Y, from those of X. When the association between the two variables is strong, regression offers us a statistical model that

Table 1. Descriptor features of mushrooms for the edible and poisonous classes.

Features	Values
Cap shape	conical (c), convex (x), flat (f), bell (b), sunken (s), knobbed (k)
Cap surface	grooves (g), scaly (y), smooth (s), fibrous (f)
Cap colour	Cap _colour buff (b), cinnamon (c), grey (g), green (r), brown (n), purple (u), red (e), white (w), yellow (y), pink (p)
Bruises	Bruises no (f), bruises (t)
Odour	anise (l), creosote (c), fishy (y), foul (f), almond (a), none (n), pungent (p), spicy (s), musty (m)
Gill attachment	descending (d), free (f), notched (n), attached (a)
Gill spacing	crowded (w), distant (d), close (c)
Gill size	narrow (n), broad (b)
Gill color	brown (n), buff (b), chocolate (h), grey (g), black (k), orange (o), pink (p), purple (u), red (e), green (r), yellow (y), white (w)
Stalk shape	tapering (t), enlarging (e)
Stalk root	club (c), cup (u), equal (e), bulbous (b), rhizomorphs (z), rooted (r), missing (?)
Stalk surface above ring	scaly (y), silky (k), smooth (s), fibrous (f)
Stalk surface below ring	scaly (y), silky (k), smooth (s), fibrous (f)
Stalk colour above ring	buff (b), cinnamon (c), grey (g), orange (o), brown (n), red (e), white (w), yellow (y), pink (p)
Stalk colour below ring	buff (b), cinnamon (c), grey (g), orange (o), brown (n), red (e), white (w), yellow (y), pink (p)
Veil type	universal (u), partial (p)
Veil colour	orange (o), white (w), yellow (y), brown (n)
Ring number	one (o), two (t), none (n)
Ring type	evanescent (e), flaring (f), large (l), cobwebby (c), none (n), pendant (p), sheathing (s), zone (z)
Spore print colour	brown (n), buff (b), chocolate (h), green (r), black (k), purple (u), white (w), yellow (y), orange (o)
Population	clustered (c), numerous (n), abundant (a), several (v), solitary (y), scattered (s)
Habitat	grasses (g), leaves (l), meadows (m), paths (p), urban (u), waste (w), woods (d)

can achieve predictive purposes. It is a statistical technique that analyzes the relationship between two quantitative variables, trying to verify whether this relationship is linear. If we have two variables is simple regression, if there are more than two variables, multiple regression. The objective is to explain the behavior of a variable Y, which we will call the explained variable (or dependent or endogenous), from another variable X, which we will call the explanatory variable (or independent or exogenous).

Randomforest. Decision trees are among the most widely used supervised classification methods. They are non-parametric, robust and easy to interpret. They work by making successive partitions in the space of variables always looking for the variable and the threshold value of the variable that maximize the homogeneity of the resulting partitions. The homogeneity of a partition can be measured in several ways, one of the most common being the Gini index:

$$G = \sum f * k * (1 - f * k)K, \tag{1}$$

where k is each of the classes present in the partition, K the total number of classes present in the partition and fk the proportion of cases in the partition that belong to the k class. To calculate the Gini index of a complete tree, one would have to sum the Gini indices of all its partitions. The partitioning process continues until all partitions are completely homogeneous. At that point the pruning process of the tree begins using a cross-validation procedure that prevents the tree from overfitting to the training data. This basically involves regrouping the smaller partitions that respond only to noise in the training data. Once the tree has been pruned, each partition of the variable space is assigned the most frequent class so that any new cases are classified according to where they are placed in the variable space.

Random Forest [8] uses several decision trees (between 500 and 2,000). Each of them is trained with a random subset of cases (obtained by bootstrapping) called in-bag, the rest of the cases form the out-of-bag. In addition, only a random subset of the predictors is considered in each split (node) of the trees. Each new case is presented to each of the trees (which have not been previously pruned) and assigned to the class most frequently chosen by the trees. The proportion of trees that have voted to each class can also be interpreted as the probability of belonging to that class. The introduced randomness decreases the correlation between trees giving more sense to the use of an ensemble of classifiers. The Random Forest algorithm uses two parameters: the number of trees and the number of predictors to be used in each partition of each of the trees. However, one of the great advantages of this algorithm is its low sensitivity to these parameters, so that the default values usually produce good results [11, 15, 16].

***K*-Nearest Neighbor.** Neighborhood classification rules are based on searching a set of prototypes for the k prototypes closest to the pattern to be classified. Predictions are made based on the examples most similar to the one to be predicted. The learning cost is 0, all cost is passed to the prediction computation.

This is known as a lazy learning mechanism. We must specify a metric in order to measure proximity. For computational reasons, the Euclidean distance is usually used, δ for this purpose.

The most general neighborhood classification rule is the k-nearest neighbor classification rule or simply k-NN. It is based on the assumption that the nearest prototypes have similar a posterior probability. If Ki(X) is the number of samples of the class present in the k nearest neighbors of X, this rule can be expressed as [10] :

$$d(X) = wcsiKc(X) = maxi = 1- > JKi(K). \tag{2}$$

Support Vector Machines. Support Vector Machines (SVM) is a classification technique. It is based on the idea of structural risk minimization (SRM). Support Vector Machines proved to have great performance, even more than traditional learning machines such as neural networks [9] and are included as an effective method for solving classification problems. An SVM first maps the input points to a higher-dimensional feature space and finds a hyperplane that separates them and maximizes the margin m between the classes in this space.

Maximizing the margin m is a quadratic programming (QP) problem and can be solved by its dual problem by introducing Lagrange multipliers. Without any knowledge of the mapping, the SVM finds the optimal hyperplane using the dot product with functions in the feature space that are called kernels. The solution of the optimal hyperplane can be written as the combination of a few input points which are called support vectors [5, 12]

2.4   Validation

Sensitivity, specificity, accuracy and area under the curve ROC, were calculated to evaluate the performance of the four proposed algorithms used for classification.

Once the classification algorithms were trained with the training data set, the validation metrics were obtained from the test data subset.

ROC Curve. The ROC curve for a binary classification problem plots the true positive rate as a function of the false positive rate. ROC curve for a binary classification problem plots the true positive rate as a function of the false positive rate. The points on the curve are obtained by sweeping the classification threshold from the most positive to the most negative classification value. For completely random classification, the ROC curve is a straight line connecting the origin to (1, 1). Any improvement over random classification results in an ROC curve at least partially above this straight line [12].

Specificity and Sensitivity. Sensitivity (SN), has the capability of test to correctly an individual as positive in disease (i.e., poisonous mushrooms). The specificity (SP) is the opposite of SN, this metric has the ability of a test to correctly classify an individual as disease-free (i.e., edible mushrooms) [21].

Accuracy. It is statistically defined as the closeness between a measured value and a true value of a measurement. Thus, the smaller the measurement error, the more accurate the measurement [3].

Area Under the Curve. It is a natural criterion or summary statistic often used to measure the classification quality of a classifier[14]. Any deviation from this classification decreases the AUC. The test is based on the observation that the AUC value is exactly the probability P $(X > Y)$ where X is the random variable corresponding to the distribution of the outputs for the positive examples and Y the one corresponding to the negative examples [13].

The AUC can be viewed as a measure based on pairwise comparisons between classifications of the two classes. With perfect ranking, all positive examples are ranked above the negative ones and area is equal to 1 [12] .

3   Experiments and Results

This section explains the experiments performed and their results. As shown in Fig. 1, first the acquisition of the dataset was done, which was downloaded from the UCI Machine Learning repository. Once the dataset was acquired, the proposed preprocessing was performed, i.e., converting the categorical variables for easier manipulation and interpretation. Once the previous process was completed, all the features with missing more than 70% of their observations were eliminated, in this case the stalk root feature, obtaining a data set of 21 features to carry out the experiment.

The data set was split into two subsets: the first represents 80% of the total data and was used as the training set, the remaining 20% corresponds to the test set.

Once the two data sets were obtained, the training of the proposed algorithms was performed using the training data subset. Finally, the performance of the algorithms was measured using the subset of data corresponding to the test (20). In the Table 2, where the AUC, specificity, sensibility and accuracy which can measure the classification performance among the different classes, are presented.

These metrics were calculated for each model; the results were reviewed, finding that the best performance is obtained by the K-NN and Random Forest algorithms, with all evaluation metrics equal to 1 (results highlighted in bold). On the other hand, linear regression and SVM obtain results close to 1 for all their metrics.

Table 2. Accuracy, sensibility, specificity according to the different classification models.

Models	AUC	Specificity	Sensibility	Accuracy
Linear regression	0.979	0.995	0.939	0.966
K-NN	1	1	1	1
Random Forest	1	1	1	1
SVM	0.978	0.971	0.984	0.978

4   Discussion and Conclusions

In this work, a public dataset was used, which contains 22 descriptive characteristics of edible and poisonous mushrooms. After preprocessing, four classification algorithms were implemented to evaluate their performance in recognizing whether the mushroom is edible or poisonous based on the features.

According to the results reported in the present work, the best performance under the AUC metric is 1, obtained by K-NN and RF and all higher than 0.978 (the minimum obtained by SVM).

It is interesting to mention that the linear regression algorithm, with a low computational cost and fast implementation, has a similar performance to those that are more complex, obtaining an AUC of 0.979, being the sensitivity metric with 0.939 the one that reduces its capability to recognize poisonous mushrooms.

To avoid overfitting and to give validity to the results obtained, a blind test approach was used, dividing the data set into 80% for training and 20% for testing. From this comparison it is concluded that, regardless of the classification model used, all the above mentioned have the ability to obtain an Area Under the Curve close to 1, which allows asserting that it is possible to develop a classification model that identifies whether a mushroom is edible or poisonous based on descriptive characteristics.

A clear advantage over other works [1] used for classification in mycology, agronomy and agriculture is the use of low computational cost algorithms, with similar performance, which allows classifying mushrooms quickly and without a constant need for feedback or high training cost requiring thousands of samples.

This work can serve as a basis for the development of systems to assist in the identification of different mushrooms, as well as for the development of mobile applications to assist producers and the general public in the identification, to avoid ingestion or handling of poisonous mushrooms.

References

1. Predicción de productos de plantas agrícolas de hongos utilizando técnicas de aprendizaje automático. In: 2022 Conferencia Internacional sobre Investigación e Innovación en Tecnologías de la Información (ICITRI)
2. Alkaromi, M.A.: Komparasi algoritma klasifikasi untuk dataset iris dengan rapid miner. IC Tech **11**(2) (2014)
3. Armenteros, A.M.R., Balboa, J.L.G., Mingorance, J.L.M.: Error, incertidumbre, precisión y exactitud, términos asociados a la calidad espacial del dato geográfico. In: Catastro: formación, investigación y empresa: Selección de ponencias del I Congreso Internacional sobre catastro unificado y multipropósito, pp. 95–102 (2010)
4. Atkinson, G.F.: Studies of American Fungi: Mushrooms. Hafner Pub (1961)
5. Betancourt, G.A.: Las máquinas de soporte vectorial (svms). Scientia et technica **1**(27) (2005)
6. Blackwell, M.: The fungi: 1, 2, 3... 5.1 million species? Am. J. Botany **98**(3), 426–438 (2011)
7. Bon, M.: Flore mycologique d'europe 3. les lépiotes. doc. mycol. mémoire hors série 3, 1–153 (1993)

8. Breiman, L.: Random forests. Mach. Learn. **45**(1), 5–32 (2001)
9. Burges, C.J.: A tutorial on support vector machines for pattern recognition. Data Min. Knowl. Disc. **2**(2), 121–167 (1998)
10. Cambronero, C.G., Moreno, I.G.: Algoritmos de aprendizaje: knn & kmeans. Intelgencia en Redes de Comunicación, Universidad Carlos III de Madrid **23** (2006)
11. Cánovas-García, F., Alonso-Sarría, F., Gomariz-Castillo, F.: Modificación del algoritmo random forest para su empleo en clasificación de imágenes de teledetección. In: Aplicaciones de las Tecnologías de la Información Geográfica (TIG) para el desarrollo económico sostenible XVII Congreso Nacional de Tecnologías de Información Geográfica, Málaga, vol. 29, pp. 359–368 (2016)
12. Cortes, C., Mohri, M.: AUC optimization vs. error rate minimization. In: Advances in Neural Information Processing Systems, vol. 16, pp. 313–320 (2003)
13. Green, D.M., Swets, J.A., et al.: Signal Detection Theory and Psychophysics, vol. 1. Wiley, New York (1966)
14. Hanley, J.A., McNeil, B.J.: The meaning and use of the area under a receiver operating characteristic (ROC) curve. Radiology **143**(1), 29–36 (1982)
15. Hastie, T., Tibshirani, R., Friedman, J.: The Elements of Statistical Learning. SSS, Springer, New York (2009). https://doi.org/10.1007/978-0-387-84858-7
16. Liaw, A., Wiener, M., et al.: Classification and regression by randomforest. R News **2**(3), 18–22 (2002)
17. Matheny, P.B., et al.: Major clades of agaricales: a multilocus phylogenetic overview. Mycologia **98**(6), 982–995 (2006)
18. Montes, B., Restrepo, A., McEwen, J.G.: Nuevos aspectos sobre la clasificación de los hongos y su posible aplicación médica. Biomedica **23**(2), 213–224 (2003)
19. Moreno, J.A.C., et al.: Los hongos: héroes y villanos de la prosperidad humana, vol. 17, no. 9 (2016)
20. Ortega, J.H.J.C., Lagman, A.C., Natividad, L.R.Q., Bantug, E.T., Resureccion, M.R., Manalo, L.: Analysis of performance of classification algorithms in mushroom poisonous detection using confusion matrix analysis. Int. J. **9**(1.3) (2020)
21. Parikh, R., Mathai, A., Parikh, S., Sekhar, G.C., Thomas, R.: Understanding and using sensitivity, specificity and predictive values. Indian J. Ophthalmol. **56**(1), 45 (2008)
22. Pérez-Silva, E.: Hongos de zonas urbanas: ciudad de méxico y estado de méxico. Scientia Fungorum **47**, 57–66 (2018)
23. Ria, N.J., Badhon, S.S.I., Khushbu, S.A., Akter, S., Hossain, S.A.: State of art research in edible and poisonous mushroom recognition. In: 2021 12th International Conference on Computing Communication and Networking Technologies (ICCCNT), pp. 01–05. IEEE (2021)
24. Schlimmer, J.: UCI Machine Learning Repository: Mushroom Data Set. University of California, School of Information and Computer Science, Irvine, CA, USA (1987)
25. UCI: Mushroom classification (1987)
26. Vellinga, E.C.: Ecology and distribution of lepiotaceous fungi (agaricaceae)-a review. Nova Hedwigia **78**(3), 273–300 (2004)
27. Wibowo, A., Rahayu, Y., Riyanto, A., Hidayatulloh, T.: Classification algorithm for edible mushroom identification. In: 2018 International Conference on Information and Communications Technology (ICOIACT), pp. 250–253. IEEE (2018)

Traceable Health Data for Consciously Trained ML Models

José R. García[1] ![ORCID], Jesús Favela[1]([✉]) ![ORCID], and Carlos E. Sánchez-Torres[2] ![ORCID]

[1] CICESE, Carretera Ensenada, Tijuana No. 3918, Ensenada, B.C., Mexico
garcia.research.mail@gmail.com, favela@cicese.mx
[2] UABC, Carretera Ensenada, Tijuana No. 3917, Ensenada, B.C., Mexico
hello@sanchezcarlosjr.com

Abstract. Advances in Machine Learning have highlighted the importance of gathering and labeling data, particularly in settings such as healthcare. Issues of ownership, privacy, data provenance, bias in trained models and informed consent, have raised an interest in the ethical use of personal data. We introduce THREAD, a Reference Architecture raised on a decentralized network to gather healthcare data, sign consent, annotate data and obtain credit from it, allowing users to trace the use of their data, and data scientist to trace their provenance, while protecting user privacy. We demonstrate the use of THREAD through an implementation of it, able to gather breast cancer risk factors data to train a machine learning model. The e-wallet implemented on THREAD is used to sign user's consent to participate in the study. Credit is given according to valid data uploaded and the validation of data from other users. A customized computational notebook is used to import the data consciously gathered by TRHEAD and train ML models. Blockchain is used as a decentralized registry for each operation performed over users data and as a repository of consent letters stored as Non Fungible Tokens (NFT) representing rights of their ownership. Thus, the data consumed can be traced, confirming that they were ethically/consciously trained.

Keywords: Consciously-trained IA · Blockchain · Data privacy · DApp

1 Introduction

Wearable devices allow the naturalistic sensing of health data from individuals during daily activities. This has led to large datasets and their analysis using machine learning [7]. However, this has brought with it a culture of carelessness towards the information we share, encouraged by the data hunger of some information consumers. This malpractice has spread to health information sensing campaigns, where loopholes in data collection are exploited for profit.

Protecting user privacy is not a trivial problem that allows for absolute solutions such as banning data consumption. The scientific and technical development that has shaped modern society depends to a large extent on the constant

J. Bravo and G. Urzáiz (Eds.): UCAmI 2023, LNNS 842, pp. 279–284, 2023.
https://doi.org/10.1007/978-3-031-48642-5_28

flow of information to adapt technology to the needs of society. Our proposal consists of a reference architecture for a system capable of handling the flow of health data from active sensing via mobile and wearable devices. Systems implemented following this reference architecture should provide data provenance without compromising patient privacy, so that even if the data is in the public domain it cannot be linked to the patient's identity. To this end, we propose the use of the Blockchain as a technology that will allow the creation of a public and immutable record of the operations performed over the data.

2   Blockchain and Web3 in Healthcare

Today's healthcare systems are highly fragmented and specialized. The use of the Blockchain, as a decentralized and distributed environment, would allow each institution to conduct its operations independently from a central authority, taking advantage of the interoperability offered by its node-based structure [4].

Decentralization. The blockchain structure is based on nodes which could represent a hospital or healthcare/wellness entity, so that their interconnection becomes a Decentralized Autonomous Organization (DAO) [1] where decisions are made based on the consensus of the majority of nodes, in this case, hospitals.

Availability. Patient medical records can be stored digitally in the cloud while their connection to the patient and the operations performed on them can be written to the Blockchain as an immutable record of them. The consensus mechanism of the network ensures that these records are correct [6] and the node-based structure allows their storage on multitude devices, so that if a technical failure occurs in one of them there will always be a backup copy. Not relying on a single storage center allows information to always be available for peer-to-peer sharing.

Smart Contracts. A smart contract is a set of promises, specified in digital form, that include protocols under which the parties carry out these premises [5]. Their terms and conditions are written on the Blockchain and run autonomously, usually inside a virtual machine such as Ethereum Virtual Machine (EVM), when the conditions for its execution are met. Smart contracts allow the execution of the operations in a transparent, tamper-proof, and conflict-free way [3].

In a medical center implementing a blockchain-based resource management system, smart contracts could be used to automate administrative and asset control tasks. Being autonomous, they are executed as long as the conditions stipulated in their code are met, thus avoiding human error and malicious actions by eliminating human intervention.

DApps. We cannot expect medical staff or patients to have technical knowledge of the blockchain, so a mediating element between the end-user and the blockchain-based infrastructure is necessary. To address this issue, Decentralized Applications (DApps) can be used, which are composed of several layers of code, the interface being the one that the user perceives and where all the user's interaction is centered. It also has the Web3 layer, where the code that consumes web3 libraries is located, allowing the interaction with the chain and consumption of its resources through the different protocols that it provides, such as HTTP, WebSockets and WebRTC. A third layer that deserves mention is the smart contract, as the main tool for DApp operations on the network, which may include the deployment of new smart contracts.

3   Proposed Approach

As shown above, a Blockchain-based architecture could draw on the capabilities of this technology to provide an environment for sensing, storing and consuming data to ensure data provenance without compromising user identity. We propose the TRHEAD reference architecture, specifically designed for the gathering and processing of mobile and wearable sensing data, in order to provide a platform where people can make their contribution, either by offering their data, their services as data curator or by exchanging goods for the above.

3.1   The TRHEAD Reference Architecture for eHealth Data Collection and Usage

The main objective of this architecture is to train Machine Learning (ML) algorithms using consciously/legally collected data, keeping the traceability of the data and the operations on it isolated from the user's real identity, so that the user's privacy is respected while ensuring the provenance of the data. The trained models will be distributed as stand-alone instances of a computational notebook, which will contain the digital consent letters that allow the consumption of the data used to train them.

3.2   Architecture

The architecture of the proposed system is composed of a DApp as a graphical interface for user interaction and in charge of executing the libraries that allow communication with the blockchain. The blockchain itself functions as a decentralized, immutable and public registry of the operations carried out on the platform and the NFTs (smart contracts) that represent the rights over the data. A cloud storage system is used to store both the health data and the metadata associated with the NFTs that are recorded on the chain. In order to test the feasibility of the proposed architecture, an implementation of the architecture was carried out, the structure of which is shown in the Fig. 1 and explained below.

The TRHEAD DApp has different interfaces that allow each of the roles involved in the data flow to perform their tasks abstracting from the interaction with the blockchain.

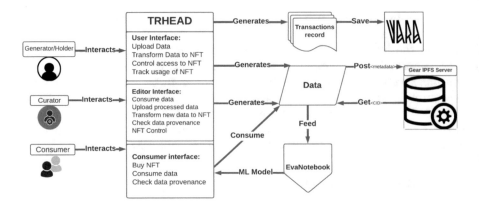

Fig. 1. TRHEAD architecture implementation

The roles mentioned are three and are described below:

- **Data generators or data holders:** in charge of sensing their own data or storing large volumes of other people's data. Usually patients with wearable sensors or companies with large databases collected from their users.
- **Data curators:** in charge of cleaning, filtering and labeling data, just to mention some of the tasks they must perform to generate volumes of data with the required quality to obtain valuable information from them.
- **Data consumers:** in charge of consuming volumes of data and generating information from it in the form of scientific articles, new services of social or economic benefit and Machine Learning models.

DApp. The web application written in ReactJS connects to the user's wallet through a browser extension to sign with it all user transactions and manage their assets, either internal network credit or NFTs. Data generators and data holders can upload the healthcare data in their possession to the platform and automatically generate NFTs that represent ownership over these assets, allowing them to control their access and maintain traceability of its consumption.

Data curators can purchase these NFTs with internal platform credit and with them a digital letter of consent for their consumption specifying the term and extent of their privileges over them. This is for the purpose of reprocessing, segmenting and tagging the data to create a new set of value-added data that can be of greater use to consumers.

Consumers, for their part, have the options of consuming the raw data from the platform, preferring data already processed and/or labelled by specialists, or directly requiring a ML model already trained using the data found on the platform. Each of these options includes corresponding digital letters of consent and payment in network credit.

Vara Network and Gear. The Vara Network is a stand-alone layer-1 decentralized network built and running on top of Gear Protocol. Gear Protocol is a Substrate-based smart-contract platform that enables anyone to develop and run a dApp in Vara Network as well as other networks powered by the Gear Protocol's runtime and technology. In our architecture the Vara Network provides a distributed, immutable and public record of the operations performed on the data handled in the platform. When a new dataset is uploaded from the dApp it will be stored in the Gear Technologies IPFS and an NFT will be created with the information and access control of it in that storage.

EvaNotebook: EvaNotebook is a computational notebook designed to operate within a browser environment with tools to manage decentralized datasets and build transparent applications for a data scientist and end-users. Its design lends itself to the ETL (Extract, Transform, Load) operations but with emphasis in Web3 and Blockchain. It achieves this with the assistance of a decentralized database and incorporates various application protocols, including WebRTC, WebSockets, libP2P, IPFS, and OrbitDB.

4   Implementation Results: BCRAT

BCRAT is a multifaceted decentralized application focused on breast cancer risk factors. It utilizes NFTs to collect data, emphasizing the novel risk factor of insomnia in patients, a condition suggested by recent research as potentially increasing the risk of the disease [2]. This data is linked to the user's digital wallet, which records their consent to data usage, ensuring transparency and respect for user's privacy.

When a data scientist gathers sufficient data, they load this information into their favorite tool, train their model, and then connect to the primary notebook via WebRTC from a secondary notebook. This secondary notebook then provides an interpretation of these data points across four levels of breast cancer risk by requesting a prediction from the primary notebook. This dual-purpose functionality of BCRAT not only aids in the assessment of breast cancer risk but also promotes transparency and accessibility in data management.

Web Interface. Was implemented in ReactJS using the NFT Marketplace example from Gear Technologies Github repository, so the health data stored as NFT becomes part of a decentralized marketplace running over the Vara network. The user has the power to upload, delete, set price and sell their digital assets, all these transactions are signed with their e-wallet and can only be carried out using their private key thanks to this business rule being set in the smart contracts of the marketplace.

EvaNotebook. It acts as a platform for deploying models in a decentralized way. We have developed two types of notebooks: a primary one that runs the model, and a replica that communicates with the primary notebook using user's risk factors. See 3.2 section.

5   Conclusions and Discussion

We have proposed a reference architecture for a system capable of maintaining medical data traceability while respecting patient privacy. We have implemented a system using this architecture that respects the principles of data provenance and consent by creating a digital identity for each patient (data generator/holder), data curator and consumer that allows them to perform operations on medical data in the form of NFTs. We offer an alternative for training machine learning models with consciously/legally collected data thanks to the digitization of consent letters. We advocate for a world in which information is not centralized in huge databanks whose security can be penetrated or its integrity compromised by human error, but distributed among those who generate it and actually care about its final destination.

References

1. Faqir, Y.E., Arroyo, J., Hassan, S.: An overview of decentralized autonomous organizations on the blockchain. In: 16th International Symposium on Open Collaboration (2020). https://doi.org/10.1145/3412569.3412579
2. Fiorentino, L., Rissling, M., Liu, L., Ancoli-Israel, S.: The symptom cluster of sleep, fatigue and depressive symptoms in breast cancer patients: severity of the problem and treatment options. Drug Discov. Today Dis. Model. **8**, 167–173 (2011). https://doi.org/10.1016/j.ddmod.2011.05.001
3. Hassanein, A.A., El-Tazi, N., Mohy, N.N.: Blockchain, smart contracts, and decentralized applications: an introduction. In: Rawal, B.S., Manogaran, G., Poongodi, M. (eds.) Implementing and Leveraging Blockchain Programming, pp. 97–114. Springer, Singapore (2022). https://doi.org/10.1007/978-981-16-3412-3_6
4. Hölbl, M., Kompara, M., Kamišalić, A., Zlatolas, L.N.: A systematic review of the use of blockchain in healthcare. Symmetry 2018 **10**, 470 (2018). https://doi.org/10.3390/SYM10100470
5. Khan, S.N., Loukil, F., Ghedira-Guegan, C., Benkhelifa, E., Bani-Hani, A.: Blockchain smart contracts: applications, challenges, and future trends. Peer-to-Peer Network. Appl. **14**(5), 2901–2925 (2021). https://doi.org/10.1007/s12083-021-01127-0
6. Liang, X., Shetty, S., Tosh, D., Kamhoua, C., Kwiat, K., Njilla, L.: Provchain: a blockchain-based data provenance architecture in cloud environment with enhanced privacy and availability. In: Proceedings - 2017 17th IEEE/ACM International Symposium on Cluster, Cloud and Grid Computing, CCGRID 2017, pp. 468–477 (2017). https://doi.org/10.1109/CCGRID.2017.8
7. Mann, S.P., Savulescu, J., Ravaud, P., Benchoufi, M.: Blockchain, consent and prosent for medical research. J. Med. Ethics **47**, 244–250 (2021). https://doi.org/10.1136/MEDETHICS-2019-105963

Viric Learning - A Novel Transfer Learning Method

Alfonso Barragán[1]([⊠])(ID), Jesús Fontecha[1](ID), Iván González[1](ID),
Esperanza Jonhson[2](ID), David Carneros-Prado[1](ID), and Laura Villa[1](ID)

[1] University of Castilla-La Mancha (MAmI Research Group),
Paseo de la Universidad, 4, 13071 Ciudad Real, Spain
`alfonso.barragan@alu.uclm.es`,
`{jesus.fontecha,ivan.gdiaz,David.Carneros,laura.villa}@uclm.es`
[2] Spillskolen - Høgskolen i Innlandet INN, Hamar, Norway
`esperanza.johnson@inn.no`

Abstract. Increasingly, information systems include intelligent components to make decisions or perform actions based on data, owing to advances in data science and artificial intelligence. Although advances have been made at the machine level to enable the execution of more complex models, due to the changing context they can quickly become outdated and require a high number of resources to adapt to new conditions, thus transfer learning is being investigated. This discipline seeks to optimize learning tasks of an intelligent model based on the learning from another model. Therefore, a method is proposed that focuses on what all models receive in a generic way: data. The objective is to elaborate a design data register that, in controlled training sequences, transfers only the quality knowledge of the model that from it was created to a new one by means of optimization algorithms. Likewise, it will also obtain clarity on the criteria used by the models to perform their functions, the development of objective mechanisms to obtain traceability of the training data and an optimal method of adaptation to the changing context to which the information systems are subjected.

Keywords: Transfer learning · Nature Inspired Algorithms · Neural Network · Decision Tree · Random Forest · Naive Bayes · Physical Activity Data

1 Introduction

The rapid proliferation of artificial intelligence systems has revolutionized various domains, facilitated by the abundance of available data, which is estimated to be approximately 64.2 zettabytes in 2020[1]. Notably, advancements in deep learning and automatic classifier metaoptimizers have yielded highly capable and accurate models, although at the cost of increased computational requirements.

---

[1] Statista. 2021. https://es.statista.com/. Last visited 10th of June, 2023.

© The Author(s), under exclusive license to Springer Nature Switzerland AG 2023
J. Bravo and G. Urzáiz (Eds.): UCAmI 2023, LNNS 842, pp. 285–291, 2023.
https://doi.org/10.1007/978-3-031-48642-5_29

A drawback of many AI models is their contextual dependence, which hinders their application in similar scenarios, necessitating updates or new models with expanded data sources. To address this issue, *transfer learning* has emerged as a discipline aimed at optimizing models (M) for changes in input and output parameters, as well as the domains (D_x and D_y) governing them. Transfer learning has been a widely explored field that has enabled the realization of numerous use cases, ranging from the detection of outbreaks of Kyasanur forest disease [1] to improving the process of designing control systems [2].

In this study, a novel approach is proposed that emphasizes data-focused knowledge transmission through artificially generated data samples and is optimized to comprehensively encapsulate the models' original knowledge. This analogy resembles the process of a virus invading an organism and replicating within host cells. The primary goal is to achieve adaptability and robustness by harnessing the full semantic knowledge of the source models.

1.1   Application of Nature Inspired Optimization Algorithms

As part of the viric learning method, the use of optimization algorithms is required. Within this proposal, nature-inspired algorithms (NIA) have been chosen for this purpose.

NIAs, also known as bio-inspired algorithms, are optimization and search algorithms that draw inspiration from the processes observed in nature. These algorithms emulate the behavior of natural systems, such as animals, plants, and other biological entities, to address complex computational problems, harnessing the efficiency and effectiveness of nature to find optimal solutions.

Given the conditions of the viric learning method, it is crucial to utilize an optimization algorithm that can modify the learning parameters with the goal of been like the original model, starting from a base set of data samples.

As a solution, the Symbiotic Organisms Search (SOS) algorithm [3] is recommended. This algorithm is based on differential vector movement [4] and boasts high robustness and stability, effectively addressing issues related to highly variable data distributions while preserving the essential characteristics of the raw data.

The SOS algorithm models interactions observed between organisms in nature, such as mutualism, commensalism, and parasitism, to guide the search for optimal solutions. Each organism represents a potential solution to the optimization problem: they cooperate and compete with each other in a manner that mimics the natural behavior of living beings.

2   Approach

2.1   Viric Learning Method

The viric learning method is based on the following idea: given an already trained classifier, meaningful data samples can be generated by means of optimization algorithms that do not modify the parameters that govern the knowledge of the model.

In other words, we have a model M that executes the inference function f on domain X to obtain result y. Before it can be used, the f function requires tuning, which, depending on M, is performed in one way or another. The virus that transmits knowledge must be a set of data samples that does not significantly modify f during the tuning procedure.

The process is depicted in Fig. 1. First, it is created the population of data to be optimized throught a random sample selection. Then, the model which has the knowledge is loaded. Next, the optimization algorithm is initiated, within which false training cycles would be performed, as they do not actually modify the model's performance, and the data samples (population) would be modified by NIAs algorithm. These algorithm would take the accuracy metrics from the model with the population data and modified its attributes (updated population).

The process concludes when these metrics are stabilized at 0.

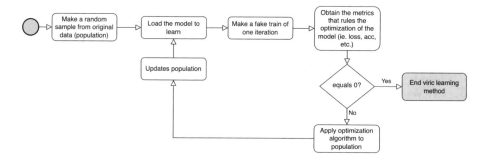

Fig. 1. Viric learning method flowchart.

Finally, the target model was trained using the resulting data (updated population) from the iterations on viric learning method.

2.2   Activity Dataset

To evaluate the proposed method's performance, a limited dataset with multiple classes was searched. This makes it too complex for a single model to cover it completely and for us to assess performance from several perspectives.

This dataset contains minute-by-minute measurements obtained from an activity bracelet [5]. These measurements are included heart rate, steps per minute, and the type of activity that subjects performed during that minute. These data do not include any personal data, so they are completely anonymous. The dataset used is a selection of 11624 samples, spread over several days and weeks.

The goal is to create a machine learning classifier that, through steps and heart rate, can distinguish between sitting, walking, sleeping, and getting up events.

2.3   Experiment Baseline

Using the Activity dataset mentioned in Sect 2.2, generic classifiers have been implemented (Decision Tree, Random Forest, Naive Bayes, Neural Network), and their accuracy rates and F-scores are compared in Table 1 throught a cross-validation method of 10 splits and 100 iterations.

Table 1. Model accuracy comparison between decision tree (DT), random forest (RF), naive bayes (NB), neural network model for general activity (NNGEN), and a specific model for sitting (NNST), sleeping (NNSL), get-up (NNG), walking (NNW)

Model	DT	RF	NB	NNGEN	NNST	NNSL	NNG	NNW
Avg. Accuracy	0.754	0.763	0.374	0.793	0.886	0.944	0.932	0.782
Avg. f1-score	0.728	0.733	0.346	0.804	0.868	0.968	0.862	0.951

After implementing these general classifiers, specific classifiers have been created for each activity class (Neural Network Model for sitting, sleeping, getting up, and walking respectevely), the results are shown in Table 1.

Overall, we observed a trend of increasing accuracy for the specific in comparison to general classifiers. The average improvement was 17%.

Through this baseline, it is compared the learning under normal conditions that this variety of models would have and the possibilities offered by the viric learning method proposed in this work. In the Sect. 3, the results are discussed in comparison to the baseline.

3   Evaluation and Results

This section presents the key elements of the evaluation. First, the generated viric data are compared with the data in its natural state. Next, a general overview of the experiments is provided. Finally, the results of each experiment are presented.

3.1   Viric Data vs. Natural Data

Figure 2 shows the comparison between the real data and the normalized real data from the datasets. As can be seen, the optimized data sample are far from the central clusters that comprise natural data.

The optimised data sample has a different distribution that resembles the outliers. This differentiation leads to an improvement of the learning rate of the models trained on viral data.

3.2   Experiments

The results of the viric-learning method are presented in the following section through the creation of three experiments: training a completely new model with viric data , training a model already performing an classification task, and training a generalist model handling several classification tasks. The evaluation of every experiment is shown by confusion matrices obtained from 10 training - test iterations with a random sampling on every iteration. The train used data from viric learning method while the test phase use the data without modification.

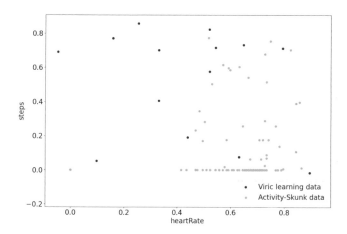

Fig. 2. Viric learning data vs. Activity data.

Figure 3 shows the confusion matrices for the first case. Three deep neural networks with the same architecture utilizing the same datasets were compared. The first neural network was entirely new and was trained to detect walking events using the standard dataset over 1000 iterations. The second neural network depicts a fine-tuned version with class balancing and training for 10,000 iterations. Finally, a neural network trained using the method proposed in this publication over 1000 iterations is presented.

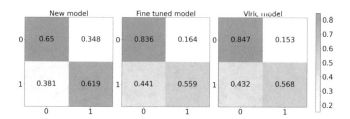

Fig. 3. Confusion matrix from neural networks models.

In the second experiment, a new neural network model was created and partially trained to detect the walking events. This training consisted of 100 iterations with 387 data samples. The model was then evaluated using a test dataset of 2325 samples, followed by another training cycle and evaluation with the same test set.

Original model
Viric model
0 0.878 0.122
1 0.829 0.171
0 0.933 0.067
1 0.788 0.212
0 1
0 1
0.8
0.6
0.4
0.2

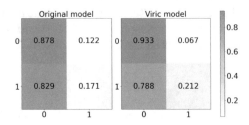

Fig. 4. Confusion matrix from neural network model trained with standard data and confusion matrix from a post-train with viric data.

Figure 4 illustrates the results of these tests, showing an improved accuracy of the model in all classes from 52.45% to 57.25%.

In the third experiment, generalist models from the Sect. 2.3 were used and trained with viric data in the same proportion as natural data. The confusion matrices shown in Fig. 5 and 6 demonstrate the similarities between the generalist models trained with 9899 samples and the models trained on viric data with 2775 data samples.

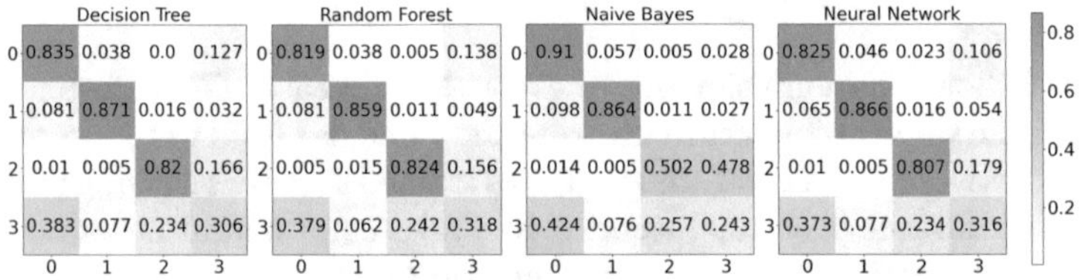

Decision Tree
0 0.835 0.038 0.0 0.127
1 0.081 0.871 0.016 0.032
2 0.01 0.005 0.82 0.166
3 0.383 0.077 0.234 0.306
Random Forest
0 0.819 0.038 0.005 0.138
1 0.081 0.859 0.011 0.049
2 0.005 0.015 0.824 0.156
3 0.379 0.062 0.242 0.318
Naive Bayes
0 0.91 0.057 0.005 0.028
1 0.098 0.864 0.011 0.027
2 0.014 0.005 0.502 0.478
3 0.424 0.076 0.257 0.243
Neural Network
0 0.825 0.046 0.023 0.106
1 0.065 0.866 0.016 0.054
2 0.01 0.005 0.807 0.179
3 0.373 0.077 0.234 0.316
0 1 2 3
0.8
0.6
0.4
0.2

Fig. 5. Confusion matrix for generalist models, with a mixed sample with viric data.

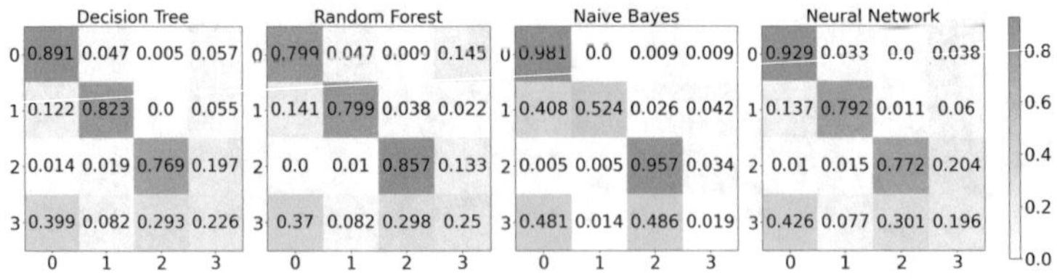

Decision Tree
0 0.891 0.047 0.005 0.057
1 0.122 0.823 0.0 0.055
2 0.014 0.019 0.769 0.197
3 0.399 0.082 0.293 0.226
Random Forest
0 0.799 0.047 0.009 0.145
1 0.141 0.799 0.038 0.022
2 0.0 0.01 0.857 0.133
3 0.37 0.082 0.298 0.25
Naive Bayes
0 0.981 0.0 0.009 0.009
1 0.408 0.524 0.026 0.042
2 0.005 0.005 0.957 0.034
3 0.481 0.014 0.486 0.019
Neural Network
0 0.929 0.033 0.0 0.038
1 0.137 0.792 0.011 0.06
2 0.01 0.015 0.772 0.204
3 0.426 0.077 0.301 0.196
0 1 2 3
0.8
0.6
0.4
0.2
0.0

Fig. 6. Confusion matrix from generalist models.

In terms of accuracy, both models are comparable, with the natural data models outperforming the viric data models by an average of 3%, considering that the viric data models utilized approximately one-third of the data samples.

4 Conclusions

In conclusion, the presented experiments demonstrate that the viric learning method effectively transfers knowledge between models through the use of viric data. This method is promising for optimizing models and for enhancing their adaptability to new contexts.

As mentioned in Sect 3, this method has been able to achieve that models with smaller samples and a reduced number of iterations perform just as well as classifiers trained and fine-tuned with the majority of the data.

This method can help establish relationships between data and models. Thanks to this, it is believed that it could enable future work to determine if a particular model has been trained with a certain dataset or new analysis techniques based on this underlying relationship between models, data, and learning.

References

1. Keshavamurthy, R., Charles, L.E.: Predicting Kyasanur forest disease in resource-limited settings using event-based surveillance and transfer learning. Sci. Rep. **13** (2023)
2. Fatehi, M., Toloei, A., Zio, E., Niaki, S., Keshtegar, B.: Robust optimization of the design of monopropellant propulsion control systems using an advanced teaching-learning-based optimization method. Eng. Appl. Artif. Intell. **126**, 106778 (2023)
3. Cheng, M.Y., Prayogo, D. :Symbiotic organisms search: a new metaheuristic optimization algorithm. Comput. Struct. **139**, 98–112 (2014)
4. Molina, D. et al.: Comprehensive taxonomies of nature- and bio-inspired optimization: Inspiration Versus Algorithmic Behavior, Critical Analysis Recommendations. Cognitive Comput. **12**, 897–939 (2020)
5. Barragán, A.: Activity Skunk Dataset version 1.0 (Zenodo, July 2023). https://doi.org/10.5281/zenodo.8191612

Granular Linguistic Model Based Multimodal Data Integration for Automated Evaluation of Core Soft Skills

Jared D.T. Guerrero-Sosa[1]([✉]) [iD], Francisco P. Romero[1] [iD],
Victor H. Menendez[2] [iD], Jesus Serrano-Guerrero[1] [iD], Jose A. Olivas[1] [iD],
and Andres Montoro-Montarroso[1] [iD]

[1] Department Information Systems and Technologies, University of Castilla La
Mancha. Paseo de la Universidad, 4, Ciudad Real, Spain
`JaredDavidtadeo.Guerrero@alu.uclm.es`,
`{FranciscoP.Romero,Jesus.Serrano,JoseA.Olivas,Andres.Montoro}@uclm.es`
[2] Mathematics School, Autonomous University of Yucatan. Anillo Periférico Norte,
Tablaje Cat. 13615, Mérida, Mexico
`mdoming@correo.uady.mx`

Abstract. This research introduces an innovative hybrid intelligence framework leveraging multimodal data to automate the evaluation of core soft skills, including decision-making, conflict resolution, and creativity. The model applies the principles of the Granular Linguistic Model of Phenomena (GLMP), a sophisticated method that delineates phenomena at varying granularity levels, ensuring a detailed analysis of the exhibited skills. The process involves mining significant behavioural features from diverse data sources, specifically video, audio, and text, employing deep learning algorithms. The extracted features are then subjected to the GLMP, representing the students' behaviour in a structured, interpretable format across multiple granularities. The GLMP application yields an exhaustive set of granular linguistic prompts that encapsulate the complexity of the identified soft skills. This multimodal information feeds into a fuzzy logic-based detector that evaluates the defined soft skills. This integrative approach merges granular linguistic modelling with multimodal data, enabling a comprehensive and accessible understanding of the students' soft skills. The implications of this approach extend beyond the academic sphere, finding utility in broader contexts such as college admissions and job recruitment, where objective skill evaluation is crucial. This research underscores the value of multimodal integration within the GLMP framework, highlighting its critical role in translating raw data into actionable insights. It further illuminates the potential of such methods in enhancing real-world decision-making processes and outcomes.

Keywords: Multimodal Data Integration · Granular Linguistic Model of Phenomena · Soft Skills Evaluation

J. Bravo and G. Urzáiz (Eds.): UCAmI 2023, LNNS 842, pp. 292–303, 2023.
https://doi.org/10.1007/978-3-031-48642-5_30

1   Introduction

Demonstrating social skills is crucial for students as they navigate their academic and professional lives. However, the traditional methods of evaluating social skills, such as face-to-face interviews and role-playing exercises, can be time-consuming and subjective. There are proposals oriented to specific tasks in the automatic evaluation of videos from the human aspect. For example, in the work of Akoa et al. [1], videos are evaluated from neural networks combined with a nonlinear regression model based on the metrics of the human visual system. Also, the work of Sun et al. [11] states that video sites provide users with tools to evaluate the content of videos. Still, the results may lack objectivity, so it is proposed to detect users' facial expressions when playing a video and link the results to the evaluation. In the educational field, the work of Qiusha et al. [7] proposes automatically evaluating didactic videos using image recognition and processing techniques whose results coincide with manual evaluations. On the other hand, the work of Nayak et al. [8] presents a system prototype based on text analysis and computer vision techniques. It is trained with data provided through audio and video by people in the range of 14 to 18 years for the preparation of skills required in the labour field whose motivation stems from the need for the lack of professional guidance in the development of soft skills in rural and low-income areas of India. Solé-Beteta et al. [10] propose a methodology and model to measure students' engagement in synchronous virtual learning environments to help instructors who feel that the connection with their students has weakened due to the virtuality of the learning environment.

This paper introduces an automated method that evaluates students' social skills demonstrated in videos using multimodal data such as video, audio, and text. As videos become an increasingly popular tool for students to showcase these skills, there is a growing need for efficient, objective evaluation methods. For this purpose, the proposed approach analyzes and extracts relevant features from the video content, which are then used to predict social abilities like communication, teamwork, and conflict resolution. This research aims to alleviate the evaluation duty on educators, enhance the quality of educational videos, and improve student learning outcomes. This research aims to contribute to education and assessment by developing a reliable and efficient system. The proposed system combines Machine Learning (ML) and Knowledge Extraction (KE) techniques for video analysis. The video data undergoes a cleaning and homogenization process, followed by deep learning techniques to discern various characteristics, such as the interviewee's mood and attitude changes. Subsequently, a Granular Linguistic Model and fuzzy logic-based methods are applied to produce a detailed report on the student's skill levels.

The organization of this article can be outlined as follows: In Sect. 2, we offer an exploration of the theoretical underpinnings that form the basis of our proposal. Section 3 introduces the multimodal analysis methodology that we employ in this research. Section 4 presents a thorough case study that assesses particular Machine Learning skills. Ultimately, in Sect. 5, we present our findings and illuminate potential directions for future research.

2   Background

To effectively describe soft skills, it is common to employ a hierarchical approach. This approach involves a process called granulation, which entails grouping input data into units called granules. Granules help emphasize the nuanced aspects of a phenomenon within its specific context. Granules can be composed of physical or abstract objects that are grouped based on their similarity, proximity, or functional criteria. The Granular Linguistic Model of Phenomena (GLMP) [3] is a versatile model used to explain phenomena across various granularity levels. It has found wide-ranging applications in numerous domains [2,12]. GLMP is built upon two fundamental concepts: Computational Perception (CP) and Perception Mappings (PM) (Fig. 1).

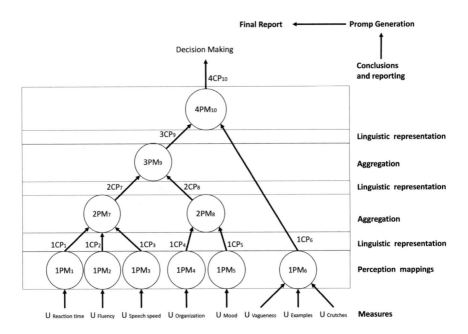

Fig. 1. Example of a GLMP.

Nevertheless, we suggest broadening this model by incorporating an extra level of structure into the perceptual maps. Our extended version, the Hierarchical Perception Mapping (HPM), embeds a hierarchical and recursive structure into the Perception Mapping (PM) tuple. This enhancement permits us to grasp and model more sophisticated phenomena, resulting in refined granularity in perception representation and increased interpretability and adaptability.

Fundamentally, CPs focus on particular facets of the system, producing information with a specific level of detail or granularity. When modelling a series of soft skills from a video, the evaluator scrutinises various elements (information units or granules) rooted in their subjective perceptions. Next, the elements of PMs and CPs will be described, drawing on the definitions from [2,3,12].

A Computational Perception (CP) is defined as a tuple (A, W, R), with the following components:

- A: $(a_1, a_2, \ldots, a_n)$ represents a vector of linguistic expressions, such as words or sentences in natural language, encompassing the entire linguistic spectrum of the CP. Each element a_i corresponds to the most appropriate linguistic value of the CP for a specific behaviour, characterized by a certain level of granularity. For instance, a student's leadership skills could be described using $A = (low, medium, high)$.
- W: $(w_1, w_2, \ldots, w_n)$ forms a vector of validity degrees, where $w_i \in [0, 1]$. Each validity degree w_i signifies the degree to which the linguistic expression a_i accurately describes the given input data. The sum of all validity degrees should equal $\sum w_i = 1$.
- R: $(r_1, r_2, \ldots, r_n)$ constitutes a vector of relevance degrees, with $r_i \in [0, 1]$, associated with each linguistic expression a_i within the specific context. It conveys the importance or significance of the linguistic expressions within $A = (low, medium, high)$. For instance, $R = (0.5, 0.5, 1)$ implies that the perception of $high$ is more pertinent than the other two options. By default, all relevance degrees are set to their maximum value $(r_i = 1)$.

Perception Mappings (PM) play a vital role in the establishment and fusion of Computational Perceptions (CPs). Each PM combines a set of input CPs to construct a unified CP. A PM is represented as a tuple (U, y, g, T), with the following elements:

- U: $(u_1, u_2, \ldots, u_n)$ forms a vector containing n input CPs, referred to as $u_i = (A_{ui}, W_{ui}, R_{ui})$. In the context of first-level Perception Mappings $(1PM)$, these inputs can also be numerical values $(z \in R)$ obtained from measurement procedures.
- y: (A_y, W_y, R_y) denotes the resulting output CP generated by the PM.
- g represents the aggregation function utilized within the PM. In Fuzzy Logic, various aggregation functions have been developed to effectively handle a wide range of linguistic expressions. For $1PM$, we construct y by employing a set of membership functions. An aggregation function can compute the degree of validity (g_W) or relevance (g_R).
- T symbolizes a text generation algorithm that facilitates the generation of all conceivable sentences related to the linguistic expressions in A_y. T serves as a linguistic template comprising a set of potential linguistic expressions.

In Hierarchical Perception Mapping (HPM), we depict a map as a tuple (F, E, V, D), in which:

– *F*: A linguistic label represents the specific phenomenon or sub-phenomenon being modeled. It's a critical descriptor that provides a meaningful name for the overall perception captured by the HPM.
– *E*: Represents the event or data being processed or evaluated. This can range from raw input data in the case of first-order HPMs to aggregated data from lower-order HPMs.
– *V*: Denotes the value associated with the HPM, typically obtained through some form of evaluation or processing of the event or data. This value essentially measures the perception related to *F*.
– *D*: Represents the set of dependent or lower-order HPMs. Each element of *D* is a tuple in the form of an HPM, thus providing a hierarchical and recursive structure. It allows us to capture and model that perceptions can be composed of sub-perceptions or sub-phenomena.

We also introduce the Prompt Generator (PG) function as an essential component that facilitates the creation of effective prompts for text generation. The PG function $PG : HPM \rightarrow Prompts$, takes an HPM as input and produces prompts that guide the text generation process. By leveraging the hierarchical and recursive structure of the HPM, the PG function is able to extract meaningful information and generate prompts that capture the essence of the perceptions at different levels of granularity. These prompts act as useful cues for the subsequent text-generation step.

3   Methodological Proposal

We've divided the process into four components, as depicted in Fig. 2.

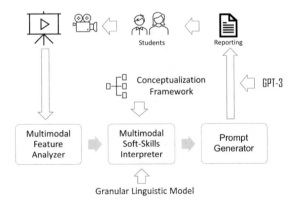

Fig. 2. Overview of the proposed approach.

- **Conceptualization Framework**: A comprehensive representation of the skill to identify is provided from a higher-level perspective. This conceptual representation encompasses distinct soft skills, each characterized by unique attributes and quantifiable aspects. A psychologist formulates the conceptualization.
- **Multimodal Feature Analyzer**: The responsibility of computing the main metrics of the process lies with a set of machine learning components. These components consist of various elements, including fundamental techniques for transcribing speech from videos, extracting audio, and segmenting the video into frames (called video preprocessing). Additionally, there are advanced tools based on Convolutional Neural Networks that are employed for audio-based emotion detection and blink and gaze detection. Transformers and autoencoders are utilized to extract topics from the text.
- **Multimodal Soft-Skill Interpreter**: It takes the data structure generated by the preceding component as input and generates a set of accurate linguistic expressions to describe the available data.
- **Prompt Generator**: It takes as input the calculated linguistic expressions and, as a result, produces the linguistic report. The report includes detailed information related to the evaluation and is enhanced by a pre-trained generative model.

3.1 Conceptualization Framework

Studying and analyzing multimedia data aims to identify essential skill assessment features. It involves experts defining dimensions and attributes, and the engineering team selecting tools for measurement. Each concept becomes a Computational Perception (CP) with a membership function. Then, an aggregation process, like a set of rules, is applied, culminating in conclusions in the form of Perception Mappings (PMs).

3.2 Multimodal Feature Analyzer

The video data will undergo a process of cleaning and standardization. The audio and video will be segmented into smaller portions and converted into WAV and MP4 formats, respectively, utilizing the FFmpeg library. Subsequently, Convolutional Neural Networks (CNN) and Autoencoder Networks, will be employed to analyse the audio and video separately. These techniques will extract significant details, such as the student's mood, attitude, and confidence level, to identify crucial content within their responses for skills assessment purposes.

Audio Transcription. We employed the open-source speech recognition system Whisper [9] trained on 680,000 h of multilingual data. It transcribed and translated a multilingual video into English using an encoder-decoder architecture. The audio was segmented, processed by a Convolutional Neural Network (CNN), and fed into a transformer model for tasks like language identification, transcription, and time stamping.

Audio Analysis. In the project, we used MyProsody[1] to analyze the audio extracted from the video. This tool measures acoustic characteristics in speech, like simultaneous speech and high entropy, and compares them to native speaker patterns. Machine learning algorithms, including Classification and Regression Trees (CART) and Multiple Regression, help establish these acoustic features. An acoustic model segments recorded speech, identifies syllable boundaries, and outlines fundamental frequency contours. MyProsody has built-in functions to recognize and measure various factors, which can be combined to derive relevant indicators for addressing the problem.

Text Analysis. In our project's Natural Language Processing (NLP) tasks, our primary tool was the Spacy library[2]. This library provides pre-trained models for multiple languages and facilitates the creation of new models or retraining existing ones using customized data tailored to specific domains. In our particular task, we examined the text extracted from the video, with a focus on extracting information and deriving context-based predictions about its meaning. We employed techniques such as Part of Speech Tagging (POS Tagging) and the identification of discursive markers for this analysis.

Text Classification. An API is used to automatically classify text based on a hierarchical taxonomy, enhancing the text extraction process. This classification is crucial for extracting topics and calculating various metrics. Texts are classified by matching specific keywords related to the text's topic against vocabularies like EUROVOC, covering a wide range of European Union-related subjects. This classification process is valuable for assessing the depth, coherence, and key elements of the text concerning the topic, quantifying its relevance to the posed questions. This approach has been employed in previous studies focused on text analysis to detect specific behaviours [6].

Video Analysis. We utilized OpenCV[3], which is a suite of software tools specifically designed for image and video processing and analysis [5]. These tools enabled us to precisely measure various aspects of the video through advanced computer vision algorithms. We harnessed video analytics capabilities to recognize and track individuals, including crucial facial features such as faces, eyes, and lips. Research has suggested that non-voluntary and semi-voluntary facial cues can provide objective estimations of emotions [4]. Parameters like eye-related events (gaze and blink), mouth activity, and head movement were quantified and have demonstrated high accuracy as indicators of stress and anxiety. To ensure efficient and accurate image processing, we employed Haar Cascades models, which rely on machine learning for visual object detection.

---

[1] https://github.com/Shahabks/myprosody.
[2] https://spacy.io/.
[3] https://opencv.org/.

3.3   Multimodal Soft Skills Interpreter

The proposed GLMP for generating linguistic reports of students' skills is structured around three fundamental components. Each of these components serves to validate and generate sentences that describe the current condition at a specific level of granularity.

The initial part focuses on denoting the process status, leveraging insights obtained from the results of the multimodal feature analyzer. Following this, these values are subjected to aggregation using fuzzy procedures to represent different attributes or dimensions of behaviour. Lastly, the concluding part of the model is dedicated to the analysis of students' skill levels.

3.4   Prompt Generator

This module utilizes the output from the preceding module to generate a comprehensive linguistic report. We employ predefined templates and enhance the report using a Generative Pre-trained model to provide more nuanced information. Our methodology integrates Hierarchical Perception Mapping (HPM) and Prompt Generator (PG) for assessing soft skills in videos.

HPM employs a recursive template for linguistic labels and associated numerical values. We fine-tune the prompt to create a comprehensive report that eliminates redundancies and emphasizes critical aspects. This methodology enables precise and relevant prompts, allowing customization to meet specific requirements.

4   Case of Study

This case study examines the effectiveness of multimodal video analysis in assessing the soft skills of undergraduate students. The experiment involves capturing and analyzing video footage of undergraduate students engaging in a presentation activity. Through the analysis of the video footage, the researchers are looking to identify factors that can accurately predict the students' soft skills, such as Decision Making, Conflict Resolution, and Creativity. The assignment was the following:

> This task consists of recording a short video (Time: Minimum 1 minute. Maximum 3 minutes) giving your name and surname, your specialization and computer science interests, and answering these questions.
>
> 1. Why are you choose the Machine Learning subject?
> 2. What is the meaning of the "Machine Learning" concept to you?
>
> It can be recorded/shared in any form, where the teacher can easily download it. Finally, copy the link in answer to the Moodle task.

4.1 Conceptualization Framework

We analyzed video footage to identify psychosociological traits predicting soft skills like Decision Making, Conflict Resolution, and Creativity (see Table 1). Each skill has dimensions and attributes provided by experts, measured as $1PM$. The GLMP was used to represent metric values with linguistic labels, defining attributes, dimensions, and skills corresponding to Perceptual Mappings and Conceptual Perceptions.

Table 1. Dimensions, attributes and measures to evaluate soft skills.

Dimension	Attribute	Measure	Source
Decision Making			
Accuracy	Speed	Reaction time, Fluency, Speech speed	Audio
	Firmness	Organization	Text
		Mood	Audio
Clearness	Ideas	Vagueness, Examples, Crutches	Text
Conflict Resolution			
Argumentation	Expression	Concreteness, Examples, Mood, Relevance	Text
	Quality	Vagueness	Text
		Fluency	Audio
Empathy	Non-Verbal Comm	Gaze, Smile	Video
Creativity			
Text Quality	Vocabulary	Redundancy, Originality, Examples, Crutches, Verbal T	Text
	Ideas	Quantity, Relevance	Text
		Reaction time	Audio
Non-Verbal	Non-Verbal	Gaze, Gesture	Video
Expression	Expression	Mood	Audio

4.2 Multimodal Feature Analyzer

This component uses machine learning techniques to calculate first-order perceptions ($1PM$), involving processes like audio retrieval, speech transcription, video segmentation, emotion detection, and text topic extraction. Audio analysis assesses reaction time, mood, and fluency, while the Whisper system transcribes speech. Transcription analysis measures various aspects of speech structure, originality, redundancy, and verb tense, and evaluates the text's relevance to the topic using MeaningCloud-trained models. Video analysis offers insights into actions, movements, and emotions by using Haar-Cascade models for detecting facial features and movements, and pre-trained models from the DeepFace library for gesture detection.

Table 2. Excerpt of results of the evaluation of the soft skills.

Student	Decision Making	Conflict Resolution	Creativity
1	Medium/Low	Low	Medium/Low
2	Low/Medium	Medium/Low	Low
3	Low/Medium	Medium	Medium/Low
4	Low/Medium	Medium/Low	Medium/High
5	Low	Low	Low/Medium
6	Medium	Medium	Medium/High
7	Medium	Low	Medium/Low
8	Low	Low	Low
9	Low/Medium	Medium	Medium/Low
10	Low	Low	Medium
11	Low/Medium	Medium/Low	Medium/Low
12	Low	Medium/Low	Low
13	Medium/Low	Medium	Low/Medium
14	Low/Medium	Low	Medium
15	High	Medium	Medium/High

4.3 Multimodal Soft-Skill Interpreter

Following the retrieval of metric results or $1PM$ from the multimodal feature analyzer, we employ fuzzy aggregations to represent skill attributes and dimensions. In our model, we utilize a total of 38 PMs, categorized across four levels: 18 at the first level (measures), 10 at the second level (attributes), 6 at the third level (dimensions), and 4 at the fourth level (soft skills). This hierarchical structure helps us capture the intricacies of the skill assessment process effectively.

Outputs CPs have predefined linguistic expressions such as:

- $A = (Bad, Normal, Good)$ for expression, vocabulary, ideas, argumentation
- $A = (Low, Medium, High)$ for reaction time, fluency, speech speed, organization, vagueness, examples, quality, non-verbal communication, accuracy, decision making, conflict resolution, creativity
- $A = (Low, High)$ for crutches
- $A = (Noemotion, Normal, Appasionate)$ for mood

4.4 Prompt Generator

Using GLMP allows us to analyse the problem by dividing its factors into smaller units and describing their relationships at different levels of granularity. Representing soft skills with various levels of detail enables transparent descriptions of metrics, attributes, and dimensions and their impact on the final output. We propose employing HPM to generate prompts that provide detailed descriptions at different hierarchy levels, resulting in a comprehensive and coherent report.

We received 15 responses and the results for the soft skills of each student (see an excerpt in Table 2). An example of the provided report for student 1 can be seen in Table 3. As can be seen, there were some differences in their skill levels. This comparison of the students' performance highlights the strengths and weaknesses of each student. This information can be used to provide targeted support and instruction for each student, to help them improve their skills in areas where they struggle.

Table 3. Example of Results.

Soft skill	Natural Language Report
Decision Making	According to the evaluation, the level of **decision making** is *medium/low*. This is due to several reasons. First, the **accuracy** is *low*, which means that the person was not able to express himself clearly and precisely. This is due to a *medium/low* **firmness** in the **organization of ideas** and the *no emotion* **mood** he presented. In addition, the **speed of speech** was *low/medium*, which affected the **fluency** of his speech.
Conflict Resolution	The evaluation shows that the level of **conflict resolution** is *low*. This is mainly due to a lack of empathy on the part of the person. It was noted that the person's **smile** was *low* and his **gaze** was *normal/diffuse*. Both of these elements are important in establishing an empathetic connection with others during a negotiation.
Creativity	The level of **creativity** was evaluated as *medium/low*. This is due to the evaluation of **text quality** and **non-verbal expression**. The **text quality** was evaluated as *high/medium*, which means that the person was able to handle the **vocabulary** and **distribute the ideas** *appropriately*. However, the **non-verbal expression** was evaluated as *poor*, since elements such as **blinks**, **gaze**, **posture changes** and **gestures** were not taken into account

5   Conclusions

We have described an approach to represent the evaluation of soft skills in students using video analysis that delivers relevant descriptions of the function of the input data. The model integrates components related to computer vision, audio and text analysis, psychology and fuzzy logic. The results of this analysis can help inform the design of more effective teaching and learning strategies in undergraduate education. The work is in an early stage, the conceptualization of several behaviours is made, and some have been detected; however, only in a few examples. In future work, a tuning process of the measures scales is needed.

Acknowledgements. The Spanish Government has partially supported this work under the grant SAFER: PID2019-104735RB-C42 (ERA/ERDF, EU), and project PLEC2021-007681 funded by MCIN/AEI /10.13039/501100011033 and by the European Union NextGenerationEU/ PRTR.

References

1. Akoa, B.E., Simeu, E., Lebowsky, F.: Using statistical analysis and artificial intelligence tools for automatic assessment of video sequences. In: Eschbach, R., Marcu, G.G., Rizzi, A. (eds.) Color Imaging XIX: Displaying, Processing, Hardcopy, and Applications, vol. 9015, p. 90150O. International Society for Optics and Photonics, SPIE (2014). https://doi.org/10.1117/12.2044797

2. de Anda-Trasviña, A., Nieto-Garibay, A., Gutiérrez, J.: Natural language report of the composting process status using linguistic perception. Appl. Soft Comput. **127**, 109357 (2022)

3. Conde-Clemente, P., Alonso, J.M., Trivino, G.: Toward automatic generation of linguistic advice for saving energy at home. Soft. Comput. **22**(2), 345–359 (2018)

4. Giannakakis, G., et al.: Stress and anxiety detection using facial cues from videos. Biomed. Signal Process. Control **31**, 89–101 (2017). https://doi.org/10.1016/j.bspc.2016.06.020

5. Kaehler, A., Bradski, G.: Learning OpenCV 3 - Computer Vision in C++ with the OpenCV Library. O'Reilly Media, Inc. (2016)

6. Losada, D.E., Gamallo, P.: Evaluating and improving lexical resources for detecting signs of depression in text. Lang. Resour. Eval. **54**(1), 1–24 (2020). https://doi.org/10.1007/s10579-018-9423-1

7. Min, Q., Zhou, Z., Li, Z.: An approach to automatic evaluation of instructional videos. In: 2021 the 5th International Conference on Management Engineering, Software Engineering and Service Sciences, pp. 63–68. ICMSS 2021, Association for Computing Machinery, New York, NY, USA (2021). https://doi.org/10.1145/3459012.3459022

8. Nayak, S., Kumar, S., Agarwal, D., Parikh, P.: AI-enabled personalized interview coach in Rural India. In: Rodrigo, M.M., Matsuda, N., Cristea, A.I., Dimitrova, V. (eds.) Artificial Intelligence in Education. Posters and Late Breaking Results, Workshops and Tutorials, Industry and Innovation Tracks, Practitioners' and Doctoral Consortium, pp. 89–93. Springer International Publishing, Cham (2022). https://doi.org/10.1007/978-3-031-11647-6_15

9. Radford, A., et al: Introducing Whisper (2022). https://openai.com/blog/whisper/

10. Solé-Beteta, X., Navarro, J., Gajšek, B., Guadagni, A., Zaballos, A.: A data-driven approach to quantify and measure students' engagement in synchronous virtual learning environments. Sensors **22**(9) (2022). https://doi.org/10.3390/s22093294

11. Sun, Y., Nomiya, H., Hochin, T.: Automatic evaluation of motion picture contents by estimation of fgacial expression intensity. In: 2019 20th IEEE/ACIS International Conference on Software Engineering, Artificial Intelligence, Networking and Parallel/Distributed Computing (SNPD), pp. 227–232 (2019). https://doi.org/10.1109/SNPD.2019.8935660

12. Trivino, G., Sanchez, A., Montemayor, A.S., Pantrigo, J.J., Cabido, R., Pardo, E.G.: Linguistic description of traffic in a roundabout. In: International Conference on Fuzzy Systems, pp. 1–8. IEEE (2010)

Understanding the Role of the User in Information Propagation on Online Social Networks: A Literature Review and Proposed User Model

Eleana Jerez-Villota[1,2](✉) ⓘ, Francisco Jurado[2] ⓘ,
and Jaime Moreno-Llorena[2] ⓘ

[1] Departamento de Ciencias de la Computación, Universidad de las Fuerzas Armadas
ESPE, Sangolquí, Ecuador
[2] Departamento de Ingeniería Informática, Universidad Autónoma de Madrid,
Madrid, Spain
eijerez@espe.edu.ec, eleana.jerez@estudiante.uam.es,
{Francisco.Jurado,Jaime.Moreno}@uam.es

Abstract. Users play a crucial role in information propagation due to their sentiments, emotions, interests, and relationships with other users. Thus, this paper details a literature review process conducted to analyze the user's role in the information propagation process. Several works in the state of art have provided valuable insights into the attributes and behaviours of nodes in the information propagation process on Online Social Networks (OSNs), contributing to a broader understanding of this complex phenomenon. However, building a model that accurately represents a user during information propagation in OSNs is a great challenge. Consequently, in this paper, we propose a user model with four sets of features, containing user attributes and actions distributed in two groups to encompass user behaviour during information propagation on Twitter. The user model captures the essence of individual behaviour within social networks and offers potential for further research and applications in the Analysis of Online Social Networks (AOSNs) field.

Keywords: Information propagation process · User model · User behaviour · Online Social Networks

1 Introduction

The Digital 2023 April Global Statshot Report [14] presents that the use of Online Social Networks (OSNs) has grown with 150 million new user identities in the last year, representing an annual growth of 3.2%. It implies that more than 4.76 billion people use social networks, meaning 59.4% of the world population.

The data presented above reflects that OSNs have become an indispensable means of communication for people who commonly use them to share information related to their daily lives or events occurring in their political, economic, or social environment. Researchers such as psychologists, physicists, and computer scientists have been interested in their analysis [2,8,13,26,41,43,45,46].

The discipline of Analysis of Online Social Networks (AOSNs) emerged in the 20th century, and it is concerned with mathematically or visually exploring the patterns formed by individuals, groups or entities that constitute the "nodes" and the relationships between them that are the "edges" [27]. Among others, one of the main topics of interest for researchers within the AOSNs discipline is Information Propagation. They study this topic to understand how information propagates, influences, and shapes the behaviour of users in the digital environment.

According to [3], one of the main factors affecting the information propagation process in OSNs, and in particular, those mediated by the Internet, is the network topology. To name a few examples, in [47], the authors study how people traverse their psychological thresholds to approve new information or adopt innovations and how to employ the topology of networks for spreading enhancement; the work presented in [32] shows an attempt to use the network topology to quantify the influence among users; in [35] authors employed a critical parameter based on network topology to study the propagation feature; and in [9], the authors mentioned that the large size and complex topology of OSNs make the conception of an accurate rumour propagation model challenging.

Nevertheless, in our opinion, not only the network's topology is an important factor in the propagation process, but also the user's features, such as social influence, relationships, and similarity of interests with other users. Additionally, their behaviour during the diffusion process will decide the direction and scope of the propagation, making it another determining factor. However, the variety and complexity observed in human behaviours in OSNs make it highly challenging for researchers to model this phenomenon.

Consequently, in this paper, we have conducted a literature review of previous works to gain a panoramic view of the user's role in the information propagation process. Furthermore, we propose a user model for analysing the information propagation in a particular social network.

Thus, the rest of the paper is structured as follows: in Sect. 2, we present the literature review; in Sect. 3, we detailed the results of the literature review; in Sect. 4 we propose a user model in the propagation process;finally, this paper ends with the conclusions, future works and limitations in Sect. 5.

2 Literature Review

This section details the literature review performed of previous works to gain a panoramic view of the users' role in the information propagation process on OSNs. During the literature review process, we followed the activities detailed in this section [12,25].

2.1   Research Question

In order to analyze and understand users' role in the information propagation process, the next research question (RQ) was formulated:

- RQ: What node's attributes affect the propagation of information in social networks?

2.2   Search of Primary Researches

We identified the primary research papers set using a search string in the following scientific databases: ACM Digital Library, Web of Science, IEEE Xplore, Science Direct, and Scopus. The search string were compound of a set of keywords defined according to the research question (see Table 1). After that, we based our selection criteria on citations, ranking and content of article titles.

Table 1. Keywords used for searching papers in the literature review.

Keywords								
	"attribute" **OR**				"dissemination" **OR**			
"node" **AND**	"feature"	**AND**	"information"	**AND**	"propagation" **OR**	**AND**	"social network"	
					"spreading" **OR**			
					"diffusion"			

2.3   Review of Articles for Inclusion and Exclusion

We applied a process of inclusion and exclusion which involved an analysis of paper titles and a critical reading of the papers' abstracts.

As inclusion criteria, we define:

- The inclusion of the papers where the abstract or title mentioned information propagation on OSNs **OR**
- The inclusion of the papers papers where the abstract or title mentioned users' attributes or features in the information propagation process.

In addition, as exclusion criteria, we established:

- The exclusion of the papers published in conferences **AND**
- The exclusion of the papers that did not have a quartile in Journal Citation Reports (JCR)[1] or Scimago Journal & Country Rank (SJR)[2] **AND**
- The exclusion of the papers that did not belong to the field of Computer Science **AND**

---

[1] https://clarivate.com/webofsciencegroup/solutions/journal-citation-reports.

[2] https://www.scimagojr.com/.

- The exclusion of the papers that did not report on information propagation on OSNs **OR**
- The exclusion of the papers that did not report users' attributes or features in the information propagation process.

After using the search string described in Sect. 2.2, we initially obtained 100 papers (see Fig. 1). We then eliminated duplicates, resulting in 82 papers. Finally, by applying the inclusion and exclusion criteria described in this section, we obtained a total of 30 relevant articles.

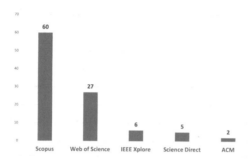

Fig. 1. Number of papers found per database.

2.4   Classification of Articles

Based on the relevant articles obtained after the inclusion and exclusion process, we identified a set of concepts that define the research context. Subsequently, we consolidated these works into different categories as follows:

- Types of information that propagate on OSNs.
- Methods to control the propagation of information on OSNs.
- Node actions in the information propagation process on OSNs.

2.5   Data Extraction and Mapping

Using the classification of articles obtained in the previous subsection and the Wieringa Classification [37] to identify the paper contributions, Table 2 lists the references of the relevant articles and their classification by area and type of contribution. From Table 2, we created Fig. 2, where the papers are categorized based on both classifications, using two axes: type of contribution (X-axis) and areas (Y-axis).

Table 2. Papers classification by area and type of contribution.

Ref.	Types of information that propagate								Methods to control the propagation								Node actions in the information propagation process							
	Tool	Model	Method	Process	Evaluation	Validation	Proposal	Experience	Tool	Model	Method	Process	Evaluation	Validation	Proposal	Experience	Tool	Model	Method	Process	Evaluation	Validation	Proposal	Experience
2																								
4			×		×																			
5		×								×								×			×			
6																×		×						×
7										×						×								
10		×	×																					
11											×								×					
15																		×						
16										×								×	×					
18																		×						
19																		×	×					
20			×					×			×							×	×					×
21										×						×								
22																		×						
23		×			×														×					
24		×										×												
28										×								×						
29											×							×	×					
30		×																×						
31											×							×						
33																			×					
36																		×						
38			×					×																
39					×														×					
40		×																×						
42																								
43		×									×							×						
44		×	×							×								×	×					
46		×						×										×						×

3 Review of the Results

In this section, we present some important aspects of the articles found in the literature review, with the focus on the role played by users in the information propagation process on OSNs.

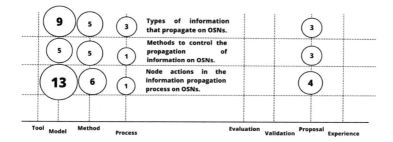

Fig. 2. The mapping of found previous works for the literature review.

The review shows how the network topology is one of the main factors affecting the propagation process, but it is not the only one. Users' attributes such as their social influence, their relationships and the similarity of interests they have with other users and their behaviour during the information propagation process are also determining factors [5,7,16,18,22,23,28,33,42,46]. For its part, social influence has been widely studied even before the emergence of OSNs. Nowadays, it seems more visible in OSNs. Scientists continue to study this phenomenon, for instance, Di Ianni and Varricchio developed a diffusion model in which some of a node's neighbours exert a negative influence on that node, i.e., they induce the node to reject the information to be diffused [6]. On the other hand, private companies, governments, and health institutions also use OSNs to influence people's decisions. They invest large sums of money in digital marketing campaigns [11,43].

In their experiments about information dissemination on the Douban Chinese social network, Niu et al. found that in most cases, information diffuses with multiple origins; they attribute this to the two kinds of influence present in a diffusion process: internal and external [22]. In [16], they used the ICWSM blog dataset [3] and measured the social pressure on each user, i.e., the fact that many different neighbours have diffused a given content; through the number of incoming neighbours that have already diffused the information. Doo and Liu [7] proposed a social-influence diffusion model with social incentives. It has three main features. The first one, they define an influenced diffusion probability for each node. Second, they categorise the nodes into two classes active and inactive. Active nodes can influence inactive but not vice versa. The third feature is that they use a diffusion threshold to control how the influence propagates.

According to [42], the best way to reflect interactions between people on social media platforms, and to show the behaviour of information propagation

---

[3] https://www.icwsm.org/data/.

on them, is the retweet. It is a feature on Twitter where users can repost or share someone else's tweet with their followers. When a user retweets a tweet, it appears on their timeline, attributing the original tweet to the original author. This sharing feature allows information, posts, or messages to be spread rapidly across the platform and reach a broader audience [1]. Li et al. claim that retweeting is the primary method of information propagation in social networks. They investigate three user activities associated with information propagation: mentions, replies, and retweets [18]. Zhao et al. [44] proposed a semi-supervised graph and embedding model based on a graph attention network for spam bot detection in social networks. They conducted several experiments and used a Twitter dataset to evaluate their model. The features that they took from the dataset were: account_age, the number of days an account was created on Twitter; no_followers, the number of followers for an account on Twitter; no_followings, the number of followings for an account on Twitter; no_userfavourites, the number of favourites received by an account on Twitter; no_statuses, the number of tweets posted by the account on Twitter, including retweets; no_tweets, the number of tweets posted by the account on Twitter; no_retweets, the number of tweets retweeted. Shang et al. [28] used Sina Weibo, which is similar to Twitter and allows users to post short messages, images, videos, and links, making it a popular platform for sharing updates, news, and opinions. They used it to study the precise prediction of information diffusion popularity based only on early repost information, given that the underlying user relation network is unknown.

Research communities take some users' behaviours to perform their studies on OSNs. Li et al. [18] used Twitter and Sina Weibo to carry out their experiments. They proposed the dynamic topic model to model user behavioural attributes using a time factor. According to [5], on social networks, nodes prefer to contact more frequently with those sharing similar interests or visit communities and have regular social behaviours. In [46], they proposed an approach to nonparametrically estimate the hidden interaction network behind the spreading information process. Their results show that Twitter users involved in the spreading process can be divided into four communities: big name users, famous active and inactive users, and nobody users. Stai et al. [30] take into account user interests and their features, i.e., interest periodicity or interest dependence on the topic of the propagated information, to develop a framework for modelling and controlling useful information diffusion in generalized networks.

4   Proposal of User Model in Information Propagation

After carrying out the literature review, we can say that the behaviour of users is a determinant factor in the propagation process, as each individual has their way of thinking, emotions, and will. Users can decide whether to share information or not, and based on their choice, the propagation process continues or stops. If the propagation process continues, it could mean that the individuals believe in the information and share it. Although modelling the behaviour of individuals within social networks is challenging, in our opinion, it is important to have a user model that represents the person in the information propagation process.

We propose a user model in the information propagation process based on some relevant aspects that we found in the results of the literature review [2, 10, 18, 20, 22, 44, 46]. Also, we based our proposed user model on the gaps that identified in the literature review, such as: the users' interactions through replies, since most of the works focus their studies on analysing tweets and retweets in the information propagation process [2, 18, 19, 42, 43, 46]; the account verification attribute, as it allows us to distinguish authentic accounts from fake or imper-sonator accounts; and the users' sentiments and emotions during actions in the information propagation process, including posting tweets, spreading informa-tion through retweets, and interacting with other users through replies.

As a consequence, the proposed user model contemplates user attributes related to their profile, the social relationships they maintain, the sentiments and emotions they involve in their interactions and the actions they perform.

Although the proposed approach could be applied to other social networks, it is used for Twitter, as an illustrative example. Despite the recent modifications the platform has undergone, and the existence of other social media platforms that are also subject to study, it can still be exploited as it remains popular, has an API that allows researchers to access public platform data, facilitates rapid information dissemination, and provides rich and structured data. Furthermore, a tweet represents a message that can be propagated in social networks with similar Twitter features.

To carry out our proposal, we incorporated specific characteristics from the six categories defined in the model proposed by Lee et al. [17]. These categories include the user profile characteristics, the user activity characteristics, the user's previous retweet characteristics, and the user availability characteristics.

For our model, we have identified four categories of attributes and actions to model a person user. We have grouped those categories in two groups: User Attributes Group and User Actions Group.

In the User Attributes Group, we have the next three categories containing attributes of the user who participates in the information propagation process:

- In the Profile Attribute Category, having the account age attribute could allow us to infer, for example, if the user has experience using the platform's functionalities [17, 44]. The user name and its length, the profile description and its length, and the profile URL and its length are also important because they could show the user's interest in the social network by the effort he has made to generate these data [17]. Finally, the verified account could guarantee us that the owner of the profile is not a robot but a person.
- In the Social Network Attributes Category, we take into account social net-work topology measures, such as the number of followers, the number of followings, and the ratio of followers to followings [2, 10, 17, 22, 44, 46].
- Researchers have found that the use of words in a person's writings, such as blogs and essays, is related to their personality [2, 17, 44, 46]. Therefore, it is essential to collect users' sentiments and emotions from the information they propagate. Hence, we have Sentiments and Emotion in the User Actions Category.

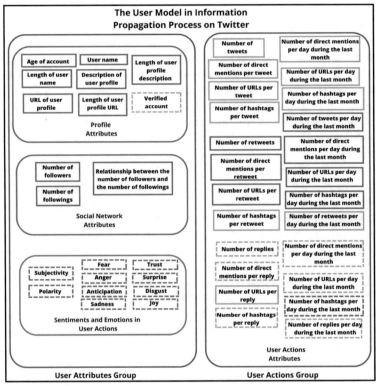

Fig. 3. The User Model in Information Propagation Process on Twitter.

For its part, the User Actions Group contains the actions performed by the user in the network, and will allow us to infer their behavior within the information propagation process [18,28,33,42,46]. Therefore, User Actions Attributes Category has attributes that describe the information that the user writes and reads based on tweets, retweets, and replies. The user can issue tweets, retweets, and replies, but they can also read all this from their followings.

As a result, we have a user model (see Fig. 3) to cover the identified challenges by introducing attributes and actions to model a person user. The attributes in the model are within rectangles. The solid line rectangles represent attributes extracted from the literature review, while the dashed line rectangles represent attributes that we didn't obtain from the literature review and are our contribution to the model.

5   Conclusions, Future Works and Limitations

Along this paper, after carried out a literature review, we can conclude that the user has a crucial role in the information propagation process on OSNs. Each individual's way of thinking, their emotions and their will influence their decision to share information, ultimately affecting whether the propagation process continues or stops.

Another conclusion we can extract after the research is that modelling the behaviour of individuals within social networks is a challenging task, but having a user model that accurately represents the person in the information propagation process is essential. The user model which we present in this work intends to tackle these challenges by having attributes and actions to model a person user, grouped as User Attributes and User Actions. These categories encompass Profile Attributes, Social Network Attributes, Sentiments and Emotions in User Actions, and User Actions Attributes.

Our future research lines include conducting several experiments to evaluate the proposed user model; developing a network model and an information model to create simulations of the information propagation process, which will include the proposed user model. In addition, we will use the user model to create user profiles by applying clustering algorithms.

A limitation of our model is that some of the proposed attributes may not be extractable from other social networks or even from Twitter, given its recent modifications. This situation could potentially pose challenges for future work. However, the presence of several well-defined attributes enables us to conduct multiple experiments, ultimately allowing us to refine the model and achieve a definitive user model for the information propagation process.

Acknowledgments. This research work has been co-funded by the Spanish National Research Agency (Agencia Estatal de Investigación) through the project Indigo! (PID2019-105951RB-I00), and the Regional Government of Madrid (Comunidad de Madrid) through the e- Madrid-CM project (P2018/TCS-4307). The e-Madrid-CM project is also supported by Structural Funds (ESF and ERDF).

References

1. Social media terms. https://www.socialpilot.co/social-media-terms
2. Brooks, H.Z., Porter, M.A.: A model for the influence of media on the ideology of content in online social networks. Phys. Rev. Res. **2**(2), 023041 (2020)
3. Chen, S., et al.: Rumor propagation model with consideration of scientific knowledge level and social reinforcement in heterogeneous network. Phys. A **559**, 125063 (2020)
4. Dang, Q., Gao, F., Zhou, Y.: Early detection method for emerging topics based on dynamic Bayesian networks in micro-blogging networks. Expert Syst. Appl. **57**, 285–295 (2016)
5. Deng, X., Chang, L., Tao, J., Pan, J.: Reducing the overhead of multicast using social features in mobile opportunistic networks. IEEE Access **7**, 50095–50108 (2019)
6. Di Ianni, M., Varricchio, G.: Latency-bounded target set selection in signed networks. Algorithms **13**(2), 32 (2020)
7. Doo, M., Liu, L.: Probabilistic diffusion of social influence with incentives. IEEE Trans. Serv. Comput. **7**(3), 387–400 (2014)
8. Gu, K., Wang, L., Yin, B.: Social community detection and message propagation scheme based on personal willingness in social network. Soft. Comput. **23**, 6267–6285 (2019)

9. Hosni, A.I.E., Li, K., Ahmad, S.: Minimizing rumor influence in multiplex online social networks based on human individual and social behaviors. Inf. Sci. **512**, 1458–1480 (2020)

10. Jing, Y., Wei, Z., Xiao-Mei, Z., Qing-Chao, Z.: Tip over community: backbone nodes detection based on community structure. IEEE Access **6**, 36050–36063 (2018)

11. Karczmarczyk, A., Jankowski, J., Wątrobski, J.: Multi-criteria seed selection for targeting multi-attribute nodes in complex networks. Symmetry **13**(4), 731 (2021)

12. Keele, S., et al.: Guidelines for performing systematic literature reviews in software engineering (2007)

13. Keikha, M.M., Rahgozar, M., Asadpour, M., Abdollahi, M.F.: Influence maximization across heterogeneous interconnected networks based on deep learning. Expert Syst. Appl. **140**, 112905 (2020)

14. Kemp, S.: Digital 2023 april global statshot report. https://datareportal.com/reports/digital-2023-april-global-statshot

15. Kong, J.H., Han, M.M.: Malware containment using weight based on incremental pagerank in dynamic social networks. KSII Trans. Internet Inform. Syst. **9**(1), 421 (2015)

16. Lagnier, C., Gaussier, E., Kawala, F.: User-centered probabilistic models for content diffusion in the blogosphere. Online Soc. Netw. Media **5**, 61–75 (2018)

17. Lee, K., Mahmud, J., Chen, J., Zhou, M., Nichols, J.: Who will retweet this? Detecting strangers from twitter to retweet information. ACM Trans. Intell. Syst. Technol. (TIST) **6**(3), 1–25 (2015)

18. Li, Q., Liu, L., Xu, M., Wu, B., Xiao, Y.: GDTM: a gaussian dynamic topic model for forwarding prediction under complex mechanisms. IEEE Trans. Comput. Soc. Syst. **6**(2), 338–349 (2019)

19. Liu, X., He, D., Yang, L., Liu, C.: A novel negative feedback information dissemination model based on online social network. Phys. A **513**, 371–389 (2019)

20. Mulders, D., de Bodt, C., Bjelland, J., Pentland, A., Verleysen, M., de Montjoye, Y.A.: Inference of node attributes from social network assortativity. Neural Comput. Appl. **32**, 18023–18043 (2020)

21. Nguyen, N.P., Yan, G., Thai, M.T.: Analysis of misinformation containment in online social networks. Comput. Netw. **57**(10), 2133–2146 (2013)

22. Niu, J., Wang, D., Stojmenovic, M.: How does information diffuse in large recommendation social networks? IEEE Network **30**(4), 28–33 (2016)

23. Niu, Y.W., Qu, C.Q., Wang, G.H., Wu, J.L., Yan, G.Y.: Information spreading with relative attributes on signed networks. Inf. Sci. **551**, 54–66 (2021)

24. Pham, D.V., Nguyen, G.L., Nguyen, T.N., Pham, C.V., Nguyen, A.V.: Multi-topic misinformation blocking with budget constraint on online social networks. IEEE Access **8**, 78879–78889 (2020)

25. Ren, R., Zapata, M., Castro, J.W., Dieste, O., Acuña, S.T.: Experimentation for chatbot usability evaluation: a secondary study. IEEE Access **10**, 12430–12464 (2022)

26. Sarkar, S., Guo, R., Shakarian, P.: Using network motifs to characterize temporal network evolution leading to diffusion inhibition. Soc. Netw. Anal. Min. **9**(1), 1–24 (2019). https://doi.org/10.1007/s13278-019-0556-z

27. Scott, J.: What is social network analysis? Bloomsbury Academic (2012)

28. Shang, J., et al.: RNe2Vec: information diffusion popularity prediction based on repost network embedding. Computing **103**, 271–289 (2021)

29. Shi, J., Salmon, C.T.: Identifying opinion leaders to promote organ donation on social media: network study. J. Med. Internet Res. **20**(1), e7 (2018)

30. Stai, E., Karyotis, V., Papavassiliou, S.: Analysis and control of information diffusion dictated by user interest in generalized networks. Comput. Soc. Netw. **2**(1), 1–31 (2015). https://doi.org/10.1186/s40649-015-0025-4
31. Tan, Z., Ning, J., Liu, Y., Wang, X., Yang, G., Yang, W.: ECRModel: an elastic collision-based rumor-propagation model in online social networks. IEEE Access **4**, 6105–6120 (2016)
32. Wang, C., Tang, J., Sun, J., Han, J.: Dynamic social influence analysis through time-dependent factor graphs. In: 2011 International Conference on Advances in Social Networks Analysis and Mining, pp. 239–246. IEEE (2011)
33. Wang, G., Zhang, X., Tang, S., Wilson, C., Zheng, H., Zhao, B.Y.: Clickstream user behavior models. ACM Trans. Web (TWEB) **11**(4), 1–37 (2017)
34. Wang, J., Jiang, C., Quek, T.Q., Wang, X., Ren, Y.: The value strength aided information diffusion in socially-aware mobile networks. IEEE Access **4**, 3907–3919 (2016)
35. Wang, L., Li, L., Chen, G., Ye, Q.: Edge instability: a critical parameter for the propagation and robustness analysis of large networks. Inf. Sci. **536**, 358–371 (2020)
36. Wang, Z., Du, C., Fan, J., Xing, Y.: Ranking influential nodes in social networks based on node position and neighborhood. Neurocomputing **260**, 466–477 (2017)
37. Wieringa, R., Maiden, N., Mead, N., Rolland, C.: Requirements engineering paper classification and evaluation criteria: a proposal and a discussion. Requirements Eng. **11**, 102–107 (2006)
38. Wu, Y., Huang, H., Zhao, J., Wang, C., Wang, T.: Using mobile nodes to control rumors in big data based on a new rumor propagation model in vehicular social networks. IEEE Access **6**, 62612–62621 (2018)
39. Xiao, Y., Yu, H., Li, Q., Liu, L., Xu, M., Xiao, H.: MPURank: a social hotspot tracking scheme based on tripartite graph and multi messages iterative driven. IEEE Trans. Comput. Soc. Syst. **6**(4), 715–725 (2019)
40. Xu, X.: Cultural communication in double-layer coupling social network based on association rules in big data. Pers. Ubiquit. Comput. **24**(1), 57–74 (2020)
41. Yang, L., Qiao, Y., Liu, Z., Ma, J., Li, X.: Identifying opinion leader nodes in online social networks with a new closeness evaluation algorithm. Soft Comput. **22**, 453–464 (2018)
42. Yin, H., Yang, S., Song, X., Liu, W., Li, J.: Deep fusion of multimodal features for social media retweet time prediction. World Wide Web **24**, 1027–1044 (2021)
43. Zhan, Q., Zhuo, W., Liu, Y.: Social influence maximization for public health campaigns. IEEE Access **7**, 151252–151260 (2019)
44. Zhao, C., Xin, Y., Li, X., Zhu, H., Yang, Y., Chen, Y.: An attention-based graph neural network for spam bot detection in social networks. Appl. Sci. **10**(22), 8160 (2020)
45. Zhao, Y., Yang, N., Lin, T., Philip, S.Y.: Deep collaborative embedding for information cascade prediction. Knowl. Based Syst. **193**, 105502 (2020)
46. Zheng, Y., Zhao, X., Zhang, X., Ye, X., Dai, Q., et al.: Mining the hidden link structure from distribution flows for a spatial social network. Complexity **2019** (2019)
47. Zhou, C., Zhao, Q., Lu, W.: Impact of repeated exposures on information spreading in social networks. PLoS ONE **10**(10), e0140556 (2015)

Author Index

J. Bravo and G. Urzáiz (Eds.): UCAmI 2023, LNNS 842, pp. 317–318, 2023.
https://doi.org/10.1007/978-3-031-48642-5

Printed in the United States
by Baker & Taylor Publisher Services